Ulrich Kulisch
Rudolf Lohner
Axel Facius (eds.)

Perspectives on Enclosure
Methods

SpringerWienNewYork

Prof. Dr. Ulrich Kulisch
Priv.-Doz. Dr. Rudolf Lohner
Dr. Axel Facius

Institut für Angewandte Mathematik, Universität Karlsruhe,
Deutschland

© 2001 Springer-Verlag/Wien
Printed in Austria

Cover illustration: Double bubble image © 1995 by John M. Sullivan, University of Illinois

Typesetting: Camera-ready by authors
Printing: Novographic, A-1238 Wien
Binding: Papyrus, A-1100 Wien

Printed on acid-free and chlorine-free bleached paper
SPIN: 10835368

With numerous Figures

CIP data applied for

ISBN 3-211-83590-3 Springer-Verlag Wien New York

To Götz Alefeld

on the occasion of his

60th birthday.

Preface

Interval mathematics and enclosure methods have been developed to a high standard during the last decades together with their side effects on the arithmetic of computers, on programming languages and compilers, on the elementary functions and the run time system. A good number of meetings, the SCAN meetings in particular, have been devoted to this area. The latest of these meetings: scan2000 – GAMM-IMACS International Symposium on Scientific Computing, Computer Arithmetic and Validated Numerics was held at Universität Karlsruhe (TH), September 19-22, 2000 jointly with INTERVAL 2000 – International Conference on Interval Methods in Science and Engineering. The conference attracted 193 participants from 33 countries from all over the world. 12 invited lectures and 153 contributed talks were given at the symposium.

This volume contains selected papers on enclosure methods and validated numerics. The majority of contributions is based on invited lectures delivered at the scan2000 conference. These contributions are supplemented by a few selected additional papers. All papers represent original work following the common goal to push the limits of enclosure methods forward. We wish to thank all authors for their contributions and for adapting their manuscripts to the goals of this volume.

The editors and authors dedicate this volume to Prof. Götz Alefeld on the occasion of his 60th birthday. He has been among the leaders in the field for decades. The development of enclosure methods owes many outstanding contributions to him. His work shaped our current understanding of interval and enclosure methods in particular for linear and nonlinear systems of equations. He participated at the scan2000 conference as an author and as the president of the Gesellschaft für Angewandte Mathematik und Mechanik (GAMM).

Karlsruhe, Germany,
February 2001

Ulrich Kulisch
Rudolf Lohner
Axel Facius

Contents

Proving Conjectures by Use of Interval Arithmetic

Andreas Frommer

Fachbereich Mathematik, Universität Wuppertal,
D-42097 Wuppertal, Germany.
Andreas.Frommer@math.uni-wuppertal.de

Abstract. Machine interval arithmetic has become an important tool in computer assisted proofs in analysis. Usually, an interval arithmetic computation is just one of many ingredients in such a proof. The purpose of this contribution is to highlight and to summarize the role of interval arithmetic in some outstanding results obtained in computer assisted analysis. 'Outstanding' is defined through the observation that the importance of a mathematical result is at least to some extent indicated by the fact that it has been formulated as a 'conjecture' prior to its proof.

1 Computer Assisted Proofs in Analysis

A computer assisted proof is a mathematical proof which is partly done by executing an algorithm on a computer. Of course, like the other parts of the proof, the algorithm as well as its implementation have to be correct. When devising the algorithm and proving its correctness, we rely on a precise *specification* on how the computer performs the elementary steps of the algorithm. And when we then run the algorithm on a given computer, we rely on the fact that the computer indeed performs these elementary steps exactly according to the specification.

In computer assisted proofs from discrete mathematics, the elementary steps of the algorithm are just arithmetic operations with integers. On virtually every computer, arithmetic operations with integers are specified to exactly match the result of the corresponding mathematical operation. This might sound trivial, but even in the integer case we have the possible exception of overflows. Consequently, the algorithm implemented will have been executed by using the mathematical definition of integer arithmetic operations only if no overflow occured at run time. But this is easy to check in practice since overflows are signaled to the user. Due to this simple direct matching between computer integer operations and their mathematical counterpart, computer assisted proofs in discrete mathematics have a long and successful history. As the most prominent example in this area, let us just mention the proof of the Four Colour Theorem by Appel and Haken [5,6]

When a computer is used to prove a result in analysis, the picture becomes quite a bit more involved. This is due to the fact that the basic quantities in analysis are the real numbers, an infinite continuum. Only a finite subset of it can correctly be represented by the floating point ('machine') numbers of the computer. Already the conversion of a real to a floating point number now becomes a tricky and cumbersome issue. Moreover, even when we stick to floating point numbers, the mathematically correct result of an arithmetic operation will in general not be representable as a floating point number. Consequently, floating point arithmetic can always be just an approximation to the arithmetic on the reals. The IEEE Standard 754 [4] specifies such a floating point arithmetic in a way that the result of the floating point operation is closest to the mathematically correct result. Note that, in addition to overflow, we may now also experience underflow.

The discrepancy between mathematical arithmetic operations and their floating point counter parts has a tremendous impact: Suppose we develop an algorithm using real numbers by assuming the 'correct' mathematical arithmetic operations. Let us call this the 'theoretical' algorithm. When we show that such an algorithm is correct, we have *not at all* proven that its computer implementation, the 'machine' algorithm will give us correct results. And indeed, the machine algorithm will for almost certain give us a different result, because it works with a different arithmetic. During the past 50 years, error analysis of (floating point) machine algorithms has thus been a topic research subject in numerical analysis. Backward error analysis [14,23,24] has proven very useful in this context. The idea is to interpret the output of the machine algorithm as the result of the theoretical algorithm with a different input and to establish bounds on the difference of the respective inputs. However, when trying to prove a theorem in analysis with the help of a computer, a backward analysis is rarely helpful because we usually need information about the computed result with respect to the original input. This is the place where interval arithmetic comes into play.

'Correct' interval arithmetic is the set theoretic extension of the arithmetic on the reals to compact intervals, see [3]. As such it also suffers from the fact that it cannot be exactly matched onto a floating point counter part. But through the judicious use of directed roundings on floating point operations with the end points of the intervals we obtain a *machine interval arithmetic* working with intervals whose end points are floating point numbers. Operations with (non interval) floating point numbers are included in this concept by viewing floating point numbers as machine intervals with identical end points. Machine interval arithmetic exhibits the very crucial property of *containment* : if $\mathbf{a}$ and $\mathbf{b}$ are two intervals and $\circ \in \{+, -, *, /\}$ is an interval arithmetic operation, then the machine arithmetic counter part $\diamondsuit$ will give a result $\mathbf{a} \diamondsuit \mathbf{b}$ which satisfies

$$\mathbf{a} \diamondsuit \mathbf{b} \supseteq \mathbf{a} \circ \mathbf{b}.$$

With the rounding modes available in the IEEE Standard 754 , the result $\mathbf{a} \diamondsuit \mathbf{b}$ can in addition be made closest possible to $\mathbf{a} \circ \mathbf{b}$, see [16] or [3, Ch. 4].

We again distinguish between the theoretical algorithm and the machine algorithm. Due to the containment property, it is possible to identify crucial relations which, when observed for some output of the machine algorithm, necessarily also hold for the corresponding output of the theoretical algorithm. The two most important ones are:

a) $\mathbf{b} \subseteq \mathbf{a}$. Here, the (machine) interval $\mathbf{a}$ is an *a priori* given, fixed input whereas the the interval $\mathbf{b}$ is an output, i.e. a quantity computed by the algorithm,

b) $\mathbf{b} \cap \mathbf{c} = \emptyset$, where both intervals $\mathbf{b}$ and $\mathbf{c}$ are outputs.

Note that b) means that all elements of one of the two intervals are strictly less than all elements of the other interval. Therefore, b) can be interpreted to contain the case where we wish to establish strict inequalities between numbers.

A computer assisted proof in analysis, using interval arithmetic, can now be worked out in the following manner:

1. Develop a theoretical interval arithmetic algorithm which checks one or several relations of type a) and/or b) for certain outputs. The algorithm must be such that the conjecture to prove is true if all these checks return 'true'.
2. Run the machine algorithm.
3. If all checks in the machine algorithm return 'true', the conjecture is proved.

Enclosure methods using interval arithmetic work this way. Let $f : \mathbb{R}^n \to \mathbb{R}^n$ be some function of which we wish to find a zero. Then the mathematical algorithm would, for example, be based on a fixed point operator $H : \mathbb{R}^n \to \mathbb{R}^n$ whose fixed points are known to be the zeros of f. Assume that H is continuous and that we know an enclosure function $\mathbf{H}$ for H, i.e. a function based on 'correct' interval arithmetic which gives an interval vector $\mathbf{H}(\mathbf{x})$ containing the range of H over a given interval vector $\mathbf{x}$. Then, if $\mathbf{H}(\mathbf{x}) \subseteq \mathbf{x}$ (this is relation a) above) we know that $H(\mathbf{x}) \subseteq \mathbf{x}$ and so H has a fixed point in $\mathbf{x}$ by Brouwer's theorem.

The above procedure is certainly a very – if not the most – important application of interval arithmetic in practical computations. One then usually focuses on making $\mathbf{x}$ as small as possible in order to get very tight bounds for the zero. But note that we have just described a general procedure which one can use to prove lots of conjectures, each conjecture saying that a certain function f has a zero z in some interval vector $\mathbf{x}$. For example, Moré in [19] conjectured that a nonlinear function arising in the modelling of chemical destillation columns (the 'methanol-8-problem') does have a zero, but was not able to prove it using analytical tools. A computer assisted proof for this conjecture was given by Alefeld, Gienger and Potra [2] in 1994 using the method just described.

Relation b) above is crucial to verifying branch and bound methods in global optimization [15]. Here, the bounding step checks strict inequalities between the range of the objective function over a given interval vector and a candidate optimum.

Machine interval arithmetic has been used in lots of other computer assisted proofs in analysis. Some of these results have even been enthusiastically welcomed by the scientific community, since they solved problems which have remained open for some time. In the following pages we describe some of these *conjectures* and how machine interval arithmetic was used to solve them. In these descriptions we make a distinction between what we call the *paper work*, i.e. the classical, theoretical part of the proof written down on paper and the *computer work*, i.e. the algorithmic, computer assisted part involving machine interval arithmetic.

2   The Kepler Conjecture

This conjecture is among the most famous and oldest mathematical conjectures. It deals with the problem of finding a densest packing for equally sized balls in three-dimensional space. It appears in a book by the astronomer Johannes Kepler in 1611 and even found its way into Hilbert's 18th problem.

Conjecture 1 (Kepler Conjecture). The face-centered cubic packing is the densest packing of equally sized balls.

The face-centered cubic packing is a grid based regular arrangement which one gets, for example, when piling oranges on a market stall. Figure 1 shows this arrangement.

Fig. 1. the face-centered cubic packing

In the last ten years there was some controversy whether the Kepler conjecture has actually been proved. The work by Hsiang, who claimed a proof in 1993 has been severely criticized by the mathematical community – this even found its way into the popular literature [22]. At the time of writing of this note, a complete, accepted proof has not yet been published as a reviewed journal article. But the work of Hales and Ferguson, as explained on Hale's web site [11], was presented at several conferences and is now about to be accepted as a proof of Kepler's conjecture. Our summary here is based on [1,10,11].

2.1   The Paper Work

The ratio of space filled with balls in the face-centered cubic packing is $\pi/\sqrt{18}$. So proving Kepler's conjecture means solving an optimization problem on the space of all possible arrangements of spheres with a known value for the maximum. The major step making this optimization problem (in principle) tractable on a computer was to show that it can be reduced to a problem involving only a finite number of variables which parametrize graphs describing arrangements of spheres. Another important ingredient for a proof is to find a particularly well suited such parametrization. A major breakthrough here were Hale's 'hybrid decompositions' mixing Voronoi cells and Delaunay tetrahedrons. Hale has then identified a five step program to achieve the proof of Kepler's conjecture. The next paragraph sketches the computer assisted work in that program.

2.2   The Computer Work

Hybrid decompositions correspond to graphs. A first computer assisted part of the proof (which does not use interval arithmetic) is to classify all possible graphs into a finite number of classes, which turned out to be approximately 5000. For each such class there is a 'scoring function' and Kepler's conjecture is proved if the score does not exceed 8. Each score is a nonlinear optimization problem in up to 150 variables with up to 2000 linear restrictions. The optimization problems are solved with variants of the simplex algorithm.

Machine interval arithmetic is used when establishing the linear restrictions. A flavour of the tremendous amount of work involved can be obtained from [10] which deals with just one (but the very most difficult) class of graphs, the pentagonal prisms. In this work, machine interval arithmetic was for example used to show that a certain polynomial in six variables is positive over a given interval vector. This allowed to reduce the number of variables in certain situations, thus shrinking the subsequent computational effort. The most important use of interval arithmetic was made to establish a lot of those linear inequality constraints involving amazingly 'odd' constants. Establishing these inequalities means enclosing the range of a six-variable function defined through a composition of arithmetic operations, the square root and the arctan. Second degree Taylor formulas were sometimes needed to establish the desired inequalities, since standard 'naive' interval arithmetic would have required far too many subdivisions of the initial six-dimensional interval vector. The final result established a score < 7.9997 for the pentagonal prism.

3   The Double Bubble Conjecture

All of us know – even if this is not so easy to prove – that the sphere is the surface of smallest area enclosing a given volume (in 3d). Assume now that we have *two* volumes of equal size, and ask again for the surface of smallest area enclosing both these volumes. Let us call such a surface a *minimizing* surface. Just taking two spheres will not yield a minimizing surface, because constellations where a part of the surface is a bounding surface for both volumes will have smaller area. A *double bubble* enclosing two volumes of equal size is a surface made up of two pieces of equally sized spheres and one flat circle, meeting at an angle of 120°. See Fig. 2 which we reproduce by kind permission of John Sullivan from the University of Illinois at Urbana-Champaign. The double bubble conjecture can be traced back to at least the year 1911, where this conjecture was formulated based on physical experiments, see [7].

Conjecture 2 (Double Bubble Conjecture). The double bubble is the surface of smallest area enclosing two equal volumes. If the area is a and v either volume, then $a^3 = 243\pi v^2$.

The proof of this conjecture was announced by Hass, Hutchings and Schlafly [12] in 1995. The complete paper recently appeared as [13] and our presentation is based on that paper.

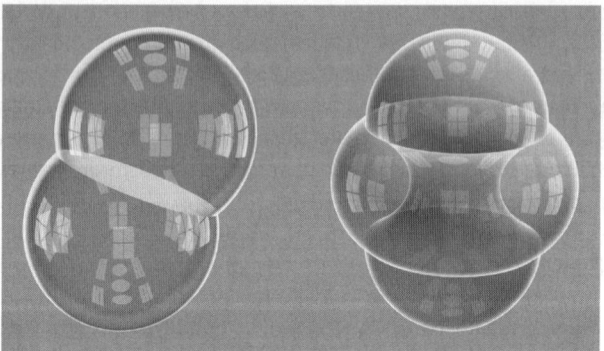

Fig. 2. double bubble and torus bubble

3.1   The Paper Work

One first has to find categories of surfaces which are known to contain a minimizing one. Classes for which this can be shown are very large and general, but one can then show that it is sufficient to consider piecewise smooth two-dimensional surfaces. In this class, a *bubble* can be rigorously defined as follows: (i) each smooth part has constant mean curvature, (ii) the singular sets are either smooth triple curves along which smooth surfaces meet at an angle of 120° or isolated vertices to which six smooth surfaces and four triple curves converge at fixed angles and (iii) the mean curvatures around an edge where three smooth surfaces have common boundary sum up to 0.

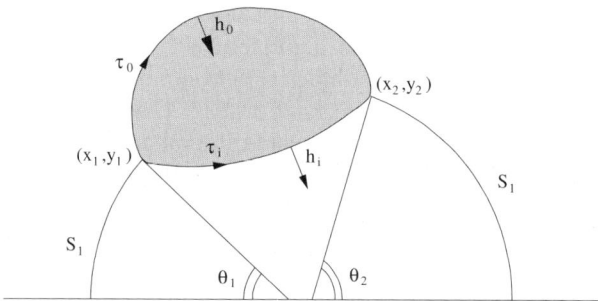

Fig. 3. cross section of a torus bubble

An important, deep result is that there exists a minimizing surface which is a *bubble*. An additional result shows that the set of possible minimizing surfaces can be further reduced to bubbles which are surfaces of revolution. Finally, it can be shown that the minimizing surface is either the double bubble or a so-called *torus bubble* (see Fig. 2). The cross section of a torus bubble (as given in Fig. 3, which is rotated by 90° compared to Fig. 2) consists of four smooth curves. Two of them, S_1 and S_2 are segments of circles. In between, we have two other curves τ_i and τ_o which, upon rotation, yield two smooth surfaces of constant mean curvature h_i and h_o, an inner and an outer one. These two surfaces enclose a volume which is topologically equivalent to a torus (grey area in Fig. 3). The other volume of the torus bubble is enclosed by the surfaces generated by S_1, S_2 and τ_i, and it is topologically equivalent to a sphere. The Delaunay curves τ_i and τ_o are completely specified by their starting point (x_1, y_1), the slope at the starting point and their mean curvature. But the slopes are defined by θ_1 since the curves have to meet at angles of 120° in (x_1, y_1). Moreover, since the mean curvatures have to add up to 0 in (x_1, y_1), we have an additional linear equality constraint relating h_o and h_i. Finally, for a minimizing torus bubble it can be shown that the radii of S_1 and S_2 have to be equal. Therefore, upon a scaling, possible minimizing torus bubbles are parametrized by just two parameters θ_1 and h_o. Hass and Schlafly showed, that we can even restrict the parameter ranges to $0 \leq \theta_1 \leq 90°$ and $0 \leq h_o \leq 10$.

3.2   The Computer Work

The algorithmic part of the proof for the double bubble conjecture consists in checking that no parameter pair (θ_1, h_o) in the interval vector $R = ([0, 90], [0, 10])$ gives us a minimizing bubble. This is a genuine interval arithmetic problem, since we have to devise an algorithm which treats *ranges* for the parameters as a whole. One will therefore subdivide R into finitely many interval vectors $\mathbf{x}$ and then show, using interval arithmetic, that $\mathbf{x}$ cannot contain a minimizing bubble. Interestingly, the algorithm in [13] does *not* rely on computing areas and volumes and then compare the ratio with that of the double bubble. Rather, the algorithm is based upon a list of other sufficient conditions for a torus bubble to *not* be minimizing. Each such sufficient condition boils down to checking a strict inequality between certain functions of θ_1 and h_o. The algorithm uses interval arithmetic enclosures for these functions and then checks whether, for a given input $\mathbf{x}$, we get strict inequality for the corresponding output intervals for at least one such function. If the procedure fails for all functions, $\mathbf{x}$ is further subdivided, and the same procedure is repeated, recursively. We cannot discuss all the sufficient conditions considered in [13] here, but as an example let us mention that one condition is to compute the volume of the torus part and that of the ball part and to check that these are not equal. This condition is very useful, because the parameter range R contains lots of torus bubbles which enclose two different volumes. The algorithm implemented avoids the use of π (since π is not a floating point number) as well as the use of trigonometric functions. The latter is achieved by taking $\cos \theta_1$ as an input parameter rather than θ_1 itself. The only non arithmetic machine interval operation necessary in the algorithm is then the interval square root. Another ingredient of the algorithm are (simple) enclosure functions for functions defined as integrals, needed to compute volumes and to determine the position of (x_2, y_2). To this purpose, the authors use interval arithmetic evaluations of Riemann sums. Division of the integration interval into 32 parts was always sufficient.

In total, 15 016 subintervals of R were considered involving the computation of more than 50 000 integrals. The computing time was about 10 seconds on a fast (1999) PC. The implementation was done in C++ and the algorithm was run on a variety of different hardware platforms.

4   The Dirac-Schwinger Conjecture

The Dirac-Schwinger conjecture arises in quantum mechanics. It owes its name to the fact that it further refines the asymptotics of a formula first stipulated by Dirac and then improved by Schwinger. The conjecture concerns the ground-state energy of an atom in the quantum mechanical, non-relativistic model. An atom with N electrons and a nucleus charge of Z has a ground-state energy $E(N, Z)$ which is the smallest eigenvalue of the associated Hamiltonian. The ground-state energy of the atom itself is

$$E(Z) = \inf_{N \geq 1} E(N, Z) .$$

The conjecture is on the asymptotic behaviour of $E(Z)$ as $Z \to \infty$.

Conjecture 3 (Dirac-Schwinger conjecture). For $Z \to \infty$ we have

$$E(Z) = -c_0 Z^{7/3} + \frac{1}{8} Z^2 - c_1 Z^{5/3} + o(Z^{5/3})$$

with constants c_0, c_1. Here $o(Z^{5/3})$ denotes a quantity which when divided by $Z^{5/3}$ tends to 0 as $Z \to \infty$.

The importance of the Dirac-Schwinger conjecture resides in the fact that it improves by 'one order' the 'Scott conjecture' $E(Z) = -c_0 Z^{7/3} + \frac{1}{8} Z^2 + \mathcal{O}(Z^\gamma), \gamma < 2$, proved (without computer assistance) in the late 80s.

The Dirac-Schwinger conjecture appeared in a paper by Schwinger in 1981 [21]; it was proved by Fefferman and Seco in a series of eight papers published in 1994/95. Their computer assisted proof even gives a more quantitative result for the $o(Z^{5/3})$-term by replacing it by $\mathcal{O}(Z^{5/3-\varepsilon_0})$ with $\varepsilon_0 = \frac{1}{2835}$. Our presentation here is based on the paper [9] by Fefferman and Seco.

4.1   The Paper Work

The asymptotic extension stated in the conjecture is not at all proved directly. It follows – after quite a bit of mathematics – from a strong inequality to be fulfilled for the second derivative F'' given below. Define $y : [0, \infty) \to \mathbb{R}$ as the solution to the boundary value problem

$$y''(x) = x^{-1/2} y^{3/2}(x), \quad y(0) = 1, \quad y(\infty) = 0$$

with $-y(x)/x$ being the 'Thomas-Fermi potential' . Then let

$$F : (0, \Omega_c) \to \mathbb{R}, \quad F(\Omega) = \int_0^\infty \left(\frac{y(x)}{x} + \frac{\Omega^2}{x^2} \right)^{1/2} dx$$

with Ω_c^2 being the unique maximum of $x \cdot y(x)$ on $(0, \infty)$. The crucial inequality to establish is

$$F''(\Omega) \leq c < 0 \quad \text{for all } \Omega \in (0, \Omega_c).$$

This inequality has an important physical interpretation related to zero-energy orbits for a certain Hamiltonian. From that, the Dirac-Schwinger conjecture can be deduced.

4.2   The Computer Work

The computational work breaks into two parts. The first part is to compute a tight enclosure for the Thomas-Fermi function y. The second part is then to use this information on y to rigorously verify $F''(\Omega) < 0$.

As it is usual in enclosure methods for ODEs, enclosures for the solution are given by a finte number of (interval) parameters on a sequence of intervals which add up to the whole time domain. On each such interval, the enclosure can be viewed as a polynomial with interval coefficients and a remainder term. Let us first consider initial value problems. Then existence of a solution within an enclosure is shown by computationally verifying that the integral operator representing the ode is contracting. To this purpose, machine interval arithmetic computations are used. As opposed to what may now be considered the standard approach (see the program AWA [18], e.g.), [9] explictly computes (and uses) a bound on the Lipschitz constant of the integral operator. Note also that due to the singularities at 0 and ∞, one has to take special care of 'end' intervals $(0, r)$ and (R, ∞). The solution to the boundary value problem is finally obtained by the shooting technique . Here, shooting is applied from both end points, 0 and ∞, so that taking intersections will give improved enclosures on any chosen time point. As an interesting detail, let us mention that when shooting from the left end point with an initial slope that is too small, the solution will *not* exist on the whole time interval but reach a finite time point at which it vanishes. This produces difficulties in the algorithm computing the enclosures. This can be fixed, however, by using a criterion which predicts this situation at an earlier time point.

With their algorithm, Fefferman and Seco succeed in rigorously computing the first 11 decimal digits of Ω_c and of the time point r_c where this maximum occurs.

To finally prove $F''(\Omega) < c \le 0$ for $\Omega \in (0, \Omega_c)$ one needs an enclosure function for F'' and then evaluates this enclosure function on a a set of subintervals covering $(0, \Omega_c)$. It can be shown that the following representation, using $u(x) = xy(x)$ holds for F'':

$$F''(\Omega) = -\lim_{\delta \to \infty} \left(\int_{r_1(\Omega)+\delta}^{r_2(\Omega)-\delta} \left(u(x) - \Omega^2\right)^{-3/2} y(x)dx + c(\Omega)\delta^{-1/2} \right).$$

Here, $r_1(\Omega), r_2(\Omega)$ are the two solutions to $u(r) = \Omega^2$ and $c(\Omega)$ is uniquely specified by requiring the limit to be finite. Split the domain of integration into the three parts $[r_1(\Omega) + \delta, a], [a, b]$ and $[b, r_2(\Omega) - \delta]$. The part $[a, b]$ is not close to poles of the integrand, so its value can be enclosed (since we already have an enclosure for u) by machine interval arithmetic computation of Riemann sums. The intervals $[r_1(\Omega) + \delta, a]$ and $[b, r_2(\Omega) - \delta]$ approach the poles when $\delta \to 0$. But we can compute the enclosure for $u(x)$ in such a manner that $u(x) - \Omega^2$ is (at least piecewise) given as a polynomial with interval coeffcients. This can be converted into a similar representation of an enclosure for $\left(u(x) - \Omega^2\right)^{-3/2}$. It is then possible to directly determine an enclosure for the integral which does not depend on δ (but on a and b, resp.). Of course, it is crucial that the whole process described will not severely overestimate the range of F'', so that the right choice of a and b as well as some delicacy in computing the Riemann sums is necessary. Moreover, for values of Ω close to 0 or Ω_c additional difficulties arise. Here, a change of variables has to be done in the representation of F''.

5 'Chaos conjectures'

This survey would be incomplete without a section on how machine interval arithmetic computations have been and are used in computer assisted proofs of chaos in dynamical systems. Maybe such results have not really been formulated as 'conjectures' in the classical sense. But often, evidence for chaos was and is first obtained by numerical integration of the system. Since these computations do not have proof quality, we can therefore regard them as conjectures as well. We will describe three pieces of work where interval arithmetic was used to turn such conjectures into theorems.

The first dates back to 1982 , where Lanford III [17] proved what he called the 'Feigenbaum conjectures' on some universal features displayed by infinite sequences of period doubling bifurcations. Among other things, Lanford proves the first seven digits in the Feigenbaum constant $\mu_\infty = 1.401155\ldots$ to be correct. This constant arises when studying the family of logistic mappings $\psi_\mu(x) = 1 - \mu x^2$ and μ_∞ is the smallest positive number such that the orbits of ψ_μ show chaotic behaviour. In his proof, Lanford computes rigorous upper bounds on the l_1-norms of certain linear operators on infinite dimensional spaces. Interval arithmetic is used as a tool for automated forward error analysis, i.e. all operations on the reals are replaced by machine interval arithmetic operations. The resulting interval is then guaranteed to contain the 'correct' result, and the right end point of the interval is a guaranteed upper bound.

Our second sample is by Neumaier and Rage [20] from 1993. The purpose of their work is to prove the existence of (and to accurately compute) a transversal homoclinic point for a given dynamical system with diffeomorphism F. They proceed as follows: First, they compute guaranteed enclosures for a hyperbolic fixed point of F. Then they compute enclosures for the stable and unstable manifold of this fixed point. These are iterated to get enclosures of a 'candidate' homoclinic point. Finally, they check for transversal intersection within this enclosure in a way that also proves existence of a homoclinic point. For the above steps to work one needs 'constructive', i.e. more than just local theoretical results whose hypotheses can be checked using machine interval arithmetic . Such results are developed in [20]. The computation of an enclosure of a fixed point of F is a fairly standard interval problem, see [3], e.g. The enclosures for the stable and unstable manifolds proposed in [20] are convex hulls of parallel, lower-dimensional interval vectors and as such amenable to computation with (machine) interval arithmetic. The conditions to check that these enclosures actually do contain the respective manifolds require to check an inequality on norms of certain matrices and that a computed interval vector is contained within another one. The existence test for the homoclinic point requires to compute an enclosure of the solution set of a linear system with interval coefficients and to then check an inequality on certain components of that enclosure. As an example, Neumaier and Rage prove the existence of a homoclinic point for the standard map

$$F(p,q) = \left(q + p - \frac{k}{2\pi} \sin(2\pi q) \bmod 1, \ p - \frac{k}{2\pi} \sin(2\pi q) \bmod 1 \right)$$

for two different values of k and compute the homoclinic point with a guaranteed accuracy of up to 10 decimal digits.

We end by briefly describing the results of Camacho and de Figueiredo [8] on the dynamics of the Jouanolou foliation. Without even explaining terminology, suffice it to say that in this article the authors prove that four interesting foliations do not have non-trivial minimal sets. The open problem is whether there are at all foliations having non-trivial minimal sets. In the proof of their results the authors use machine interval arithmetic in two major steps. After having identified two small sectors in the complex projective plane which a minimal set has to cross (this is the 'paper work'), they find a sphere centered at a singular point which is transversal to the foliation. The radius of this sphere is determined in a computer assisted proof. It has to be such that the normal to the sphere is nowhere parallel to the field defined by the foliation. This can be formulated as proving that the global maximum of some function (depending on four real parameters) does not reach 0. To get an upper bound for the maximum the authors use a machine interval arithmetic branch and bound algorithm as described in [15], e.g. Several such optimization problems are solved to get a sphere with largest radius possible. The last step is to follow all possible trajectories starting in the two sectors identified in the first step and show that those intersect the sphere. To achieve this, the sectors are covered with interval vectors $\mathbf{x}$. Then, the machine interval arithmetic based ODE solver AWA of Lohner [18] is used to enclose all trajectories starting in $\mathbf{x}$. If at some time point the enclosure is contained in the sphere, the interval $\mathbf{x}$ is deleted from the computing list, otherwise it is bisected and its two parts are again stored in the list. The computational proof is then achieved if the list has become empty, which eventually happened for all four foliations considered in [8].

References

1. M. Aigner and E. Behrends. *Alles Mathematik*. Vieweg, Braunschweig, 2000.
2. G. Alefeld, A. Gienger, and F. A. Potra. Efficient numerical validation of solutions of nonlinear systems. *SIAM J. Numer. Anal.*, 31:252–260, 1994.

3. G. Alefeld and J. Herzberger. *Introduction to Interval Computations*. Academic Press, New York, 1983.
4. American National Standard Institute. IEEE standard for binary floating point arithmetic IEEE/ANSI 754-1985. Technical report, New York, 1985.
5. K. Appel and W. Haken. Every planar graph is four colorable, part I: Discharging. *Illinois J. of Mathematics*, 21:429–490, 1977.
6. K. Appel and W. Haken. Every planar graph is four colorable, part II: Reducibility. *Illinois J. of Mathematics*, 21:491–567, 1977.
7. C. V. Boys. *Soap Bubbles*. Dover Publ. Inc., New York, 1959. (first edition 1911).
8. C. Camacho and L. de Figueiredo. The dynamics of the Jouanolou foliation on the complex projective 2-space. *Ergodic Theory Dyn. Sys.*, to appear.
9. C. L. Fefferman and L. A. Seco. Interval arithmetic in quantum mechanics. In Kearfott, B. R. et al., editor, *Applications of Interval Computations*, volume 3 of *Appl. Optim.*, pages 145–167. Kluwer, Dordrecht, 1995. Proceedings of an international workshop.
10. S. P. Ferguson. *Sphere Packings, V.* PhD thesis, Department of Mathematics, University of Michigan, 1997.
11. J. Hales. The Kepler conjecture. Technical report, 1998. `http://www.math.lsa.umich.edu/~hales/countdown/`.
12. J. Hass, M. Hutchings, and R. Schlafly. The double bubble conjecture. *Electron. Res. Announc. Am. Math. Soc.*, 1:98–102, 1995.
13. J. Hass and R. Schlafly. Double bubbles minimize. *Ann. Math. (2)*, 151:459–515, 2000.
14. N. J. Higham. *Accuracy and Stability of Numerical Algorithms*. SIAM, Philadelphia, 1996.
15. R. B. Kearfott. *Rigorous Global Search*. Kluwer Academic Publishers, 1996.
16. U. Kulisch and W. Miranker. *Computer Arithmetic in Theory and Practice*. Academic Press, New York, 1981.
17. O. E. Lanford III. A computer-assisted proof of the feigenbaum conjectures. *Bull. Am. Math. Soc. New Ser.*, 6:427–434, 1982.
18. R. Lohner. AWA: Software for the computation of guaranteed bounds for solutions of ordinary initial value problems. Technical report, Institut für Angewandte Mathematik, Universität Karlsruhe, 1994. Software available at `ftp://ftp.iam.uni-karlsruhe.de/pub/awa/`.
19. J. J. Moré. A collection of nonlinear model problems. In E. L. Allgower and K. Georg, editors, *Computational Solution of Nonlinear Systems of Equations*, volume 26 of *Lectures in Applied Mathematics*. American Mathematical Society, Providence, 1990.
20. A. Neumaier and T. Rage. Rigorous verification in discrete dynamical systems. *Physica D*, 67:327–346, 1993.
21. J. Schwinger. Thomas-Fermi model: The second correction. *Physical Review*, A24:2253–2361, 1981.
22. S. Singh. *Fermat's Enigma: The Quest to Solve the World's Greatest Mathematical Problem*. Walker & Company, 1997.
23. J. H. Wilkinson. *The Algebraic Eigenvalue Problem*. Oxford University Press, Oxford, 1965.
24. J. H. Wilkinson. *Rounding Errors in Algebraic Processes*. Dover, New York, 1994. Originally published as Notes on Applied Science No. 32, Her Majesty's Stationery Office, London, 1963.

Advanced Arithmetic for the Digital Computer – Interval Arithmetic Revisited

Ulrich W. Kulisch*

Universität Karlsruhe (TH), Institut für Angewandte Mathematik,
D-76128 Karlsruhe, Germany.
ulrich.kulisch@math.uni-karlsruhe.de

Abstract. This paper deals with interval arithmetic and interval mathematics. Interval mathematics has been developed to a high standard during the last few decades. It provides methods which deliver results with guarantees. However, the arithmetic available on existing processors makes these methods extremely slow. The paper reviews a number of basic methods and techniques of interval mathematics in order to derive and focus on those properties which by today's knowledge could effectively be supported by the computer's hardware, by basic software, and by the programming languages. The paper is not aiming for completeness. Unnecessary mathematical details, formalisms and derivations are left aside whenever possible. Particular emphasis is put on an efficient implementation of interval arithmetic on computers.

Interval arithmetic is introduced as a shorthand notation and automatic calculus to add, subtract, multiply, divide, and otherwise deal with inequalities. Interval operations are also interpreted as special powerset or set operations. The inclusion isotony and the inclusion property are central and important consequences of this property. The basic techniques for enclosing the range of function values by centered forms or by subdivision are discussed. The Interval Newton Method is developed as an always (globally) convergent technique to enclose zeros of functions.

Then extended interval arithmetic is introduced. It allows division by intervals that contain zero and is the basis for the development of the extended Interval Newton Method. This is the major tool for computing enclosures at all zeros of a function or of systems of functions in a given domain. It is also the basic ingredient for many other important applications like global optimization, subdivision in higher dimensional cases or for computing error bounds for the remainder term of definite integrals in more than one variable. We also sketch the techniques of differentiation arithmetic, sometimes called automatic differentiation, for the computation of enclosures of derivatives, of Taylor coefficients, of gradients, of Jacobian or Hessian matrices.

The major final part of the paper is devoted to the question of how interval arithmetic can effectively be provided on computers. This is an essential prerequisite for its superior and fascinating properties to be more widely used in the scientific computing community. With more appropriate processors, rigorous methods based on interval arithmetic could be comparable in speed with today's "approximate" methods. At processor speeds of gigaFLOPS there remains no alternative

---

* This Article was prepared while the author was staying at the Electrotechnical Laboratory, Agency of Industrial Science and Technology, MITI, at Tsukuba, Ibaraki 305-8568, Japan, November 1999 - March 2000.

but to furnish future computers with the capability to control the accuracy of a computation at least to a certain extent.

1   Introduction and Historical Remarks

In 1958 the Japanese mathematician Teruo Sunaga published a paper entitled "Theory of an Interval Algebra and its Application to Numerical Analysis" [61]. Sunaga's paper was intended to indicate a method of rigorous error estimation alternative to the methods and ideas developed in J. v. Neumann and H. H. Goldstine's paper on "Numerical Inverting of Matrices of High Order". [48]

Sunaga's paper is not the first one using interval arithmetic in numerical computing. However, several ideas which are standard techniques in interval mathematics today are for the first time mentioned there in rudimentary form. The structure of interval arithmetic is studied in Sunaga's paper. The possibility of enclosing the range of a rational function by interval arithmetic is discussed. The basic idea of what today is called the Interval Newton Method can be found there, and also the methods of obtaining rigorous bounds in the cases of numerical integration of definite integrals or of initial value problems of ordinary differential equations by evaluating the remainder term of the integration routine in interval arithmetic are indicated in Sunaga's paper. Under "Conclusion" Sunaga's paper ends with the statement "that a future problem will be to revise the structure of the automatic digital computer from the standpoint of interval calculus".

Today Interval Analysis or Interval Mathematics appears as a mature mathematical discipline. However, the last statement of Sunaga's paper still describes a "future problem". The present paper is intended to help close this gap.

This paper is supposed to provide an informal, easily readable introduction to basic features, properties and methods of interval arithmetic. In particular it is intended to deepen the understanding and clearly derive those properties of interval arithmetic which should be supported by computer hardware, by basic software, and by programming languages. The paper is not aiming for completeness. Unnecessary mathematical details, formalisms and derivations are put aside, whenever possible.

Interval mathematics has been developed to a high level during the last decades at only a few academic sites. Problem solving routines which deliver validated results are actually available for all the standard problems of numerical analysis. Many applications have been solved using these tools. Since all these solutions are mathematically proven to be correct, interval mathematics has occasionally been called the Mathematical Numerics in contrast to Numerical Mathematics, where results are sometimes merely speculative. Interval mathematics is not a trivial subject which can just be applied naively. It needs education, training and practice. The author is convinced that with the necessary skills interval arithmetic can be useful, and can be successfully applied to any serious scientific computing problem.

In spite of all its advantages it is a fact that interval arithmetic is not widely used in the scientific computing community as a whole. The author sees several reasons for this which should be discussed briefly. A broad understanding of these reasons is an essential prerequisite for further progress.

Forty years of nearly exclusive use of floating-point arithmetic in scientific computing has formed and now dominates our thinking. Interval arithmetic requires a much higher level of abstraction than languages like Fortran-77, Pascal or C provide. If every single interval operation requires a procedure call, the user's energy and attention are forced down to the level of coding, and are dissipated there.

The development and implementation of adequate and powerful programming environments like PASCAL-XSC [26,27,17] or ARITH-XSC [76] requires a large body of experienced and devoted scientists (about 20 man years for each) which is not easy to muster. In such environments interval arithmetic, the elementary functions for the data types real and interval, a long real and a long real interval arithmetic including the corresponding elementary functions, vector and matrix arithmetic, differentiation and Taylor arithmetic both for real and interval data are provided by the run time system of the compiler. All operations can be called by the usual mathematical operator symbols and are of maximum accuracy. This releases the user from coding drudgery. This means, for instance, that an enclosure of a high derivative of a function over an interval — needed for step size control and to guarantee the value of a definite integral or a differential equation within close bounds — can be computed by the same notation used to compute the real function value. The compiler interprets the operators according to the type specification of the data. This level of programming is essential indeed. It opens a new era of conceptual thinking for mathematical numerics.

A second reason for the low acceptance of interval arithmetic in the scientific computing community is simply the prejudices which are often the result of superficial experiments. Sentences like the following appear again and again in the literature: "The error bounds are overly conservative; they quickly grow to the computer representation of $[-\infty, +\infty]$", "Interval arithmetic is expensive because it takes twice the storage and at least twice the work of ordinary arithmetic."

Such sentences are correct for what is called "naive interval arithmetic". Interval arithmetic, however, should not be applied naively. Its properties must be studied and understood first, before it can be applied successfully. Many program packages have been developed using interval arithmetic, which deliver close bounds for their solutions. In no case are these bounds obtained by substituting intervals in a conventional floating-point algorithm. Interval arithmetic is an extension of floating-point arithmetic, not a replacement for it. Sophisticated use of interval arithmetic often leads to safe and better results. There are many applications where the extended tool delivers a guaranteed answer faster than the restricted tool of floating-point arithmetic delivers an "approximation". Examples are numerical integration (because of automatic step size control) and global optimization (intervals bring the continuum on the computer). One interval evaluation of a function over an interval may suffice to prove that the function definitively has no zero in that interval, while 1000 floating-point evaluations of the function in the interval could not provide a safe answer. Interval methods that have been developed for systems of ordinary differential and integral equations may be a bit slower. But they deliver not just unproven numbers. Interval methods deliver close bounds and prove existence and uniqueness of the solution within the computed bounds. The bounds include both discretization and rounding errors. This can save a lot of computing time by avoiding experimental reruns.

The main reason why interval methods are sometimes slow is already expressed in the last statement of Sunaga's early article. It's not that the methods are slow. It is the missing hardware support which makes them slow. While conventional floating-point arithmetic nowadays is provided by fast hardware, interval arithmetic has to be simulated by software routines based on integer arithmetic. The IEEE arithmetic standard, adopted in 1985, seems to support interval arithmetic. It requires the basic four arithmetic operations with rounding to nearest, towards zero, and with rounding downwards and upwards. The latter two are needed for interval arithmetic. But IEEE arithmetic separates the rounding from the operation, which proves to be a severe drawback. In a conventional floating-point computation this does not cause any difficulties. The rounding mode is set only once. Then a large number of operations is performed with this rounding mode. However, when interval arithmetic is performed the rounding mode has to be switched very frequently. The lower bound of the result of every interval operation has to be rounded downwards and the upper bound rounded upwards. Thus, the rounding mode has to be reset for every arithmetic operation. If setting the rounding mode and the arithmetic operation are equally fast this slows down interval arithmetic unnecessarily by a factor of two in comparison to conventional floating-point arithmetic. On all existing commercial processors, however, setting the rounding mode takes a multiple (three, ten, twenty and even more) of the time that is needed for the arithmetic operation. Thus an interval operation is unnecessarily at least eight (or twenty and even more) times slower than the corresponding floating-point operation. The rounding should be part of the arithmetic operation as required by the theory of computer arithmetic [33,34]. Every one of the rounded operations $\boxdot, \triangledown, \triangle, \circ \in \{+, -, *, /\}$ with rounding to nearest, downwards or upwards should be equally fast and executed in a single cycle.

The IEEE arithmetic standard requires that these 12 operations for floating-point numbers give computed results that coincide with the rounded exact result of the operation for any operands [77]. The standard was developed around 1980 as a standard for microprocessors at a time when the typical microprocessor was the 8086 running at 2 MHz and serving a memory space of 64 KB. Since that time the speed of microprocessors has been increased by a factor of more than 1000. IEEE arithmetic is now even provided by supercomputers, the speed of which is still faster by magnitudes. Advances in computer technology are now so profound that the arithmetic capability and repertoire of computers can and should be expanded. In contrast to IEEE arithmetic a general theory of advanced computer arithmetic requires that all arithmetic operations in the usual product spaces of computation: the complex numbers, real and complex vectors, real and complex matrices, real and complex intervals as well as real and complex interval vectors and interval matrices are provided on the computer by a general mathematical mapping principle which is called a semimorphism. For definition see [33,34]. This guarantees, among other things, that all arithmetic operations in all these spaces deliver a computed result which differs from the exact result of the operation by (no or) only a single rounding.

A careful analysis within the theory of computer arithmetic shows that the arithmetic operations in the computer representable subsets of these spaces can be realized on the computer by a modular technique provided fifteen fundamental operations are made available on a low level, possibly by fast hardware routines. These fifteen operations are

$$\boxplus, \ \boxminus, \ \boxtimes, \ \boxslash, \ \boxdot,$$
$$\triangledown, \ \triangledown, \ \triangledown, \ \triangledown, \ \triangledown,$$
$$\triangle, \ \triangle, \ \triangle, \ \triangle, \ \triangle.$$

Here $\boxdot, \circ \in \{+, -, *, /\}$ denotes operations using a monotone and antisymmetric rounding $\square$ from the real numbers onto the subset of floating-point numbers, such as rounding to the nearest floating-point number. Likewise $\triangledown$ and $\triangle$, $\circ \in \{+, -, *, /\}$ denote the operations using the monotone rounding downwards $\triangledown$ and upwards $\triangle$ respectively. $\square, \triangledown$ and $\triangle$ denote scalar products with only a single rounding. That is, if $a = (a_i)$ and $b = (b_i)$ are vectors with floating-point components a_i, b_i, then $a \bigcirc b := \bigcirc(a_1 * b_1 + a_2 * b_2 + \ldots + a_n * b_n)$, $\bigcirc \in \{\square, \triangledown, \triangle\}$. The multiplication and addition signs on the right hand side of the assignment denote exact multiplication and summation in the sense of real numbers.

Of these 15 fundamental operations above, traditional numerical methods use only the four operations $\boxplus, \boxminus, \boxtimes$ and $\boxslash$. Conventional interval arithmetic employs the eight operations $\triangledown, \triangledown, \triangledown, \triangledown$ and $\triangle, \triangle, \triangle, \triangle$. These eight operations are computer equivalents of the operations for real intervals; they provide interval arithmetic. The IEEE arithmetic standard requires 12 of these 15 fundamental operations: $\boxdot, \triangledown, \triangle$, $\circ \in \{+, -, *, /\}$. Generally speaking, interval arithmetic brings guarantees into computation, while the three scalar products $\square, \triangledown$ and $\triangle$ bring high accuracy.

A detailed discussion of the implementation of the three scalar products on all kinds of computers is given in [40]. Basically the products $a_i * b_i$ are accumulated in fixed-point arithmetic with or without a single rounding at the very end of the accumulation. In contrast to accumulation in floating-point arithmetic, fixed-point accumulation is error free. Apart from this important property it is simpler than accumulation in floating-point and it is even faster. Accumulations in floating-point arithmetic are very sensitive with respect to cancellation.

So accumulations should be done in fixed-point arithmetic whenever possible whether the data are integers, floating-point numbers or products of two floating-point numbers. An arithmetic operation which can always be performed correctly on a digital computer should not be simulated by a routine which can easily fail in critical situations. Many real life and expensive accidents have been attributed to loss of numeric accuracy in a floating-point calculation or to other arithmetic failures. Examples are: bursting of a large turbine under test due to wrongly predicted eigenvalues; failure of early space shuttle retriever arms under space conditions; disastrous homing failure on ground to air missile missions; software failure in the Ariane 5 guidance program.

Advanced computer arithmetic requires a correct implementation of all arithmetic operations in the usual product spaces of computations. This includes interval arithmetic and in particular the three scalar products $\square$, ∇ and $\triangle$. This confronts us with another severe slowdown of interval arithmetic.

All commercial processors that provide IEEE arithmetic only deliver a rounded product to the outside world in the case of multiplication. Computation of an accurate scalar product requires products of the full double length. So these products have to be simulated on the processor. This slows down the multiplication by a factor of up to 10 in comparison to a rounded hardware multiplication. In a software simulation of the accurate scalar product the products of double length then have to be accumulated in fixed-point mode. This process is again slower by a factor of about 5 in comparison to a possibly wrong hardware accumulation of the products in floating-point arithmetic. Thus in summary a factor of at least 50 is the penalty for an accurate computation of the scalar product on existing processors. This is too much to be readily accepted by the user. In contrast to this a hardware implementation of the optimal scalar product could even be faster than a conventional implementation in floating-point arithmetic [40].

Another severe shortcoming which makes interval arithmetic slow is the fact that no reasonable interface to the programming languages has been accepted by the standardization committees so far. Operator overloading is not adequate for calling all fifteen operations $\square$, ∇, $\triangle$, $\circ \in \{+, -, *, /, \cdot\}$, in a high level programming language. A general operator concept is necessary for ease of programming (three real operations for $+, -, *, /$ and the dot product with three different roundings) otherwise clumsy and slow function calls have to be used to call different rounded arithmetic operations.

All these factors which make interval arithmetic on existing processors slow are quite well known. Nevertheless, they are generally not taken into account when the speed of interval methods is judged. It is, however, important that these factors are well understood. Real progress depends critically on an understanding of their details. Interval methods are not slow per se. It is the actual available arithmetic on existing processors which makes them slow. With better processor and language support, rigorous methods could be comparable in speed to today's "approximate" methods. Interval mathematics or mathematical numerics has been developed to a level where already today library routines could speedily deliver validated bounds instead of just approximations for small and medium size problems. This would ease the life of many users dramatically.

Future computers must be equipped with fast and effective interval arithmetic. At processor speeds of gigaFLOPS it is almost the only way to check the accuracy of a computation. Computer-generated graphics requires validation techniques in many cases.

After Sunaga's early paper the publication of Ramon E. Moore's book on interval arithmetic in 1966 [44] certainly was another milestone in the development of interval arithmetic. Moore's book is full of unconventional ideas which were out of the mainstream of numerical analysis of that time. To many colleagues the book appeared as a utopian dream. Others tried to carry out his ideas with little success in general. Computers were very very slow at that time. Today Moore's book appears as an exposition of extraordinary intellectual and creative power. The basic ideas of a great many well established methods of validation numerics can be traced back to Moore's book.

We conclude this introduction with a brief sketch of the development of interval arithmetic at the author's institute. Already by 1967 an ALGOL-60 extension implemented on a Zuse Z 23 computer provided operators and a number of elementary functions for a new data type *interval* [68,69]. In 1968/69 this language was implemented on a more powerful computer, an Electrologica X8. To speed up the arithmetic, the hardware of the processor was extended by the four arithmetic operations with rounding downwards ∇, $\circ \in \{+, -, *, /\}$. Operations with rounding upwards were produced by use of the relation $\triangle(a) = -\nabla(-a)$. Many early interval methods have been developed using these tools. Based on this experience a book [5] was written by two collaborators of that time. The English translation which appeared in 1983 is still a standard monograph on interval arithmetic [6].

At about 1969 the author became aware that interval and floating-point arithmetic basically follow the same mathematical mapping principles, and can be subsumed by a general mathematical theory of what is called advanced computer arithmetic in this paper. The basic assumption is that all arithmetic operations on computers (for real and complex numbers, real and complex intervals as well as for vectors and matrices over these four basic data types) should be defined by four simple rules which are called a semimorphism. This guarantees the best possible answers for all these arithmetic operations. A book on the subject was published in 1976 [33] and the German company Nixdorf funded an implementation of the new arithmetic. At that time a Z-80 microprocessor with 64 KB main memory had to be used. The result was a PASCAL extension called PASCAL-SC, published in [37,38]. The language provides about 600 predefined arithmetic operations for all the data types mentioned above and a number of elementary functions for the data types real and interval. The programming convenience of PASCAL-SC allowed a small group of collaborators to implement a large number of problem solving routines with automatic result verification within a few months. All this work was exhibited at the Hannover fair in March 1980 with the result that Nixdorf donated a number of computers to the Universität Karlsruhe. This allowed the programming education at Universität Karlsruhe to be decentralized from the summer of 1980. PASCAL-SC was the proof that advanced computer arithmetic need not be restricted to the very large computers. It had been realized on a microprocessor. When the PC appeared on the scene in 1982 it looked poor compared with what we had already two years earlier. But the PASCAL-SC system was never marketed.

In 1978 an English version of the theoretical foundation of advanced computer arithmetic was prepared during a sabbatical of the author jointly with W. L. Miranker at the IBM Research Center at Yorktown Heights. It appeared as a book in 1981 [34].

In May 1980 IBM became aware of the decentralized programming education with PASCAL-SC at the Universität Karlsruhe. This was the beginning of nearly ten years of close cooperation with IBM. We jointly developed and implemented a Fortran extension corresponding to the PASCAL extension with a large number of problem solving routines with automatic result verification [74–76].

In 1980 IBM had only the /370 architecture on the market. So we had to work for this architecture. IBM supported the arithmetic on an early processor (4361 in 1983) by microcode and later by VLSI design. Everything we developed for IBM was offered on the market as IBM program products in several versions between 1983 and 1989. But the products did not sell in the quantities IBM had expected. During the 1980s scientific computing had moved from the old mainframes to workstations and supercomputers. So the final outcome of these wonderful products was the same as for all the other earlier attempts to establish interval arithmetic effectively. With the next processor generation or a new language standard work for a particular processor loses its attraction and its value.

Nevertheless all these developments have contributed to the high standard attained by interval mathematics or mathematical numerics today. What we have today is a new version of PASCAL-SC, called PASCAL-XSC [26,27,29], with fast elementary functions and a corresponding C++ extension called C-XSC [28]. Both languages are translated into C so that they can be used on nearly all platforms. The arithmetic is implemented in software in C with all the regrettable consequences with respect to speed discussed earlier. Toolbox publications with problem solving routines are available for both languages [17,18].

Of course, much valuable work on the subject had been done at other places as well. International Conferences where new results can be presented and discussed are held regularly.

After completion of this paper Sun Microsystems announced an interval extension of Fortran 95 [82]. With this new product and compiler, interval arithmetic is now available on computers which are wide spread.

Acknowledgement: I acknowledge with gratitude the support of the Electrotechnical Laboratory (ETL) at Tsukuba, Japan, and the Agency of Industrial Science and Technology, MITI. I am very grateful for the opportunity to prepare this article in a pleasant scientific atmosphere far away from the usual university business. In particular I would like to thank Satoshi Sekiguchi for being a wonderful host personally and scientifically.

I also wish to thank many former and present collaborators and students who helped clarify many details which may look obvious in this article. In particular I thank Gerd Bohlender and Neville Holmes for proofreading the paper.

As Teruo Sunaga did in 1958 and many others after him, I am looking forward to, expect, and eagerly await a revision of the structure of the digital computer for better support of interval arithmetic.

2   Interval Arithmetic, a Powerful Calculus to Deal with Inequalities

Problems in technology and science are often described by an equation or a system of equations. Mathematics is used to manipulate these equations in order to obtain a solution. The Gauss algorithm, for instance, is used to compute the solution of a system of linear equations by adding, subtracting, multiplying and dividing equations in a systematic manner. Newton's method is used to compute approximately the location of a zero of a non linear function or of a system of such functions.

Data are often given by bounds rather than by simple numbers. Bounds are expressed by inequalities. To compute bounds for problems derived from given data requires a systematic calculus to deal with inequalities. Interval arithmetic provides this calculus. It supplies the basic rules for how to add, subtract, multiply, divide, and manipulate inequalities in a systematic manner: Let bounds for two real numbers a and b be given by the inequalities

$a_1 \leq a \leq a_2$ and $b_1 \leq b \leq b_2$. Addition of these inequalities leads to bounds for the sum $a + b$:

$$a_1 + b_1 \leq a + b \leq a_2 + b_2.$$

The inequality for b can be reversed by multiplication with -1: $-b_2 \leq -b \leq -b_1$. Addition to the inequality for a then delivers the rule for the subtraction of two inequalities:

$$a_1 - b_2 \leq a - b \leq a_2 - b_1.$$

Interval arithmetic provides a shorthand notation for these rules suppressing the $\leq$ symbols. We simply identify the inequality $a_1 \leq a \leq a_2$ with the closed and bounded real interval $[a_1, a_2]$. The rules for addition and subtraction for two such intervals now read:

$$[a_1, a_2] + [b_1, b_2] = [a_1 + b_1, a_2 + b_2], \tag{1}$$

$$[a_1, a_2] - [b_1, b_2] = [a_1 - b_2, a_2 - b_1]. \tag{2}$$

The rule for multiplication of two intervals is more complicated. Nine cases are to be distinguished depending on whether a_1, a_2, b_1, b_2, are less or greater than zero. For division the situation is similar. Since we shall build upon these rules later they are cited here. For a detailed derivation see [33,34]. In the tables the order relation $\leq$ is used for intervals. It is defined by

$$[a_1, a_2] \leq [b_1, b_2] :\Longleftrightarrow a_1 \leq b_1 \wedge a_2 \leq b_2.$$

Table 1. The 9 cases for the multiplication of two intervals or inequalities

Nr.	$A = [a_1, a_2]$	$B = [b_1, b_2]$	$A * B$
1	$A \geq [0,0]$	$B \geq [0,0]$	$[a_1 b_1, a_2 b_2]$
2	$A \geq [0,0]$	$B \leq [0,0]$	$[a_2 b_1, a_1 b_2]$
3	$A \geq [0,0]$	$0 \in \overset{\circ}{B}$	$[a_2 b_1, a_2 b_2]$
4	$A \leq [0,0]$	$B \geq [0,0]$	$[a_1 b_2, a_2 b_1]$
5	$A \leq [0,0]$	$B \leq [0,0]$	$[a_2 b_2, a_1 b_1]$
6	$A \leq [0,0]$	$0 \in \overset{\circ}{B}$	$[a_1 b_2, a_1 b_1]$
7	$0 \in \overset{\circ}{A}$	$B \geq [0,0]$	$[a_1 b_2, a_2 b_2]$
8	$0 \in \overset{\circ}{A}$	$B \leq [0,0]$	$[a_2 b_1, a_1 b_1]$
9	$0 \in \overset{\circ}{A}$	$0 \in \overset{\circ}{B}$	$[\min(a_1 b_2, a_2 b_1), \max(a_1 b_1, a_2 b_2)]$

$$\tag{3}$$

Table 2. The 6 cases for the division of two intervals or inequalities

Nr.	$A = [a_1, a_2]$	$B = [b_1, b_2]$	A/B	
1	$A \geq [0, 0]$	$0 < b_1 \leq b_2$	$[a_1/b_2, a_2/b_1]$	
2	$A \geq [0, 0]$	$b_1 \leq b_2 < 0$	$[a_2/b_2, a_1/b_1]$	
3	$A \leq [0, 0]$	$0 < b_1 \leq b_2$	$[a_1/b_1, a_2/b_2]$	(4)
4	$A \leq [0, 0]$	$b_1 \leq b_2 < 0$	$[a_2/b_1, a_1/b_2]$	
5	$0 \in \overset{\circ}{A}$	$0 < b_1 \leq b_2$	$[a_1/b_1, a_2/b_1]$	
6	$0 \in \overset{\circ}{A}$	$b_1 \leq b_2 < 0$	$[a_2/b_2, a_1/b_2]$	

In Tables 1 and 2 $\overset{\circ}{A}$ denotes the interior of A, i.e. $c \in \overset{\circ}{A}$ means $a_1 < c < a_2$. In the cases $0 \in B$ division A/B is not defined.

As a result of these rules it can be stated that in the case of real intervals the result of an interval operation $A \circ B$, for all $\circ \in \{+, -, *, /\}$, can be expressed in terms of the bounds of the interval operands (with the A/B exception above). In order to get each of these bounds, typically only one real operation is necessary. Only in case 9 of Table 1, $0 \in \overset{\circ}{A}$ and $0 \in \overset{\circ}{B}$, do two products have to be calculated and compared.

Whenever in the Tables 1 and 2 both operands are comparable with the interval $[0, 0]$ with respect to $\leq, \geq, <$ or $>$, the result of the interval operation $A * B$ or A/B contains both bounds of A and B. If one or both of the operands A or B, however, contains zero as an interior point, then the result $A * B$ and A/B is expressed by only three of the four bounds of A and B. In all these cases (3, 6, 7, 8, 9) in Table 1, the bound which is missing in the expression for the result can be shifted towards zero without changing the result of the operation $A * B$. Similarly, in cases 5 and 6 in Table 2, the bound of B, which is missing in the expression for the resulting interval, can be shifted toward ∞ (resp. $-\infty$) without changing the result of the operation. This shows a certain lack of sensitivity of interval arithmetic or computing with inequalities whenever in the cases of multiplication and division one of the operands contains zero as an interior point.

In all these cases — 3, 6, 7, 8, 9, of Table 1 and 5, 6 of Table 2 — the result of $A * B$ or A/B also contains zero, and the formulas show that the result tends toward the zero interval if the operands that contain zero do likewise. In the limit case when the operand that contains zero has become the zero interval, no such imprecision is left. This suggests that within arithmetic expressions interval operands that contain zero as an interior point should be made as small in diameter as possible.

We illustrate the efficiency of this calculus for inequalities by a simple example. See [4]. Let $x = Ax + b$ be a system of linear equations in fixed point form with a contracting real matrix A and a real vector b, and let the interval vector X be a rough initial enclosure of the solution $x^* \in X$. We can now formally write down the Jacobi method, the Gauss-Seidel method, a relaxation method or some other iterative scheme for the solution of the linear system. In these formulas we then interpret all components of the vector x as being intervals. Doing so we obtain a number of iterative methods for the computation of enclosures of linear systems of equations. Further iterative schemes then can be obtained by taking the intersection of two successive approximations. If we now decompose all these methods in formulas for the bounds of the intervals we obtain a major number of methods for the computation of bounds for the solution of linear systems which have been derived by well-known mathematicians painstakingly about 40 years ago, see [14]. The calculus of interval arithmetic reproduces these and other methods in the simplest way. The user does not have to take care of the many case distinctions occurring in the matrix vector multiplications. The computer executes them automatically by the preprogrammed calculus. Also the rounding errors are enclosed. The calculus evolves its own dynamics.

3   Interval Arithmetic as Executable Set Operations

The rules (1), (2), (3), and (4) also can be interpreted as arithmetic operations for sets. As such they are special cases of general set operations. Further important properties of interval arithmetic can immediately be obtained via set operations. Let M be any set with a dyadic operation $\circ : M \times M \to M$ defined for its elements. The powerset $\mathbb{P}M$ of M is defined as the set of all subsets of M. The operation $\circ$ in M can be extended to the powerset $\mathbb{P}M$ by the following definition

$$A \circ B := \{a \circ b | a \in A \wedge b \in B\} \text{ for all } A, B \in \mathbb{P}M. \tag{5}$$

The least element in $\mathbb{P}M$ with respect to set inclusion as an order relation is the empty set. The greatest element is the set M. We denote the empty set by the character string $[\,]$. The empty set is subset of any set. Any arithmetic operation on the empty set produces the empty set.

The following properties are obvious and immediate consequences of (5):

$$A \subseteq B \wedge C \subseteq D \Rightarrow A \circ C \subseteq B \circ D \text{ for all } A, B, C, D \in \mathbb{P}M, \tag{6}$$

and in particular

$$a \in A \wedge b \in B \Rightarrow a \circ b \in A \circ B \text{ for all } A, B \in \mathbb{P}M. \tag{7}$$

(6) is called the *inclusion isotony* (or *inclusion monotony*). (7) is called the *inclusion property*.

By use of parentheses these rules can immediately be extended to expressions with more than one arithmetic operation, e.g.

$$A \subseteq B \wedge C \subseteq D \wedge E \subseteq F \Rightarrow A \circ C \subseteq B \circ D \Rightarrow (A \circ C) \circ E \subseteq (B \circ D) \circ F,$$

and so on. Moreover, if more than one operation is defined in M this chain of conclusions also remains valid for expressions containing several different operations.

If we now replace the general set M by the set of real numbers, (5), (6), and (7) hold in particular for the powerset $I\!\!PI\!\!R$ of the real numbers $I\!\!R$. This is the case for all operations $\circ \in \{+, -, *, /\}$, if we assume that in case of division 0 is not an element of the denominator, for instance, $0 \notin B$ in (5).

The set $I\!\!R$ of closed and bounded intervals over $I\!\!R$ is a subset of $I\!\!PI\!\!R$. Thus (5), (6), and (7) are also valid for elements of $I\!\!R$. The set $I\!\!R$ with the operations (5), $\circ \in \{+, -, *, /\}$, is an algebraically closed[1] subset within $I\!\!PI\!\!R$. That is, if (5) is performed for two intervals $A, B \in I\!\!R$ the result is always an interval again. This holds for all operations $\circ \in \{+, -, *, /\}$ with $0 \notin B$ in case of division. This property is a simple consequence of the fact that for all arithmetic operations $\circ \in \{+, -, *, /\}$, $a \circ b$ is a continuous function of both variables. $A \circ B$ is the range of this function over the product set $A \times B$. Since A and B are closed intervals, $A \times B$ is a simply connected, bounded and closed subset of $I\!\!R^2$. In such a region the continuous function $a \circ b$ takes a maximum and a minimum as well as all values in between. Therefore

$$A \circ B = [\min_{a \in A, b \in B} (a \circ b), \ \max_{a \in A, b \in B} (a \circ b)], \ \text{for all} \ \circ \in \{+, -, *, /\},$$

provided that $0 \notin B$ in case of division.

Consideration of (5), (6), and (7) for intervals of $I\!\!R$ leads to the crucial properties of all applications of interval arithmetic. Because of the great importance of these properties we repeat them here explicitly. Thus we obtain for all operations $\circ \in \{+, -, *, /\}$:

The *set definition* of interval arithmetic:

$$A \circ B := \{a \circ b | a \in A \wedge b \in B\} \quad \text{for all } A, B \in I\!\!R, \atop 0 \notin B \text{ in case of division,} \tag{8}$$

the *inclusion isotony* (or *inclusion monotony*):

$$A \subseteq B \wedge C \subseteq D \Rightarrow A \circ C \subseteq B \circ D \quad \text{for all } A, B, C, D \in I\!\!R, \atop 0 \notin C, D \text{ in case of division,} \tag{9}$$

and in particular the *inclusion property*

$$a \in A \wedge b \in B \Rightarrow a \circ b \in A \circ B \quad \text{for all } A, B \in I\!\!R, \atop 0 \notin B \text{ in case of division.} \tag{10}$$

If for $M = I\!\!R$ in (5) the number of elements in A or B is infinite, the operations are effectively not executable because infinitely many real operations would have to be performed. If A and B are intervals of $I\!\!R$, however, the

---

[1] as the integers are within the reals for $\circ \in \{+, -, *\}$

situation is different. In general A or B or both will again contain infinitely many real numbers. The result of the operation (8), however, can now be performed by a finite number of operations with real numbers, with the bounds of A and B. For all operations $\circ \in \{+, -, *, /\}$ the result is obtained by the explicit formulas (1), (2), (3), and (4), [33,34].

For intervals $A = [a_1, a_2]$ and $B = [b_1, b_2]$ the formulas (1), (2), (3), and (4) can be summarized by

$$A \circ B = [\min_{i,j=1,2} (a_i \circ b_j), \max_{i,j=1,2} (a_i \circ b_j)] \text{ for all } \circ \in \{+, -, *, /\}, \qquad (11)$$

with $0 \notin B$ in case of division.

Since interval operations are particular powerset operations, the inclusion isotony and the inclusion property also hold for expressions with more than one arithmetic operation.

In programming languages the concept of an arithmetic expression is usually defined to be a little more general. Besides constants and variables elementary functions (sometimes called standard functions) like sqr, sqrt, sin, cos, exp, ln, tan, ... may also be elementary ingredients. All these are put together with arithmetic operators and parentheses into the general concept of an arithmetic expression. This construct is illustrated by the syntax diagram of Fig. 1. Therein solid lines are to be traversed from left to right and from top to bottom. Dotted lines are to be traversed oppositely, i.e. from right to left and from bottom to top. In Fig. 1 the syntax variable REAL FUNCTION merely represents a real arithmetic expression hidden in a subroutine.

Now we define the general concept of an arithmetic expression for the new data type interval by exchanging the data type *real* in Fig. 1 for the new data type *interval*. This results in the syntax diagram for INTERVAL EXPRESSION shown in Fig. 2. In Fig. 2 the syntax variable INTERVAL FUNCTION represents an interval expression hidden in a subroutine.

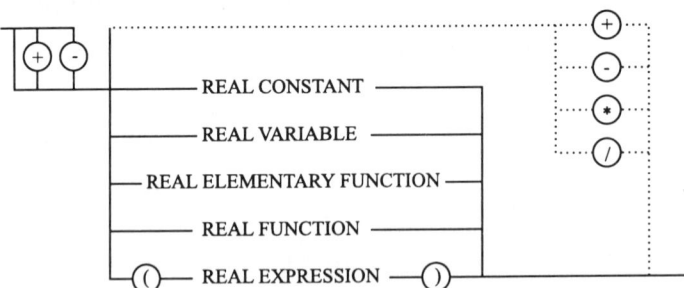

Fig. 1. Syntax diagram for REAL EXPRESSION

In the syntax diagram for INTERVAL EXPRESSION in Fig. 2 the concept of an interval elementary function is not yet defined. We simply define

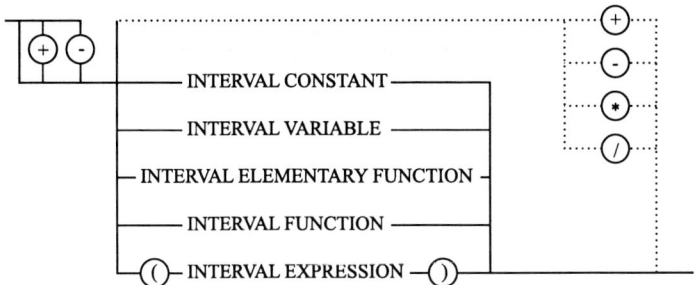

Fig. 2. Syntax diagram for INTERVAL EXPRESSION

it as the range of function values taken over an interval (out of the domain of definition $D(f)$ of the function). In case of a real function f we denote the range of values over the interval $[a_1, a_2]$ by

$$f([a_1, a_2]) := \{f(a) \mid a \in [a_1, a_2]\} \text{ with } [a_1, a_2] \in D(f).$$

For instance:

$$e^{[a_1,a_2]} = [e^{a_1}, e^{a_2}],$$

$$[a_1, a_2]^{2n} = \begin{cases} [\min(a_1^{2n}, a_2^{2n}), \max(a_1^{2n}, a_2^{2n})] & \text{for } 0 \notin [a_1, a_2], \\ [0, \max(a_1^{2n}, a_2^{2n})] & \text{for } 0 \in [a_1, a_2], \end{cases}$$

$$\sin[-\frac{\pi}{4}, \frac{\pi}{4}] = [-\frac{1}{2}\sqrt{2}, \frac{1}{2}\sqrt{2}],$$

$$\cos[0, \frac{\pi}{4}] = [0, 1].$$

For non monotonic functions the computation of the range of values over an interval $[a_1, a_2]$ requires the determination of the global minimum and maximum of the function in the interval $[a_1, a_2]$. For the usual elementary functions, however, these are known. With this definition of elementary functions for intervals the key properties of interval arithmetic, the inclusion monotony (7) and the inclusion property (8) extend immediately to elementary functions and with this to interval expressions as defined in Fig. 2:

$$A \subseteq B \Rightarrow f(A) \subseteq f(B), \text{ with } A, B \in I\!R \qquad \text{inclusion isotone,}$$

and in particular for $a \in I\!R$ and $A \in I\!R$:

$$a \in A \Rightarrow f(a) \in f(A) \qquad \text{inclusion property.}$$

We summarize the development so far by stating that interval arithmetic expressions are generally inclusion isotone and that the inclusion property holds. These are the key properties of interval arithmetic. They give interval arithmetic its raison d'être. To start with, they provide the possibility of enclosing imprecise data within bounds and then continuing the computation with these bounds. This always results in guaranteed enclosures.

As the next step we define a (computable) real function simply by a real arithmetic expression. We need the concept of an *interval evaluation of a real function*. It is defined as follows: In the arithmetic expression for the function all operands are replaced by intervals and all operations by interval operations (where all intervals must be within the domain of definition of the real operands). This is just the step from Fig. 1 to Fig. 2. What is obtained is an interval expression. Then all arithmetic operations are performed in interval arithmetic. For a real function $f(a)$ we denote the interval evaluation over the interval A by $F(A)$.

With this definition we can immediately conclude that interval evaluations of (computable) real functions are inclusion isotone and that the inclusion property holds in particular:

$$A \subseteq B \Rightarrow F(A) \subseteq F(B) \qquad \text{inclusion isotone,} \qquad (12)$$

$$a \in A \Rightarrow f(a) \in F(A) \qquad \text{inclusion property.} \qquad (13)$$

These concepts immediately extend in a natural way to functions of several real variables. In this case in (13) a is an n-tuple, $a = (a_1, a_2, \ldots, a_n)$, and A and B are higher dimensional intervals, e.g. $A = (A_1, A_2, \ldots, A_n)$.

Remark: Two different real arithmetic expressions can define equivalent real functions, for instance:

$$f(x) = x(x-1) \quad \text{and} \quad g(x) = x^2 - x.$$

Evaluation of the two expressions for a real number always leads to the same real function value. In contrast to this, interval evaluation of the two expressions may lead to different intervals. In the example we obtain for the interval $A = [1, 2]$:

$$
\begin{aligned}
F(A) &= [1,2]([1,2] + [-1,-1]) & G(A) &= [1,2][1,2] - [1,2] \\
&= [1,2][0,1] = [0,2], & &= [1,4] - [1,2] = [-1,3].
\end{aligned}
$$

Although an interval evaluation of a real function is very naturally defined via the arithmetic expression of the function, a closer look at the syntax diagram in Fig. 2 reveals major problems that appear when such evaluations are to be coded. The widely used programming languages do not provide the necessary ease of programming. An interval evaluation of a real function should be performable as easily as an execution of the corresponding expression in real arithmetic. For that purpose the programming language

1. must allow an operator notation $A \circ B$ for the basic interval operations $\circ \in \{+, -, *, /\}$, i.e. operator overloading must be provided,
2. the concept of a function subroutine must not be restricted to the data types *integer* and *real*, i.e. subroutine functions with general result type should be provided by the programming language, and

3. the elementary functions must be provided for interval arguments.

While 1. and 2. are challenges for the designer of the programming language, 3. is a challenge for the mathematician. In a conventional call of an elementary function the computer provides a result, the accuracy of which cannot easily be judged by the user. This is no longer the case when the elementary functions are provided for interval arguments. Then, if called for a point interval (where the lower and upper bound coincide), a comparison of the lower and upper bound of the result of the interval evaluation of the function reveals immediately the accuracy with which the elementary function has been implemented. This situation has forced extremely careful implementation of the elementary functions and since interval versions of the elementary functions have been provided on a large scale [37,38,26–29,76] the conventional real elementary functions on computers also had to be and have been improved step by step by the manufacturers. A most advanced programming environment in this respect is a decimal version of PASCAL-XSC [10] where, besides the usual 24 elementary functions, about the same number of special functions are provided for real and interval arguments with highest accuracy.

1., 2. and 3. are minimum requirements for any sophisticated use of interval arithmetic. If they are not met, coding difficulties absorb all the attention and capacity of users and prevent them from developing deeper mathematical ideas and insight. So far none of the widespread programming languages like Fortran, C, and even Fortran 95 and C++ provide the necessary programming ease. This is the basic reason for the slow progress in the field. It is a matter of fact that a great deal of the existing and established interval methods and algorithms have originally been developed in PASCAL-XSC even if they have been coded afterwards in other languages. Programming ease is essential indeed. The typical user, however, is reluctant to leave the programming environment he is used to, just to apply interval methods.

We summarize this discussion by stating that it does not suffice for an adequate use of interval arithmetic on computers that only the four basic arithmetic operations $+, -, *$ and $/$ for intervals are somehow supported by the computer hardware. An appropriate language support is absolutely necessary. So far this has been missing. This is the basic dilemma of interval arithmetic. Experience has shown that it cannot be overcome via slow moving standardization committees for programming languages. Two things seem to be necessary for the great breakthrough. A major vendor has to provide the necessary support and the body of numerical analysts must acquire a broader insight and skills in order to use this support.

4   Enclosing the Range of Function Values

The interval evaluation of a real function f over the interval A was denoted by $F(A)$. We now compare it with the range of function values over the interval

A which was denoted by

$$f(A) := \{f(a) \mid a \in A\}. \tag{14}$$

We have observed that interval evaluation of an arithmetic expression and of real functions is inclusion isotone (9), (12) and that the inclusion property (10), (13) holds. Since (10) and (13) hold for all $a \in A$ we can immediately state that

$$f(A) \subseteq F(A), \tag{15}$$

i.e. that the interval evaluation of a real function over an interval delivers a superset of the range of function values over that interval. If A is a point interval $[a, a]$ this reduces to:

$$f(a) \in F([a, a]). \tag{16}$$

These are basic properties of interval arithmetic. Computing with inequalities always aims for bounds for function values, or for bounds for the range of function values. Interval arithmetic allows this computation in principle.

The range of function values over an interval is needed for many applications. Its computation is a very difficult task. It is equivalent to the computation of the global minimum and maximum of the function in that interval. An interval evaluation of the arithmetic expression on the other hand is very easy to perform. It requires about twice as many real arithmetic operations as an evaluation of the function in real arithmetic. Thus interval arithmetic provides an easy means to compute upper and lower bounds for the range of function values.

In the end a complicated algorithm just performs an arithmetic expression. So an interval evaluation of the algorithm would compute bounds for the result from given bounds for the data. However, it is observed that in doing so, in general, the diameters of the intervals grow very fast and for large algorithms the bounds quickly become meaningless in particular if the bounds for the data are already large. This raises the question whether measures can be taken to keep the diameters of the intervals from growing too fast. Interval arithmetic has developed such measures and we are going to sketch these now.

If an enclosure for a function value is computed by (16), the quality of the computed result $F([a, a])$ can be judged by the diameter of the interval $F([a, a])$. This possibility of easily judging the quality of the computed result, is not available in (15). Even if $F(A)$ is a large interval, it can be a good approximation for the range of function values $f(A)$ if the latter is large also. So some means to measure the deviation between $f(A)$ and $F(A)$ in (15) is desirable.

It is well known that the set $I\!\!R$ of real intervals becomes a metric space with the so called Hausdorff metric, where the distance q of two intervals $A = [a_1, a_2]$ and $B = [b_1, b_2]$ is defined by

$$q(A, B) := \max\{|a_1 - b_1|, |a_2 - b_2|\}. \tag{17}$$

See, for instance, [6].

With this distance function q the following relation can be proved to hold under natural assumptions on f:

$$q(f(A), F(A)) \leq \alpha \cdot d(A), \text{ with a constant } \alpha \geq 0. \tag{18}$$

Here $d(A)$ denotes the diameter of the interval A:

$$d(A) := |a_2 - a_1|. \tag{19}$$

In case of functions of several real variables the maximum of the diameters $d(A_i)$ appears on the right hand side of (18).

The relation (18) shows that the distance between the range of values of the function f over the interval A and the interval evaluation of the expression for f tends to zero linearly with the diameter of the interval A. So the overestimation of $f(A)$ by $F(A)$ decreases with the diameter of A and in the limit $d(A) = 0$ no such overestimation is left.

Because of this result subdivision of the interval A into subintervals A_i, $i = 1(1)n$ with $A = \bigcup_{i=1}^{n} A_i$ is a frequently applied technique to obtain better approximations for the range of function values. Then (18) holds for each subinterval:

$$q(f(A_i), F(A_i)) \leq \alpha_i \cdot d(A_i), \text{ with } \alpha_i \geq 0 \text{ and } i = 1(1)n,$$

and, in general, the union of the interval evaluations over all subintervals

$$\bigcup_{i=1}^{n} F(A_i)$$

is a much better approximation for the range $f(A)$ than is $F(A)$.

There are yet other methods to obtain better enclosures for the range of function values $f(A)$. We have already observed that the interval evaluation $F(A)$ of a function f depends on the expression used for the representation of f. So by choosing appropriate representations for f the overestimation of $f(A)$ by the interval evaluation $F(A)$ can often be reduced. Indeed, if f allows a representation of the form

$$f(x) = f(c) + (x - c) \cdot h(x), \text{ with } c \in A, \tag{20}$$

then under natural assumptions on h the following inequality holds

$$q(f(A), F(A)) \leq \beta \cdot (d(A))^2, \text{ with a constant } \beta \geq 0. \tag{21}$$

(20) is called a centered form of f. In (20) c is not necessarily the center of A although it is often chosen as the center. (21) shows that the distance between the range of values of the function f over the interval A and the interval evaluation of a centered form of f tends toward zero quadratically

with the diameter of the interval A. In practice, this means that for small intervals the interval evaluation of the centered form leads to a very good approximation of the range of function values over an interval A. Again, subdivision is a method that can be applied in the case of a large interval A. It should be clear, however, that in general only for small intervals is the bound in (21) better than in (18).

The decrease of the overestimation of the range of function values by the interval evaluation of the function with the diameter of the interval A, and the method of subdivision, are reasons why interval arithmetic can successfully be used in many applications. Numerical methods often proceed in small steps. This is the case, for instance, with numerical quadrature or cubature, or with numerical integration of ordinary differential equations. In all these cases an interval evaluation of the remainder term of the integration formula (using differentiation arithmetic) controls the step size of the integration, and anyhow because of the small steps, overestimation is practically negligible.

We now mention briefly how centered forms can be obtained. Usually a centered form is derived via the mean-value theorem. If f is differentiable in its domain D, then $f(x) = f(c) + f'(\xi)(x - c)$ for fixed $c \in D$ and some ξ between x and c. If x and c are elements out of the interval $A \subseteq D$, then also $\xi \in A$. Therefore

$$f(x) \in F(A) := f(c) + F'(A)(A - c), \text{ for all } x \in A.$$

Here $F'(A)$ is an interval evaluation of $f'(x)$ in A.

In (20) the slope

$$h(x) = \frac{f(x) - f(c)}{x - c}$$

can be used instead of the derivative for the representation of $f(x)$. Slopes often lead to better enclosures for $f(A)$ than do derivatives. For details see [32,53,7].

Derivatives and enclosures of derivatives can be computed by a process which is called automatic differentiation or differentiation arithmetic. Slopes and enclosures of slopes can be computed by another process which is very similar to automatic differentiation. In both cases the computation of the derivative or slope or enclosures of these is done together with the computation of the function value. For these processes only the expression or algorithm for the function is required. No explicit formulas for the derivative or slope are needed. The computer interprets the arithmetic operations in the expression by differentiation or slope arithmetic. The arithmetic is hidden in the runtime system of the compiler. It is activated by type specification of the operands. For details see [17,18,53], and Section 8. Thus the computer is able to produce and enclose the centered form via the derivative or slope automatically.

Without going into further details we mention once more, that all these considerations are not restricted to functions of a single real variable. Subdivision in higher dimensions, however, is a difficult task which requires additional tools and strategies. Typical of such problems are the computation of the bounds of the solution of a system of nonlinear equations, and global optimization or numerical integration of functions of more than one real variable. In all these and other cases, zero finding is a central task. Here the extended Interval Newton Method plays an extraordinary role so we are now going to review this method, which is also one of the requirements that have to be met when interval arithmetic is implemented on the computer.

5   The Interval Newton Method

Traditionally Newton's method is used to compute an approximation of a zero of a nonlinear real function $f(x)$, i.e. to compute a solution of the equation

$$f(x) = 0. \tag{22}$$

The method approximates the function $f(x)$ in the neighborhood of an initial value x_0 by the linear function (the tangent)

$$t(x) = f(x_0) + f'(x_0)(x - x_0) \tag{23}$$

the zero of which can easily be calculated by

$$x_1 := x_0 - \frac{f(x_0)}{f'(x_0)}. \tag{24}$$

x_1 is used as new approximation for the zero of (22). Continuation of this method leads to the general iteration scheme:

$$x_{\nu+1} := x_\nu - \frac{f(x_\nu)}{f'(x_\nu)}, \quad \nu = 0, 1, 2, \ldots . \tag{25}$$

It is well known that if $f(x)$ has a single zero x^* in an interval X and $f(x)$ is twice continuously differentiable, then the sequence

$$x_0, x_1, x_2, \ldots, x_\nu, \ldots$$

converges quadratically towards x^* if x_0 is sufficiently close to x^*. If the latter condition does not hold the method may well fail.

The interval version of Newton's method computes an enclosure of the zero x^* of a continuously differentiable function $f(x)$ in the interval X by the following iteration scheme:

$$X_{\nu+1} := (m(X_\nu) - \frac{f(m(X_\nu))}{F'(X_\nu)}) \cap X_\nu, \quad \nu = 0, 1, 2, \ldots, \tag{26}$$

with $X_0 = X$. Here $F'(X_\nu)$ is the interval evaluation of the first derivative $f'(x)$ of f over the interval X_ν and $m(X_\nu)$ is the midpoint of the interval X_ν.

Instead of $m(X_\nu)$ another point within X_ν could be chosen. The method can only be applied if $0 \notin F'(X_0)$. This guarantees that $f(x)$ has only a single zero in X_0.

In contrast to (25) the method (26) can never diverge (fail). Because of the intersection with X_ν the sequence

$$X_0 \supseteq X_1 \supseteq X_2 \supseteq \cdots \tag{27}$$

is bounded. It can be shown that under natural conditions on the function f the sequence converges quadratically to x^* [6,47].

The operator

$$N(X) := x - \frac{f(x)}{F'(X)}, \quad x \in X \in I\!I\!R \tag{28}$$

is called the Interval Newton Operator. It has the following properties:

I. If $N(X) \subseteq X$, then $f(x)$ has exactly one zero x^* in X.
II. If $N(X) \cap X = [\,]$ then $f(x)$ has no zero in X.

Thus, $N(X)$ can be used to prove the existence or absence of a zero x^* of $f(x)$ in X. Since in the case of existence of a zero x^* in X the sequence (26), (27) converges, in the case of absence the situation $N(X) \cap X = [\,]$ must occur in (27).

The interval version of Newton's method (26) can also be derived via the mean value theorem. If $f(x)$ is continuously differentiable and has a single zero x^* in the interval X, and $f'(x) \neq 0$ for all $x \in X$, then

$$f(x) = f(x^*) + f'(\xi)(x - x^*) \text{ for all } x \in X \text{ and some } \xi \text{ between } x \text{ and } x^*$$

Since $f(x^*) = 0$ and $f'(\xi) \neq 0$ this leads to

$$x^* = x - \frac{f(x)}{f'(\xi)}.$$

If $F'(X)$ denotes the interval evaluation of $f'(x)$ over the interval X, we have $f'(\xi) \in F'(X)$ and therefore

$$x^* = x - \frac{f(x)}{f'(\xi)} \in x - \frac{f(x)}{F'(X)} = N(X) \text{ for all } x \in X,$$

i.e. $x^* \in N(X)$ and thus

$$x^* \in (x - \frac{f(x)}{F'(X)}) \cap X = N(X) \cap X.$$

Now we obtain by setting $X_0 := X$ and $x = m(X_0)$

$$X_1 := (m(X_0) - \frac{f(m(X_0))}{F'(X_0)}) \cap X_0,$$

and by continuation (26).

In close similarity to the conventional Newton method the Interval Newton Method also allows some geometric interpretation. For that purpose let be $X = [x_1, x_2]$ and $N(X) = [n_1, n_2]$. $F'(X)$ is the interval evaluation of $f'(x)$ over the interval X. As such it is a superset of all slopes of tangents that can occur in X. (24) computes the zero of the tangent of $f(x)$ in $(x_0, f(x_0))$. Similarly $N(X)$ is the interval of all zeros of straight lines through $(x, f(x))$ with slopes within $F'(X)$, see Fig. 3. Of course, $f'(x) \in F'(X)$.

The straight line through $f(x)$ with the least slope within $F'(X)$ cuts the real axis in n_1, and the one with the greatest slope in n_2. Thus the Interval Newton Operator $N(X)$ computes the interval $[n_1, n_2]$ which in the sketch of Fig. 3 is situated on the left hand side of x. The intersection of $N(X)$ with X then delivers the new interval X_1. In the example in Fig. 3, $X_1 = [x_1, n_2]$.

Newton's method allows some visual interpretation. From the point $(x, f(x))$ the conventional Newton method sends a beam along the tangent. The search is continued at the intersection of this beam with the x-axis. The Interval Newton Method sends a set of beams like a floodlight from the point $(x, f(x))$ to the x-axis. This set includes the directions of all tangents that occur in the entire interval X. The interval $N(X)$ comprises all cuts of these beams with the x-axis.

It is a fascinating discovery that the Interval Newton Method can be extended so that it can be used to compute all zeros of a real function in a given interval. The basic idea of this extension is already old [3]. Many scientists have worked on details of how to use this method, of how to define the necessary arithmetic operations, and of how to bring them to the computer. But inconsistencies have occurred again and again. However, it seems that understanding has now reached a point which allows a consistent realization of the method and of the necessary arithmetic. The extended Interval Newton Method is the most powerful and most frequently used tool for subdivision in higher dimensional spaces. It requires an extension of interval arithmetic which we are now going to discuss.

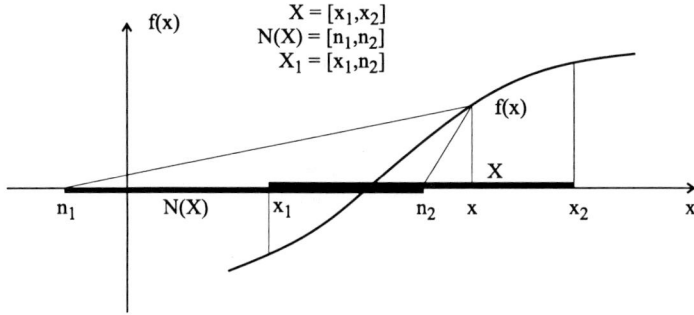

Fig. 3. Geometric interpretation of the Interval Newton Method

6   Extended Interval Arithmetic

In the definition of interval arithmetic, division by an interval which includes zero was excluded. We are now going to eliminate this exception.

The real numbers $I\!\!R$ are defined as a conditionally complete, linearly ordered field.[2] With respect to the order relation $\leq$ they can be completed by adjoining a least element $-\infty$ and a greatest element $+\infty$. We denote the resulting set by $I\!\!R^* := I\!\!R \cup \{-\infty\} \cup \{+\infty\}$. $\{I\!\!R^*, \leq\}$ is a complete lattice.[3] This completion is frequently applied in mathematics and it is well known that the new elements $-\infty$ and $+\infty$ fail to satisfy several of the algebraic properties of a field. $-\infty$ and $+\infty$ are not real numbers! For example $a + \infty = b + \infty$ even if $a < b$, so that the cancellation law is not valid. For the new elements $-\infty$ and $+\infty$ the following operations with elements $x \in I\!\!R$ are defined in analysis:

$$
\begin{aligned}
\infty + x &= \infty, & \infty * x &= \infty \text{ for } x > 0, \\
-\infty + x &= -\infty, & \infty * x &= -\infty \text{ for } x < 0, \\
\tfrac{x}{\infty} &= \tfrac{x}{-\infty} = 0, & & \\
\infty + \infty &= \infty * \infty = \infty, & & \\
-\infty + (-\infty) &= (-\infty) * \infty = -\infty.
\end{aligned}
\tag{29}
$$

together with variants obtained by applying the sign rules and the law of commutativity. Not defined are the terms $\infty - \infty$ and $0 * \infty$, again with variants obtained by applying the sign rules and the law of commutativity. These rules are well established in real analysis and there is no need to extend them for the purposes of interval arithmetic in $I\!\!I\!\!R$.

$I\!\!R$ is a set with certain arithmetic operations. These operations can be extended to the powerset $I\!\!P\!I\!\!R$ in complete analogy to (5):

$$
A \circ B := \{a \circ b \mid a \in A \wedge b \in B\} \quad \text{for all } \circ \in \{+, -, *, /\}, \\
\text{and all } A, B \in I\!\!P\!I\!\!R.
\tag{30}
$$

As a consequence of (30) again the *inclusion isotony* (6) and the *inclusion property* (7) hold for all operations and arithmetic expressions in $I\!\!P\!I\!\!R$. In particular, this is the case if (30) is restricted to operands of $I\!\!I\!\!R$. $I\!\!I\!\!R$ is a subset of $I\!\!P\!I\!\!R$.

We are now going to define division by an interval B of $I\!\!I\!\!R$ which contains zero. It turns out that the result is no longer an interval of $I\!\!I\!\!R$. But we can apply the definition of the division in the powerset as given by (30). This leads to

$$
A/B := \{a/b \mid a \in A \wedge b \in B\} \quad \text{for all } A, B \in I\!\!I\!\!R.
\tag{31}
$$

---

[2] An ordered set is called conditionally complete if every non empty, bounded subset has a greatest lower bound (infimum) and a least upper bound (supremum).
[3] In a complete lattice every subset has an infimum and a supremum.

In order to interpret the right hand side of (31) we remember that the quotient a/b is defined as the inverse operation of multiplication, i.e. as the solution of the equation $b \cdot x = a$. Thus (31) can also be written in the form

$$A/B := \{x \mid bx = a \wedge a \in A \wedge b \in B\} \quad \text{for all } A, B \in I\!R. \tag{32}$$

Now we have to interpret the right hand side of (32). We are interested in obtaining simply executable, explicit formulas for the right hand side of (32). The case $0 \notin B$ was already dealt with in Table 2. So we assume here generally that $0 \in B$. For $A = [a_1, a_2]$ and $B = [b_1, b_2] \in I\!R$, $0 \in B$ the following eight distinct cases can be set out:

1. $0 \in A$, $0 \in B$.
2. $0 \notin A$, $B = [0, 0]$.
3. $a_1 \le a_2 < 0$, $b_1 < b_2 = 0$.
4. $a_1 \le a_2 < 0$, $b_1 < 0 < b_2$.
5. $a_1 \le a_2 < 0$, $0 = b_1 < b_2$.
6. $0 < a_1 \le a_2$, $b_1 < b_2 = 0$.
7. $0 < a_1 \le a_2$, $b_1 < 0 < b_2$.
8. $0 < a_1 \le a_2$, $0 = b_1 < b_2$.

The list distinguishes the cases $0 \in A$ (case 1) and $0 \notin A$ (cases 2 to 8). Since it is generally assumed that $0 \in B$ these eight cases indeed cover all possibilities.

We are now going to derive simple formulas for the result of the interval division A/B for these eight cases:

1. $0 \in A \wedge 0 \in B$. Since every $x \in I\!R$ fulfils the equation $0 \cdot x = 0$, we have $A/B = (-\infty, +\infty)$. Here $(-\infty, +\infty)$ denotes the open interval between $-\infty$ and $+\infty$ which just consists of all real numbers $I\!R$, i.e. $A/B = I\!R$.
2. In case $0 \notin A \wedge B = [0, 0]$ the set defined by (32) consists of all elements which fulfil the equation $0 \cdot x = a$ for $a \in A$. Since $0 \notin A$, there is no real number which fulfils this equation. Thus A/B is the empty set $A/B = [\;]$.

In all other cases $0 \notin A$ also. We have already observed under 2. that in this case the element 0 in B does not contribute to the solution set. So it can be excluded without changing the set A/B.

So the general rule for computing A/B by (32) is to punch out zero of the interval B and replace it by a small positive or negative number ϵ as the case may be. The so changed interval B is denoted by $\overline{B}$ and represented in column 4 of Table 3. With this $\overline{B}$ the solution set $A/\overline{B}$ can now easily be computed by applying the rules of Table 2. The results are shown in column 5 of Table 3. Now the desired result A/B as defined by (32) is obtained if in column 5 ϵ tends to zero. Thus in cases 3 to 8 the results are obtained by the limit process $A/B = \lim_{\epsilon \to 0} A/\overline{B}$. The solution set A/B is shown in the last column of Table 3 for all the 8 cases. There, as usual in mathematics parentheses denote open interval ends, i.e. the bound is excluded. In contrast to this brackets denote closed interval ends, i.e. the bound is included.

In Table 3 the operands A and B of the division A/B are intervals of $I\!R$! The results of the division A/B shown in the last column, however, are no longer intervals of $I\!R$ nor are they intervals of $I\!R^*$ which is the set of intervals over R^*. This is logically correct and should not be surprising, since the division has been defined as an operation in $I\!P\!R$ by (30).

Table 4 shows the result of the division A/B of two intervals $A = [a_1, a_2]$ and $B = [b_1, b_2]$ in the case $0 \in B$ in a more convenient layout.

Table 3. The 8 cases of the division of two intervals A/B, with $A, B \in I\!R$ and $0 \in B$.

case	$A = [a_1, a_2]$	$B = [b_1, b_2]$	$\overline{B}$	$A/\overline{B}$	A/B
1	$0 \in A$	$0 \in B$			$(-\infty, +\infty)$
2	$0 \notin A$	$B = [0, 0]$			$[\,]$
3	$a_2 < 0$	$b_1 < b_2 = 0$	$[b_1, -\epsilon]$	$[a_2/b_1, a_1/(-\epsilon)]$	$[a_2/b_1, +\infty)$
4	$a_2 < 0$	$b_1 < 0 < b_2$	$[b_1, -\epsilon]\cup$ $[\epsilon, b_2]$	$[a_2/b_1, a1/(-\epsilon)]\cup$ $[a_1/\epsilon, a_2/b_2]$	$(-\infty, a_2/b_2]\cup$ $[a_2/b_1, +\infty)$
5	$a_2 < 0$	$0 = b_1 < b_2$	$[\epsilon, b_2]$	$[a_1/\epsilon, a_2/b_2]$	$(-\infty, a_2/b_2]$
6	$a_1 > 0$	$b_1 < b_2 = 0$	$[b_1, -\epsilon]$	$[a_2/(-\epsilon), a_1/b_1]$	$(-\infty, a_1/b_1]$
7	$a_1 > 0$	$b_1 < 0 < b_2$	$[b_1, -\epsilon]\cup$ $[\epsilon, b_2]$	$[a_2/(-\epsilon), a1/b_1]\cup$ $[a_1/b_2, a_2/\epsilon]$	$(-\infty, a_1/b_1]\cup$ $[a_1/b_2, +\infty)$
8	$a_1 > 0$	$0 = b_1 < b_2$	$[\epsilon, b_2]$	$[a_1/b_2, a_2/\epsilon]$	$[a_1/b_2, +\infty)$

Table 4. The result of the division A/B, with $A, B \in I\!R$ and $0 \in B$.

A/B	$B = [0, 0]$	$b_1 < b_2 = 0$	$b_1 < 0 < b_2$	$0 = b_1 < b_2$
$a_2 < 0$	$[\,]$	$[a_2/b_1, +\infty)$	$(-\infty, a_2/b_2] \cup [a_2/b_1, +\infty)$	$(-\infty, a_2/b_2]$
$a_1 \leq 0 \leq a_2$	$(-\infty, +\infty)$	$(-\infty, +\infty)$	$(-\infty, +\infty)$	$(-\infty, +\infty)$
$a_1 > 0$	$[\,]$	$(-\infty, a_1/b_1]$	$(-\infty, a_1/b_1] \cup [a_1/b_2, +\infty)$	$[a_1/b_2, +\infty)$

For completeness we repeat at the end of this section the results of the basic arithmetic operations for intervals $A = [a_1, a_2]$ and $B = [b_1, b_2]$ of $I\!R$ which have already been given in Section 2. In the cases of multiplication and division we use different representations. We also list the basic rules of the order relations $\leq$ and $\subseteq$ for intervals of $I\!R^*$.

I.   Equality: $[a_1, a_2] = [b_1, b_2] :\Leftrightarrow a_1 = b_1 \wedge a_2 = b_2$.
II. Addition: $[a_1, a_2] + [b_1, b_2] = [a_1 + b_1, a_2 + b_2]$.
III. Subtraction: $[a_1, a_2] - [b_1, b_2] = [a_1 - b_2, a_2 - b_1]$.
IV. Negation: $A = [a_1, a_2]$, $-A = [-a_2, -a_1]$.
V.   Multiplication:

$A \cdot B$	$b_1 \geq 0$	$b_1 < 0 < b_2$	$b_2 \leq 0$
$a_2 \leq 0$	$[a_1 b_2, a_2 b_1]$	$[a_1 b_2, a_1 b_1]$	$[a_2 b_2, a_1 b_1]$
$a_1 < 0 < a_2$	$[a_1 b_2, a_2 b_2]$	$[\min(a_1 b_2, a_2 b_1),$ $\max(a_1 b_1, a_2 b_2)]$	$[a_2 b_1, a_1 b_1]$
$a_1 \geq 0$	$[a_1 b_1, a_2 b_2]$	$[a_2 b_1, a_2 b_2]$	$[a_2 b_1, a_1 b_2]$

VI. Division, $0 \notin B$:

A/B	$b_1 > 0$	$b_2 < 0$
$a_2 \leq 0$	$[a_1/b_1, a_2/b_2]$	$[a_2/b_1, a_1/b_2]$
$a_1 < 0 < a_2$	$[a_1/b_1, a_2/b_1]$	$[a_2/b_2, a_1/b_2]$
$a_1 \geq 0$	$[a_1/b_2, a_2/b_1]$	$[a_2/b_2, a_1/b_1]$

The closed intervals over the real numbers $I\!R^*$ are ordered with respect to two different order relations, the comparison $\leq$ and the set inclusion $\subseteq$. With respect to both order relations they are complete lattices. The basic properties are:

VII. $\{I\!I\!R^*, \leq\} : [a_1, a_2] \leq [b_1, b_2] :\Leftrightarrow a_1 \leq b_1 \wedge a_2 \leq b_2$.
The least element of $I\!I\!R^*$ with respect to $\leq$ is the interval $[-\infty, -\infty]$, the greatest element is $[+\infty, +\infty]$. The infimum and supremum respectively of a subset $S \subseteq I\!I\!R^*$ are:

$$\inf_{\leq} S = [\inf_{A \in S} a_1, \inf_{A \in S} a_2], \quad \sup_{\leq} S = [\sup_{A \in S} a_1, \sup_{A \in S} a_2].$$

VIII. The interval $[-\infty, +\infty]$ is the greatest element in $\{I\!I\!R^*, \subseteq\}$, i.e. for all intervals $A \in I\!I\!R^*$ we have $A \subseteq [-\infty, +\infty]$. But a least element is missing in $I\!I\!R^*$. So we adjoin the empty set $[\,]$ as the least element of $I\!I\!R^*$. The empty set $[\,]$ is a subset of any set, thus for all $A \in I\!I\!R^*$ we have $[\,] \subseteq A$. We denote the resulting set by $\overline{I\!I\!R}^* := I\!I\!R^* \cup \{[\,]\}$. With this completion $\{\overline{I\!I\!R}^*, \subseteq\}$ is a complete lattice. The infimum and supremum respectively of a subset $S \subseteq I\!I\!R^*$ are [33,34]:

$$\inf_{\subseteq} S = [\sup_{A \in S} a_1, \inf_{A \in S} a_2], \quad \sup_{\subseteq} S = [\inf_{A \in S} a_1, \sup_{A \in S} a_2],$$

i.e. the infimum is the intersection and the supremum is the interval (convex) hull of all intervals out of S. For $\inf_{\subseteq} S$ we shall also use the usual symbol $\cap S$. $\sup_\subset S$ is occasionally written as $\bigcup S$. If in particular S just consists of two intervals A, B, this reads:

intersection: $A \cap B$, interval hull: $A \cup B$.

Since $\mathbb{R}$ is a linearly ordered set with respect to $\leq$, the interval hull is the same as the convex hull. The intersection may be empty.

In the following section we shall generalize the Interval Newton Method in such a way that for the Interval Newton Operator

$$N(X) := x - f(x)/F'(X), \quad x \in X \in I\mathbb{R} \tag{33}$$

the case $0 \in F'(X)$ is no longer excluded. The result of the division then can be taken from Tables 3 and 4. It is no longer an element out of $I\mathbb{R}$, but an element of the powerset $I\!P\mathbb{R}$. Thus the subtraction that occurs in (33) is also an operation in $I\!P\mathbb{R}$. As such it is defined by (29) and (30). As a consequence of this, of course, the operation is inclusion isotone and the inclusion property holds. We are interested in the evaluation of an expression of the form

$$Z := x - a/B, \text{ with } x, a \in \mathbb{R} \text{ and } 0 \in B \in I\mathbb{R}. \tag{34}$$

(34) can also be written as $Z = x + (-a/B)$. Multiplication of the set a/B by -1 negates and exchanges all bounds (see IV. above). Corresponding to the eight cases of Table 3, eight cases are again to be distinguished. The result is shown in the last column of Table 5.

Table 5. Evaluation of $Z = x - a/B$ for $x, a \in \mathbb{R}$, and $0 \in B \in I\mathbb{R}$.

	a	$B = [b_1, b_2]$	$-a/B$	$Z := x - a/B$
1	$a = 0$	$0 \in B$	$(-\infty, +\infty)$	$(-\infty, +\infty)$
2	$a \neq 0$	$B = [0,0]$	$[\,]$	$[\,]$
3	$a < 0$	$b_1 < b_2 = 0$	$(-\infty, -a/b_1]$	$(-\infty, x - a/b_1]$
4	$a < 0$	$b_1 < 0 < b_2$	$(-\infty, -a/b_1] \cup$	$(-\infty, x - a/b_1] \cup$
			$[-a/b_2, +\infty)$	$[x - a/b_2, +\infty)$
5	$a < 0$	$0 = b_1 < b_2$	$[-a/b_2, +\infty)$	$[x - a/b_2, +\infty)$
6	$a > 0$	$b_1 < b_2 = 0$	$[-a/b_1, +\infty)$	$[x - a/b_1, +\infty)$
7	$a > 0$	$b_1 < 0 < b_2$	$(-\infty, -a/b_2] \cup$	$(-\infty, x - a/b_2] \cup$
			$[-a/b_1, +\infty)$	$[x - a/b_1, +\infty)$
8	$a > 0$	$0 = b_1 < b_2$	$(-\infty, -a/b_2]$	$(-\infty, x - a/b_2]$

The general rules for subtraction of the type of sets which occur in column 4 of Table 5, from a real number x are:

$$
\begin{aligned}
x - (-\infty, +\infty) &= (-\infty, +\infty), \\
x - (-\infty, +y] &= [x - y, +\infty), \\
x - [y, +\infty) &= (-\infty, x - y], \\
x - (-\infty, y] \cup [z, +\infty) &= (-\infty, x - z] \cup [x - y, +\infty), \\
x - [\,] &= [\,].
\end{aligned}
$$

If in any arithmetic operation an operand is the empty set the result of the operation is also the empty set.

At the end of this Section we briefly summarize what has been developed so far.

In Section 3 we have considered the powerset $I\!\!P I\!\!R$ of real numbers and the subset $I\!\!I\!\!R$ of closed and bounded intervals over $I\!\!R$. Arithmetic operations have been defined in $I\!\!P I\!\!R$ by (5). We have seen that with these operations $I\!\!I\!\!R$ is an algebraically closed subset of $I\!\!P I\!\!R$ if division by an interval which contains zero is excluded.

With respect to the order relation $\leq$, $\{I\!\!R, \leq\}$ is a linearly ordered set. With respect to the order relation $\subseteq$, $\{I\!\!I\!\!R, \subseteq\}$ is an ordered set. Both sets $\{I\!\!R, \leq\}$ and $\{I\!\!I\!\!R, \subseteq\}$ are conditionally complete lattices (i.e. every non empty, bounded subset has an infimum and a supremum).

In this section we have completed the set $\{I\!\!R, \leq\}$ by adjoining a least element $-\infty$ and a greatest element $+\infty$. This leads to the set $I\!\!R^* := I\!\!R \cup \{-\infty\} \cup \{+\infty\}$. $\{I\!\!R^*, \leq\}$ then is a complete lattice (i.e. every subset has an infimum and a supremum). Similarly we have completed the set $\{I\!\!I\!\!R^*, \subseteq\}$ by adjoining the empty set $[\,]$ as a least element. This leads to the set $\overline{I\!\!I\!\!R^*} := I\!\!I\!\!R^* \cup \{[\,]\}$. $\{\overline{I\!\!I\!\!R^*}, \subseteq\}$ then is a complete lattice also.

Then we have extended interval division A/B, with $A, B \in I\!\!I\!\!R$ to the case $0 \in B$. We have seen, that division by an interval of $I\!\!I\!\!R$ which contains zero is well defined in $I\!\!P I\!\!R$ and that the result always is an element of $I\!\!P I\!\!R$, i.e. a set of real numbers.

This is an important result. We stress the fact that the result of division by an interval of $I\!\!I\!\!R$ which contains zero is not an element of $I\!\!I\!\!R$ nor of $\overline{I\!\!I\!\!R^*}$. Thus definition by an interval that contains zero does not require definition of arithmetic operations in the completed set of intervals $\overline{I\!\!I\!\!R^*} := I\!\!I\!\!R^* \cup \{[\,]\}$. Although this is often done in the literature, we have not done so here. Thus complicated definitions and rules for computing with intervals like $[-\infty, -\infty], [-\infty, 0], [0, +\infty], [+\infty, +\infty]$, and $[-\infty, +\infty]$ need not be considered. Putting aside such details makes interval arithmetic more friendly for the user.

Particular and important applications in the field of enclosure methods and validated numerics may require the definition of arithmetic operations in the complete lattice $I\!\!I\!\!R^*$. The infrequent occurrence of such applications certainly justifies leaving their realization to software and to the user. This paper aims to set out those central properties of interval arithmetic which should effectively be supported by the computer's hardware, by basic software, and by the programming languages.

∞ takes on a more subtle meaning in complex arithmetic. We defer consideration of complex arithmetic and complex interval arithmetic on the computer to a follow up paper.

7   The Extended Interval Newton Method

The extended Interval Newton Method can be used to compute enclosures of all zeros of a continuously differentiable function $f(x)$ in a given interval X. The iteration scheme is identical to the one defined by (26) in Section 5:

$$X_{\nu+1} := (m(X_\nu) - \frac{f(m(X_\nu))}{F'(X_\nu)}) \cap X_\nu = N(X_\nu) \cap X_\nu, \quad \nu = 0, 1, 2, \ldots,$$

with $X_0 := X$. Here again $F'(X_\nu)$ is the interval evaluation of the first derivative $f'(x)$ of the function f over the interval X_ν, and $m(X_\nu)$ is any point out of X_ν, the midpoint for example. If $f(x)$ has more than one zero in X, then the derivative $f'(x)$ has at least one zero (horizontal tangent of $f(x)$) in X also, and the interval evaluation $F'(X)$ of $f'(x)$ contains zero. Thus extended interval arithmetic has to be used to execute the Newton operator

$$N(X) = x - \frac{f(x)}{F'(X)}, \text{ with } x \in X.$$

As shown by Tables 3 and 4 the result is no longer an interval of $I\!R$. It is an element of the powerset $I\!P\!R$ which, with the exception of case 2, stretches continuously to $-\infty$ or $+\infty$ or both. The intersection $N(X) \cap X$ with the finite interval X then produces a finite set again. It may consist of a finite interval of $I\!R$, or of two separate such intervals, or of the empty set. These sets are now the starting values for the next iteration. This means that in the case where two separate intervals have occurred, the iteration has to be continued with two different starting values. This situation can occur repeatedly. On a sequential computer where only one iteration can be performed at a time all intervals which are not yet dealt with are collected in a list. This list then is treated sequentially. If more than one processor is available different subintervals can be dealt with in parallel.

Again, we illustrate this process by a simple example. The starting interval is denoted by $X = [x_1, x_2]$ and the result of the Newton operator by $N = [n_1, n_2]$. See Fig. 4.

Now $F'(x)$ is again a superset of all slopes of tangents of $f(x)$ in the interval $X = [x_1, x_2]$. $0 \in F'(X)$. $N(X)$ again is the set of zeros of straight lines through $(x, f(x))$ with slopes out of $F'(x)$. Let be $F'(x) = [s_1, s_2]$. Since $0 \in F'(x)$ we have $s_1 \leq 0$ and $s_2 \geq 0$. The straight lines through $(x, f(x))$

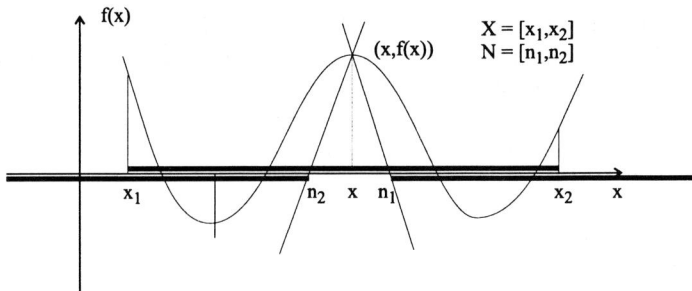

Fig. 4. Geometric interpretation of the extended Interval Newton Method.

with the slopes s_1 and s_2 cut the real axis in n_1 and n_2. Thus the Newton operator produces the set

$$N(X) = (-\infty, n_2] \cup [n_1, +\infty).$$

Intersection with the original set X (the former iterate) delivers the set

$$X_1 = N(X) \cap X = [x_1, n_2] \cup [n_1, x_2]$$

consisting of two finite intervals of $I\!R$. From this point the iteration has to be continued with the two starting intervals $[x_1, n_2]$ and $[n_1, x_2]$.

Remark: In case of division of a finite interval $A = [a_1, a_2]$ by an interval $B = [b_1, b_2]$ which contains zero, 8 non overlapping cases of the result were distinguished in Table 3 and its context. Applied to the Newton operator these 8 cases resulted in the 8 cases in Table 5. So far we have discussed the behaviour of the Interval Newton Method in the cases 3 to 8 of Table 5. We are now going to consider and interpret the particular cases 1 and 2 of Table 5 which, of course, may also occur. In Table 5 a stands for the function value and B is the enclosure of all derivatives of $f(x)$ in the interval X.

Case 2 in Table 5 is easy to interpret. If $B \equiv 0$ in the entire interval X then $f(x)$ is a constant in X and its value is $f(x) = a \neq 0$. So the occurrence of the empty interval in the Newton iteration indicates that the function $f(x)$ is a constant.

In case 1 of Table 5 the result of the Newton operator is the interval $(-\infty, +\infty)$. In this case the intersection with the former iterate X does not reduce the interval and delivers the interval X again. The Newton iteration does not converge! In this case the function value a is zero (or the numerator A in case 1 of Table 3 contains zero) and a zero has already been found.

In order to avoid rounding noise and to obtain safe bounds for the solution the value x may be shifted by a small ϵ to the left or right. This may transform case 1 into one of the other cases 2 to 8.

However, since $0 \in B$ in case 1, normally case 1 will indicate a multiple zero at the point x. This case can be further evaluated by applying the Interval

Newton Method to the derivative f' of f. The values of f' as well as enclosures $F''(X)$ for the second derivative $f''(x)$ can be obtained by differentiation arithmetic (automatic differentiation) which will be dealt with in the next section.

8   Differentiation Arithmetic, Enclosures of Derivatives

For many applications in scientific computing it is necessary to compute the value of the derivative of a function. The Interval Newton Method requires the computation of an enclosure of the first derivative of the function over an interval. The typical "school method" first computes a formal expression for the derivative of the function by applying well known rules of differentiation. Then this expression is evaluated for a point or an interval. Differentiation arithmetic avoids the computation of a formal expression for the derivative. It computes values or enclosures of derivatives just by computing with numbers or intervals. We are now going to sketch this method for the simplest case where the value of the first derivative is to be computed. If $u(x)$ and $v(x)$ are differentiable functions then the following rules for the computation of the derivative of the sum, difference, product, and quotient of the functions are well known:

$$
\begin{aligned}
(u(x) \ \pm \ v(x))' \ &= \ u'(x) \pm v'(x), \\
(u(x) \ * \ v(x))' \ &= \ u'(x)v(x) + u(x)v'(x), \\
(u(x) \ / \ v(x))' \ &= \ \tfrac{1}{v^2(x)}(u'(x)v(x) - u(x)v'(x)) \\
&= \ \tfrac{1}{v(x)}(u'(x) - \tfrac{u(x)}{v(x)}v'(x)).
\end{aligned}
\tag{35}
$$

These rules can be used to define an arithmetic for ordered pairs of numbers, similar to complex arithmetic or interval arithmetic. The first component of the pair consists of a function value $u(x_0)$ at a point x_0. The second component consists of the value of the derivative $u'(x_0)$ of the function at the point x_0. For brevity we simply write for the pair of numbers (u, u'). Then the following arithmetic for pairs follows immediately from (35):

$$
\begin{aligned}
(u, u') \ + \ (v, v') \ &= \ (u + v, u' + v'), \\
(u, u') \ - \ (v, v') \ &= \ (u - v, u' - v'), \\
(u, u') \ * \ (v, v') \ &= \ (u * v, u'v + uv'), \\
(u, u') \ / \ (v, v') \ &= \ (u/v, \tfrac{1}{v}(u' - (\tfrac{u}{v})v')), \quad v \neq 0.
\end{aligned}
\tag{36}
$$

The set of rules (36) is called differentiation arithmetic. It is an arithmetic which deals just with numbers. The rules (36) are easily programmable and are executable by a computer. These rules are now used to compute simultaneously the value and the value of the derivative of a real function at a point x_0. For brevity we call these values the function-derivative-value-pair. Why and how can this computation be done?

Earlier in this paper we have defined a (computable) real function by an arithmetic expression in the manner that arithmetic expressions are usually defined in a programming language. Apart from the arithmetic operators $+$, $-$, $*$, and $/$, arithmetic expressions contain only three kinds of operands as basic ingredients. These are constants, variables and certain differentiable elementary functions as, for instance, $\exp, \ln, \sin, \cos, \mathrm{sqr}, \ldots$. The derivatives of these functions are also well known.

If for a function $f(x)$ a function-derivative-value-pair is to be computed at a point x_0, all basic ingredients of the arithmetic expression of the function are replaced by their particular function-derivative-value-pair by the following rules:

$$
\begin{aligned}
\text{a constant:} &\quad c &&\longrightarrow (c, 0), \\
\text{the variable:} &\quad x_0 &&\longrightarrow (x_0, 1), \\
\text{the elementary functions:} &\quad \exp(x_0) &&\longrightarrow (\exp(x_0), exp(x_0)), \\
&\quad \ln(x_0) &&\longrightarrow (\ln(x_0), 1/x_0), \\
&\quad \sin(x_0) &&\longrightarrow (\sin(x_0), \cos(x_0)), \\
&\quad \cos(x_0) &&\longrightarrow (\cos(x_0), -\sin(x_0)), \\
&\quad \mathrm{sqr}(x_0) &&\longrightarrow (\mathrm{sqr}(x_0), 2x_0), \\
&\quad \text{and so on.}
\end{aligned}
\tag{37}
$$

Now the operations in the expression are executed following the rules (36) of differentiation arithmetic. The result is the function-derivative-value-pair $(f(x_0), f'(x_0))$ of the function f at the point x_0.

Example: For the function $f(x) = 25(x - 1)/(x^2 + 1)$ the function value and the value of the first derivative are to be computed at the point $x_0 = 2$. Applying the substitutions (37) and the rules (36) we obtain

$$
(f(2), f'(2)) = \frac{(25, 0)((2, 1) - (1, 0))}{(2, 1)(2, 1) + (1, 0)} = \frac{(25, 0)(1, 1)}{(4, 4) + (1, 0)} = \frac{(25, 25)}{(5, 4)} = (5, 1).
$$

Thus $f(2) = 5$ and $f'(2) = 1$.

If in the arithmetic expression for the function $f(x)$ elementary functions occur in composed form, the chain rule has to be applied, for instance

$$
\begin{aligned}
\exp(u(x_0)) &\longrightarrow (\exp(u(x_0)), \exp(u(x_0)) \cdot u'(x_0)) = (\exp u, u' \exp u), \\
\sin(u(x_0)) &\longrightarrow (\sin(u(x_0)), \cos(u(x_0)) \cdot u'(x_0)) = (\sin u, u' \cos u),
\end{aligned}
$$

and so on.

Example: For the function $f(x) = \exp(\sin(x))$ the value and the value of the first derivative are to be computed for $x_0 = \pi$. Applying the above rules we obtain

$$
\begin{aligned}
(f(\pi), f'(\pi)) &= (\exp(\sin(\pi)), \exp(\sin(\pi)) \cdot \cos(\pi)) \\
&= (\exp(0), -\exp(0)) = (1, -1).
\end{aligned}
$$

Thus $f(\pi) = 1$ and $f'(\pi) = -1$.

Differentiation arithmetic is often called automatic differentiation. All operations are performed with numbers. A computer can easily and safely execute these operations though people cannot.

Automatic differentiation is not restricted to real functions which are defined by an arithmetic expression. Any real algorithm in essence evaluates a real expression or the value of one or several real functions. Substituting for all constants, variables and elementary functions their function-derivative-value-pair, and performing all arithmetic operations by differentiation arithmetic, computes simultaneously the function-derivative-value-pair of the result. Large program packages have been developed which do just this in particular for problems in higher dimensions.

Automatic differentiation or differentiation arithmetic simply uses the arithmetic expression or the algorithm for the function. A formal arithmetic expression or algorithm for the derivative does not explicitly occur. Of course an arithmetic expression or algorithm for the derivative is evaluated indirectly. However, this expression remains hidden. It is evaluated by the rules of differentiation arithmetic. Similarly if differentiation arithmetic is performed for a fixed interval X_0 instead of for a real point x_0, an enclosure of the range of function values and an enclosure of the range of values of the derivative over that interval X_0 are computed simultaneously. Thus, for instance, neither the Newton Method nor the Interval Newton Method requires that the user provide a formal expression for the derivatives. The derivative or an enclosure for it are computed just by use of the expression for the function itself.

Automatic differentiation allows many generalizations which all together would fill a thick book. We mention only a few of these.

If the value or an enclosure of the second derivative is needed one would use triples instead of pairs and extend the rules (36) for the third component by corresponding rules: $u''+v'', u''-v'', uv''+2u'v'+vu'', \ldots$. In the arithmetic expression a constant c would now have to be replaced by the triple $(c, 0, 0)$, the variable x by $(x, 1, 0)$ and the elementary functions also by a triple with the second derivative as the third component.

Another generalization is Taylor arithmetic. It works with tuples where the first component represents the function value and the following components represent the successive Taylor coefficients. The remainder term of an integration routine for a definite integral or for an initial value problem of an ordinary differential equation usually contains a derivative of higher order. Interval Taylor arithmetic can be used to compute a safe enclosure of the remainder term over an interval. This enclosure can serve as an indicator for automatic step size control.

In arithmetics like complex arithmetic, rational arithmetic, matrix or vector arithmetic, interval arithmetic, differentiation arithmetic and Taylor arithmetic, the arithmetic itself is predefined and can be hidden in the runtime system of the compiler. The user calls the arithmetic operations by the usual operator symbols. The desired arithmetic is activated by type specification of the operands.

```
program sample;
use itaylor;
function f(x: itaylor): itaylor[lb(x)..ub(x)];
begin f := exp(5000/(sin(11+sqr(x/100))+30));
end;
var a: interval; b, fb: itaylor[0..40];
begin
        read(a);
        expand(a,b);
        fb := f(b);
        writeln ('36th Taylor coefficient: ', fb[36]);
        writeln ('40th Taylor coefficient: ', fb[40]);
end.
Test results: a = [1.001, 1.005]
36th Taylor coefficient: [-2.4139E+002, -2.4137E+002]
40th Taylor coefficient: [ 1.0759E-006, 1.0760E-006]
```

Fig. 5. Computation of enclosures of Taylor coefficients

As an example the PASCAL-XSC program shown in Fig. 5 computes and prints enclosures of the 36th and the 40th Taylor-coefficient of the function

$$f(x) = \exp(5000/(\sin(11 + sqr(x/100)) + 30))$$

over the interval $a = [1.001, 1.005]$.

First the interval a is read. Then it is expanded into the 41-tuple of its Taylor coefficients $(a, 1, 0, 0, \ldots, 0)$ which is kept in b. Then the expression for $f(x)$ is evaluated in interval Taylor arithmetic and enclosures of the 36th and the 40th Taylor coefficient over the interval a are printed.

Automatic differentiation develops its full power in the case of differentiable functions of several real variables. For instance, values or enclosures of the gradient

$$\nabla f = (\frac{\partial f}{\partial x_1}, \frac{\partial f}{\partial x_2}, \ldots, \frac{\partial f}{\partial x_n})$$

of a function $f : I\!R^n \to I\!R$ or the Jacobian or Hessian matrix can be computed directly from the expression for the function f. No formal expressions for the derivatives are needed. A particular mode, the so called reverse mode, allows a considerable acceleration for many algorithms of automatic differentiation. In the particular case of the computation of the gradient the following inequality can be shown to hold:

$$A(f, \nabla f) \leq 5A(f).$$

Here $A(f, \nabla f)$ denotes the number of operations for the computation of the gradient including the function evaluation, and $A(f)$ the number of operations for the function evaluation. For more details see [49,15,16].

9   Interval Arithmetic on the Computer

So far the basic set of all our considerations was the set of real numbers $I\!R$ or the set of extended real numbers $I\!R^* := I\!R \cup \{-\infty\} \cup \{+\infty\}$. Actual computations, however, can only be carried out on a computer. The elements of $I\!R$ and $I\!I\!R$ are in general not representable and the arithmetic operations defined for them are not executable on the computer. So we have to map these spaces and their operations onto computer representable subsets. Typical such subsets are floating-point systems, for instance, as defined by the IEEE arithmetic standard. However, in this article we do not assume any particular number representation and data format of the computer representable subsets. The considerations should apply to other data formats as well. Nevertheless, all essential properties of floating-point systems are covered.

We assume that R is a finite subset of computer representable elements of $I\!R$ with the following properties:

$$0, 1 \in R \text{ and for all } a \in R \text{ also } -a \in R.$$

The least positive non zero element of R will be denoted by L and the greatest positive element of R by G. Let be $R^* := R \cup \{-\infty\} \cup \{+\infty\}$.

Now let $\nabla : I\!R^* \to R^*$ and $\triangle : I\!R^* \to R^*$ be mappings of $I\!R^*$ onto R^* with the property that for all $a \in I\!R^*$, ∇a is the greatest lower bound of a in R^* and $\triangle a$ is the least upper bound of a in R^*. These mappings have the following three properties which also define them uniquely [33,34]:

(R1)	$\nabla a = a$ for all $a \in R^*$,	$\triangle a = a$ for all $a \in R^*$,
(R2)	$a \leq b \Rightarrow \nabla a \leq \nabla b$ for $a, b \in I\!R^*$,	$a \leq b \Rightarrow \triangle a \leq \triangle b$ for $a, b \in I\!R^*$,
(R3)	$\nabla a \leq a$ for all $a \in I\!R^*$,	$a \leq \triangle a$ for all $a \in I\!R^*$.

Because of these properties ∇ is called the monotone rounding downwards and $\triangle$ is called the monotone rounding upwards. The mappings ∇ and $\triangle$ are not independent of each other. The following equalities hold for them:

$$\nabla a = -\triangle(-a) \wedge \triangle a = -\nabla(-a) \text{ for all } a \in I\!R^*.$$

With the roundings ∇ and $\triangle$ arithmetic operations $\underline{\nabla}$ and $\underline{\triangle}, \circ \in \{+, -, *, /\}$ can be defined in R by:

(RG)	$a \underline{\nabla} b := \nabla(a \circ b)$	for all $a, b \in R$ and all $\circ \in \{+, -, *, /\}$,
	$a \underline{\triangle} b := \triangle(a \circ b)$	for all $a, b \in R$ and all $\circ \in \{+, -, *, /\}$,
		with $b \neq 0$ in case of division.

For elements $a, b \in R$ (floating-point numbers, for instance) these operations approximate the correct result $a \circ b$ in $I\!R$ by the greatest lower bound $\nabla(a \circ b)$ and the least upper bound $\triangle(a \circ b)$ in R^* for all operations $\circ \in \{+, -, *, /\}$.

In the particular case of floating-point numbers, the IEEE arithmetic standard, for instance, requires the roundings ∇ and $\triangle$, and the corresponding operations defined by (RG). As a consequence of this all processors that provide IEEE arithmetic are equipped with the roundings ∇ and $\triangle$ and the eight operations $\underline{\nabla}, \overline{\nabla}, \underline{\nabla}, \underline{\nabla}, \underline{\triangle}, \underline{\triangle}, \underline{\triangle}$, and $\underline{\triangle}$. On any computer each one of these roundings and operations should be provided by a single instruction which is directly supported by the computer hardware.

The IEEE arithmetic standard [77], however, separates the rounding from the arithmetic operation. First the rounding mode has to be set then one of the operations $\underline{\nabla}, \overline{\nabla}, \underline{\nabla}, \underline{\nabla}, \underline{\triangle}, \underline{\triangle}, \underline{\triangle}$, and $\underline{\triangle}$ may be called. This slows down these operations and interval arithmetic unnecessarily and significantly.

In the preceding sections we have defined and studied the set of intervals $I\!I\!R^*$. We are now going to approximate intervals of $I\!I\!R^*$ by intervals over R^*. We consider intervals over $I\!R^*$ with endpoints in R^* of the form

$$[a_1, a_2] = \{x \in I\!R^* \mid a_1, a_2 \in R^*, a_1 \leq x \leq a_2\}.$$

The set of all such intervals is denoted by $I\!R^*$. The empty set $[\,]$ is assumed to be an element of $I\!R^*$ also. Then $I\!R^* \subseteq I\!I\!R^*$. Note that an interval of $I\!R^*$ represents a continuous set of real numbers. It is not just a set of elements of R^*! Only the bounds of intervals of $I\!R^*$ are restricted to be elements of R^*.

With $I\!I\!R^*$ also the subset $I\!R^*$ is an ordered set with respect to both order relations $\leq$ and $\subseteq$. It can be shown that $I\!R^*$ is a complete sublattice of $I\!I\!R^*$ with respect to both order relations. For a complete proof of these properties see [33,34]. For completeness we list the order and lattice operations in both cases. We assume that $A = [a_1, a_2]$, and $B = [b_1, b_2]$ are elements of $I\!R^*$.

$\{I\!R^*, \leq\}$: $[a_1, a_2] \leq [b_1, b_2] :\Leftrightarrow a_1 \leq b_1 \wedge a_2 \leq b_2$.
The least element of $I\!R^*$ with respect to $\leq$ is the interval $[-\infty, -\infty]$. The greatest element is $[+\infty, +\infty]$. The infimum and supremum respectively of a subset $S \subseteq I\!R^*$ with respect to $\leq$ are with $A = [a_1, a_2] \in S$:

$$\inf_{\leq} S = [\inf_{A \in S} a_1, \inf_{A \in S} a_2], \quad \sup_{\leq} S = [\sup_{A \in S} a_1, \sup_{A \in S} a_2].$$

Since R^* and $I\!R^*$ only contain a finite number of elements these can also be written

$$\inf_{\leq} S = [\min_{A \in S} a_1, \min_{A \in S} a_2], \quad \sup_{\leq} S = [\max_{A \in S} a_1, \max_{A \in S} a_2].$$

$\{I\!R^*, \subseteq\}$: $[a_1, a_2] \subseteq [b_1, b_2] :\Leftrightarrow b_1 \leq a_1 \wedge a_2 \leq b_2$.

The least element of IR^* with respect to $\subseteq$ is the empty set $[\ ]$. The greatest element is the interval $[-\infty, +\infty]$. The infimum and supremum respectively of a subset $S \in IR^*$ with respect to $\subseteq$ are with $A = [a_1, a_2] \in S$

$$\inf_{\subseteq} S = [\sup_{A \in S} a_1, \inf_{A \in S} a_2], \quad \sup_{\subseteq} S = [\inf_{A \in S} a_1, \sup_{A \in S} a_2].$$

Because of the finiteness of R^* and IR^* these can also be written

$$\inf_{\subseteq} S = [\max_{A \in S} a_1, \min_{A \in S} a_2], \quad \sup_{\subseteq} S = [\min_{A \in S} a_1, \max_{A \in S} a_2],$$

i.e. the infimum is the intersection and the supremum is the interval (convex) hull of all intervals of S. As in the case of $I\!I\!R^*$ we shall use the usual mathematical symbols $\bigcap S$ for $\inf_{\subseteq} S$ and $\bigcup S$ for $\sup_{\subseteq} S$. The intersection may be empty. If in particular S consists of just two elements $A = [a_1, a_2]$ and $B = [b_1, b_2]$ this reads:

$$
\begin{aligned}
A \cap B &= [\max(a_1, b_1), \min(a_2, b_2)] & \text{intersection,} \\
A \cup B &= [\min(a_1, b_1), \max(a_2, b_2)] & \text{interval hull.}
\end{aligned}
$$

Thus, for both order relations $\leq$ and $\subseteq$ for any subset S of IR^* the infimum and supremum are the same as taken in $I\!I\!R^*$. This is by definition the criterion for a subset of a complete lattice to be a complete sublattice. So we have the results:

$$
\begin{aligned}
\{IR^*, \leq\}, & \quad \text{is a complete sublattice of} & \{I\!I\!R^*, \leq\}, & \quad \text{and} \\
\{IR^*, \subseteq\}, & \quad \text{is a complete sublattice of} & \{I\!I\!R^*, \subseteq\}.
\end{aligned}
$$

In many applications of interval arithmetic, it has to be determined whether an interval A is strictly included in an interval B. This is formally expressed by the notation:

$$A \subset \overset{\circ}{B}. \tag{38}$$

Here $\overset{\circ}{B}$ denotes the interior of B. With $A = [a_1, a_2]$ and $B = [b_1, b_2]$ (38) is equivalent to

$$A \subset \overset{\circ}{B} :\Leftrightarrow b_1 < a_1 \wedge a_2 < b_2.$$

In general, interval calculations are employed to determine sets that include the solution to a given problem. Since the arithmetic operations in $I\!I\!R$ cannot in general be executed on the computer, they have to be approximated by corresponding operations in IR. These approximations are required to have the following properties:

(a) The result of any computation in IR always has to include the result of the corresponding computation in $I\!I\!R$.
(b) The result of the computation in IR should be as close as possible to the result of the corresponding computation in $I\!I\!R$.

For all arithmetic operations $\circ \in \{+, -, *, /\}$ in $I\!\!R$ (a) means that the computer approximation $\odot$ in IR must be defined in a way that the following inequality holds:

$$A \circ B \subseteq A \odot B \text{ for } A, B \in IR \text{ and all } \circ \in \{+, -, *, /\}. \tag{39}$$

Similar requirements must hold for the elementary functions. Earlier in this paper we have defined the interval evaluation of an elementary function f over an interval $A \in I\!\!R$ by the range of function values $f(A) = \{f(a) \mid a \in A\}$. So (a) requires that for the computer evaluation $\bigcirc f(A)$ of f the following inequality holds:

$$f(A) \subseteq \bigcirc f(A) \text{ with } A \text{ and } \bigcirc f(A) \in IR. \tag{40}$$

(39) and (40) are necessary consequences of (a). There are reasonably good realizations of interval arithmetic on computers which only fulfil property (a).

(b) is an independent additional requirement. In the cases (39) and (40) it requires that $A \odot B$ and $\bigcirc f(A)$ should be the smallest interval in IR^* that includes the result $A \circ B$ and $f(A)$ in IR^* respectively. It turns out that interval arithmetic on any computer is uniquely defined by this requirement. Realization of it actually is the easiest way to support interval arithmetic on the computer by hardware. To establish this is the aim of this paper.

We are now going to discuss this arithmetic in detail. First we define the mapping $\diamondsuit : I\!\!R^* \to IR^*$ which approximates each interval A of $I\!\!R^*$ by its least upper bound $\diamondsuit A$ in IR^* with respect to the order relation $\subseteq$. This mapping has the property that for each interval $A = [a_1, a_2] \in I\!\!R^*$ its image in IR^* is

(R) $\diamondsuit A = \diamondsuit[a_1, a_2] = [\nabla a_1, \triangle a_2]$.

This mapping $\diamondsuit$ has the following properties which also define it uniquely [33,34]:

(R1) $\diamondsuit A = A$ for all $A \in IR^*$,

(R2) $A \subseteq B \Rightarrow \diamondsuit A \subseteq \diamondsuit B$ for $A, B \in I\!\!R^*$,   (monotone)

(R3) $A \subseteq \diamondsuit A$ for all $A \in I\!\!R^*$.   (upwardly directed)

We call this mapping $\diamondsuit$ the interval rounding. It has the additional property

(R4) $\diamondsuit(-A) = -\diamondsuit(A)$,   (antisymmetry)

since with $A = [a_1, a_2]$, $-A = [-a_2, -a_1]$ and $\diamondsuit(-A) = [\nabla(-a_2), \triangle(-a_1)] = [-\triangle a_2, -\nabla a_1] = -[\nabla a_1, \triangle a_2] = -\diamondsuit A$.

The interval rounding $\lozenge : I\!R^* \to I\!R^*$ is now employed in order to define arithmetic operations $\lozenge\!\!\!\!\diamond, \circ \in \{+, -, *, /\}$ in $I\!R$, i.e. on the computer, by

$$(RG) \quad A \lozenge\!\!\!\!\diamond B := \lozenge(A \circ B) \quad \text{for all } A, B \in I\!R \text{ and } \circ \in \{+, -, *, /\},$$
$$\text{with } 0 \neq B \text{ in case of division.}$$

For intervals $A, B \in I\!R$ (for instance intervals the bounds of which are floating-point numbers) these operations approximate the correct result of the interval operation $A \circ B$ in $I\!R$ by the least upper bound $\lozenge(A \circ B)$ in $I\!R^*$ with respect to the order relation $\subseteq$ for all operations $\circ \in \{+, -, *, /\}$.

Now we proceed similarly with the elementary functions. The interval evaluation $f(A)$ of an elementary function f over an interval $A \in I\!R$ is approximated on the computer by its image under the interval rounding $\lozenge$. Consequently the following inequality holds:

$$f(A) \subseteq \lozenge f(A) \text{ with } A \in I\!R \text{ and } \lozenge f(A) \in I\!R^*.$$

Thus $\lozenge f(A)$ is the least upper bound of $f(A)$ in $I\!R^*$ with respect to the order relation $\subseteq$.

If the arithmetic operations for elements of $I\!R$ are defined by (RG) with the rounding (R) the inclusion isotony and the inclusion property hold for the computer approximations of all interval operations $\circ \in \{+, -, *, /\}$. These are simple consequences of (R2) and (R3) respectively:

Inclusion isotony:

$$A \subseteq B \wedge C \subseteq D \quad \Rightarrow \quad A \circ C \subseteq B \circ D$$
$$\overset{(R2)}{\Rightarrow} \quad \lozenge(A \circ C) \subseteq \lozenge(B \circ D)$$
$$\overset{(RG)}{\Rightarrow} \quad A \lozenge\!\!\!\!\diamond C \subseteq B \lozenge\!\!\!\!\diamond D, \text{ for all } A, B, C, D \in I\!R.$$

Inclusion property:

$$a \in A \wedge b \in B \quad \Rightarrow \quad a \circ b \in A \circ B \overset{(R3)}{\Rightarrow} a \circ b \in \lozenge(A \circ B)$$
$$\overset{(RG)}{\Rightarrow} \quad a \circ b \in A \lozenge\!\!\!\!\diamond B, \text{ for } a, b \in I\!R, A, B \in I\!R.$$

Both properties also hold for the interval evaluation of the elementary functions:

Inclusion isotony: $A \subseteq B \Rightarrow f(A) \subseteq f(B) \overset{(R2)}{\Rightarrow} \lozenge f(A) \subseteq \lozenge f(B),$
for $A, B \in I\!R$.

Inclusion property: $a \in A \Rightarrow f(a) \in f(A) \overset{(R3)}{\Rightarrow} f(a) \in \lozenge f(A),$
for $a \in I\!R, A \in I\!R$.

Note that these two properties for the elementary functions are simple consequences of (R2) and (R3) respectively only. The optimality of the rounding

$\diamondsuit : I\!\!R^* \to I\!\!R^*$ which requires that the image of an interval $A \in I\!\!R^*$ is the least upper bound in $I\!\!R^*$ is not necessarily required!

With these results we can define the computer evaluation of general arithmetic expressions and of real functions in interval arithmetic for an interval $X \in I\!\!R$. If $f(x)$ is an arithmetic expression (consisting of constants, variables, and elementary functions connected by arithmetic operations and parentheses) an interval evaluation on the computer for an interval $X \in I\!\!R$ (out of the domain of definition $D(f)$) is obtained by the following rules:

- Every constant $a \in I\!\!R$ is replaced by the interval $[\nabla a, \triangle a]$.
- Every occurrence of the variable x in the expression for $f(x)$ is replaced by the interval X.
- An elementary function $\varphi(x)$ is replaced by its computer evaluation $\diamondsuit\varphi(X)$.
- Every real operation $\circ \in \{+, -, *, /\}$ is replaced by the corresponding interval operation $\diamondsuit, \circ \in \{+, -, *, /\}$.
- The interval expression thus defined in interval arithmetic is evaluated on the computer.

This procedure extends the central properties of interval arithmetic — the inclusion isotony and the inclusion property — to computer evaluations of arithmetic expressions and of real functions in interval arithmetic.

(11) in Section 3 summarizes the explicit formulas (1), (2), (3), and (4) for the operations with intervals $A = [a_1, a_2]$ and $B = [b_1, b_2] \in I\!\!R$ by

$$A \circ B = [\min_{i,j=1,2}(a_i \circ b_j), \max_{i,j=1,2}(a_i \circ b_j)] \quad \text{for all } \circ \in \{+, -, *, /\} \text{ with}$$
$$0 \notin B \text{ in case of division.}$$

Thus, the definition of the operations in $I\!\!R$ by (RG) and of the interval rounding $\diamondsuit : I\!\!R^* \to I\!\!R^*$ by (R) leads directly to the following formula for the operations for intervals $A = [a_1, a_2]$ and $B = [b_1, b_2] \in I\!\!R$:

$$A \diamondsuit B := \diamondsuit(A \circ B) = [\nabla \min_{i,j=1,2}(a_i \circ b_j), \triangle \max_{i,j=1,2}(a_i \circ b_j)], \tag{41}$$
$$\circ \in \{+, -, *, /\}, \text{ with } 0 \notin B \text{ in case of division.}$$

Since $\nabla : I\!\!R^* \to R^*$ and $\triangle : I\!\!R^* \to R^*$ are monotone mappings (R2), we obtain

$$A \diamondsuit B := \diamondsuit(A \circ B) = [\min_{i,j=1,2}(a_i \nabla b_j), \max_{i,j=1,2}(a_i \triangle b_j)], \tag{42}$$
$$\circ \in \{+, -, *, /\}, \text{ with } 0 \notin B \text{ in case of division.}$$

Employing this equation and the explicit formulas for the arithmetic operations in $I\!\!R$ listed under I, II, III, IV, V, VI, in Section 6 leads to the following formulas for the execution of the arithmetic operations $\diamondsuit, \circ \in \{+, -, *, /\}$, in $I\!\!R$ on the computer for intervals $A = [a_1, a_2]$ and $B = [b_1, b_2] \in I\!\!R$:

I. Equality: $[a_1, a_2] = [b_1, b_2] :\Leftrightarrow a_1 = b_1, a_2 = b_2.$

II. Addition: $\quad\quad\quad [a_1, a_2] \diamondplus [b_1, b_2] := [a_1 \triangledown b_1, a_2 \triangle b_2]$.

III. Subtraction: $\quad\quad [a_1, a_2] \diamondminus [b_1, b_2] := [a_1 \triangledown b_2, a_2 \triangle b_1]$.

IV. Negation: $\quad\quad\; A = [a_1, a_2], -A = [-a_2, -a_1]$.

V. Multiplication: $\;$ see Table 6.

VI. Division, $0 \notin B$: see Table 7.

Table 6. Multiplication of two intervals $A, B \in IR$ on the computer.

$A \diamondtimes B$	$b_1 \geq 0$	$b_1 < 0 < b_2$	$b_2 \leq 0$
$a_1 \geq 0$	$[a_1 \triangledown b_1, a_2 \triangle b_2]$	$[a_2 \triangledown b_1, a_2 \triangle b_2]$	$[a_2 \triangledown b_1, a_1 \triangle b_2]$
$a_1 < 0 < a_2$	$[a_1 \triangledown b_2, a_2 \triangle b_2]$	$[\min(a_1 \triangledown b_2, a_2 \triangledown b_1),$ $\max(a_1 \triangle b_1, a_2 \triangle b_2)]$	$[a_2 \triangledown b_1, a_1 \triangle b_1]$
$a_2 \leq 0$	$[a_1 \triangledown b_2, a_2 \triangle b_1]$	$[a_1 \triangledown b_2, a_1 \triangle b_1]$	$[a_2 \triangledown b_2, a_1 \triangle b_1]$

Table 7. Division of two intervals $A, B \in IR$ with $0 \notin B$ on the computer.

$A \diamonddivide B$	$b_1 > 0$	$b_2 < 0$
$a_1 \geq 0$	$[a_1 \triangledown b_2, a_2 \triangle b_1]$	$[a_2 \triangledown b_2, a_1 \triangle b_1]$
$a_1 < 0 < a_2$	$[a_1 \triangledown b_1, a_2 \triangle b_1]$	$[a_2 \triangledown b_2, a_1 \triangle b_2]$
$a_2 \leq 0$	$[a_1 \triangledown b_1, a_2 \triangle b_2]$	$[a_2 \triangledown b_1, a_1 \triangle b_2]$

These formulas show, in particular, that the operations $\diamond, \circ \in \{+, -, *, /\}$, in IR are executable on a computer if the operations $\triangledown$ and $\triangle, \circ \in \{+, -, *, /\}$, for elements of R are available. These operations have been defined earlier in this Section by

(RG) $\quad a \triangledown b := \triangledown(a \circ b)$ and $a \triangle b := \triangle(a \circ b)$ for $a, b \in R$ and $\circ \in \{+, -, *, /\}$

$$\text{with } b \neq 0 \text{ in case of division.}$$

This in turn shows the importance of the roundings $\triangledown : IR^* \to R^*$ and $\triangle : IR^* \to R^*$.

In case of division by an interval B which contains zero, eight cases had to be distinguished in Table 3. On the computer these cases have to be performed as shown in Table 8. With $A = [a_1, a_2]$ and $B = [b_1, b_2]$ Table 9 shows the same cases as Table 8 in another representation.

The generalized Newton operator requires the subtraction of a set which tends to plus or minus infinity or both or which is the empty set from a real

Table 8. The 8 cases of the division of two intervals $A \lozenge B$, with $A, B \in I\!R$ and $0 \in B$.

case	$A = [a_1, a_2]$	$B = [b_1, b_2]$	$A \lozenge B$
1	$0 \in A$	$0 \in B$	$[-\infty, +\infty]$
2	$0 \notin A$	$B = [0, 0]$	$[\,]$
3	$a_2 < 0$	$b_1 < b_2 = 0$	$[a_2 \triangledown b_1, +\infty]$
4	$a_2 < 0$	$b_1 < 0 < b_2$	$[-\infty, a_2 \triangle b_2] \cup [a_2 \triangledown b_1, +\infty]$
5	$a_2 < 0$	$0 = b_1 < b_2$	$[-\infty, a_2 \triangle b_2]$
6	$a_1 > 0$	$b_1 < b_2 = 0$	$[-\infty, a_1 \triangle b_1]$
7	$a_1 > 0$	$b_1 < 0 < b_2$	$[-\infty, a_1 \triangle b_1] \cup [a_1 \triangledown b_2, +\infty]$
8	$a_1 > 0$	$0 = b_1 < b_2$	$[a_1 \triangledown b_2, +\infty]$

Table 9. The result of the division $A \lozenge B$, with $A, B \in I\!R$ and $0 \in B$.

$A \lozenge B$	$B = [0, 0]$	$b_1 < b_2 = 0$	$b_1 < 0 < b_2$	$0 = b_1 < b_2$
$a_2 < 0$	$[\,]$	$[a_2 \triangledown b_1, +\infty]$	$[-\infty, a_2 \triangle b_2] \cup$ $[a_2 \triangledown b_1, +\infty]$	$[-\infty, a_2 \triangle b_2]$
$a_1 \leq 0 \leq a_2$	$[-\infty, +\infty]$	$[-\infty, +\infty]$	$[-\infty, +\infty]$	$[-\infty, +\infty]$
$a_1 > 0$	$[\,]$	$[-\infty, a_1 \triangle b_1]$	$[-\infty, a_1 \triangle b_1] \cup$ $[a_1 \triangledown b_2, +\infty]$	$[a_1 \triangledown b_2, +\infty]$

number x. On the computer the corresponding rules now appear in the form

$$
\begin{aligned}
x \lozenge [-\infty, +\infty] &= [-\infty, +\infty], \\
x \lozenge [-\infty, y] &= [x \triangledown y, +\infty], \\
x \lozenge [y, +\infty] &= [-\infty, x \triangle y], \\
x \lozenge ([-\infty, y] \cup [z, +\infty]) &= [-\infty, x \triangle z] \cup [x \triangledown y, +\infty], \\
x \lozenge [\,] &= [\,].
\end{aligned}
$$

After the computation of the Interval Newton Operator the intersection with a finite interval $[c_1, c_2]$ still has to be taken in the generalized Interval Newton Method. The result may be one or two finite intervals or the empty interval $[\,]$. These cases are expressed by the following explicit formulas:

$$
\begin{aligned}
(x \lozenge [-\infty, +\infty]) \cap [c_1, c_2] &= [c_1, c_2], \\
(x \lozenge [-\infty, y]) \cap [c_1, c_2] &= [x \triangledown y, c_2] \text{ or } [\,], \\
(x \lozenge [y, +\infty]) \cap [c_1, c_2] &= [c_1, x \triangle y] \text{ or } [\,], \\
x \lozenge ([-\infty, y] \cup [z, +\infty]) \cap [c_1, c_2] &= [c_1, x \triangle z] \cup [x \triangledown y, c_2] \text{ or } [\,], \\
(x \lozenge [\,]) \cap [c_1, c_2] &= [\,] \cap [c_1, c_2] = [\,].
\end{aligned}
$$

For geometric reasons $[c_1, c_2]$ can only occur as the result of the intersection in the first case.

For interval arithmetic the roundings $\nabla : \mathbb{R}^* \to R^*$ and $\triangle : \mathbb{R}^* \to R^*$ are of particular interest. They can be defined by the following properties:

$$\nabla a := \max\{x \in R^* \mid x \leq a\}, \quad \text{monotone rounding downwards, and}$$
$$\triangle a := \min\{x \in R^* \mid x \geq a\}, \quad \text{monotone rounding upwards.}$$

The following equalities hold for ∇ and $\triangle$:

$$\nabla a = -\triangle(-a), \quad \text{and} \quad \triangle a = -\nabla(-a),$$

i.e. they can be expressed by one another.

For completeness we give an explicit description of the rounding $\nabla : \mathbb{R}^* \to R^*$ in the case that R^* is a floating-point system. A floating-point number is a real number of the form $x = m \cdot b^e$. Here m is the mantissa, b is the base of the number system in use and e is the exponent. b is an integer greater than one. The exponent is an integer between two fixed integer bounds $e1$, $e2$, and in general $e1 \leq 0 \leq e2$. The mantissa is of the form $m = \circ \sum_{i=1}^{r} d[i] \cdot b^{-i}$. Here $\circ \in \{+, -\}$ is the sign of the number. The $d[i]$ are the digits of the mantissa. In a normalized floating-point system they have the property $d[i] \in \{0, 1, \ldots b-1\}$, for all $i = 1(1)r$ and $d[1] \neq 0$. Thus $|m| < 1$. Without the condition $d[1] \neq 0$, floating-point numbers are said to be unnormalized. The set of normalized floating-point numbers does not contain zero. So zero is adjoined to R^*. For a unique representation of zero it is often assumed that $m = 0,00....0$ and $e = e1$. A floating-point system thus depends on the constants $b, r, e1$, and $e2$. Here we denote it by $R = R(b, r, e1, e2)$, then $R^* := R \cup \{-\infty\} \cup \{+\infty\}$.

In the following description of $\nabla : \mathbb{R}^* \to R^*$ we use the abbreviation $G := 0.(b-1)(b-1)\ldots(b-1) \cdot b^{e2}$ for the greatest positive floating-point number. Then we obtain for ∇a:

$$\nabla a = \begin{cases} +\infty & \text{for } a = +\infty, \\ +G & \text{for } +G \leq a < +\infty, \\ +0.a[1]a[2]\ldots a[r] \cdot b^e & \text{for } b^{e1-1} \leq a < +G, \\ +0.000\ldots 0 \cdot b^{e1} & \text{for } 0 \leq a < b^{e1-1}, \\ -0.100\ldots 0 \cdot b^{e1} & \text{for } -b^{e1-1} \leq a < 0, \\ -0.a[1]a[2]\ldots a[r] \cdot b^e & \text{for } -G \leq a < -b^{e1-1} \wedge \\ & a[r+i] = 0 \text{ for all } i \geq 1, \\ -0.100\ldots 0 \cdot b^{e+1} & \text{for } -G \leq a < -b^{e1-1} \wedge \\ & a[i] = b-1 \text{ for all } i = 1(1)r \wedge \\ & a[r+i] \neq 0 \text{ for any } i \geq 1, \\ -(0.a[1]a[2]\ldots a[r] + b^{-r}) \cdot b^e & \text{for } -G \leq a < -b^{e1-1} \wedge \\ & a[i] \neq b-1 \text{ for any } i \in \{1, 2, \ldots, r\} \wedge \\ & a[r+i] \neq 0 \text{ for any } i \geq 1, \\ -\infty & \text{for } -\infty \leq a < -G. \end{cases}$$

Using the function $[a]$ (the greatest integer less than or equal to a) the description of ∇a can be shortened:

$$\nabla a = \begin{cases} +\infty & \text{for } a = +\infty, \\ +G & \text{for } +G \le a < +\infty, \\ [m_\infty \cdot b^r] \cdot b^{-r} \cdot b^e & \text{for } b^{e1-1} \le |a| \le +G, \\ 0.00\ldots0 \cdot b^{e1} & \text{for } 0 \le a < b^{e1-1}, \\ -0.100\ldots0 \cdot b^{e1} & \text{for } -b^{e1-1} \le a < 0, \\ -\infty & \text{for } -\infty \le a < -G. \end{cases}$$

The more detailed description of ∇a above shows that a normalization may still be necessary.

A few additional but very similar cases occur if in the case $e < e1$, e is set to $e1$ and unnormalized mantissas are permitted.

In these representations for ∇a we have assumed that a floating-point number is represented by the so called sign-magnitude representation. For real numbers $a \ge 0$ the rounded value ∇a is obtained by truncation of a after the rth digit of the normalized mantissa m_∞ of a. If we denote this process by $t(a)$ (truncation), we have

$$\nabla a = t(a) \quad \text{for} \quad a \ge 0.$$

This is very easy to execute. Truncation can also be used to perform the rounding ∇a in case of negative numbers $a < 0$ if negative numbers are represented by their b-complement. Then the rounded value ∇a can be obtained by truncation of the b-complement $a + x$ of a via the process:

$$\nabla a = t(a + x) - x \quad \text{for} \quad a \ge 0, \tag{43}$$

with a suitable x. See Fig. 6.

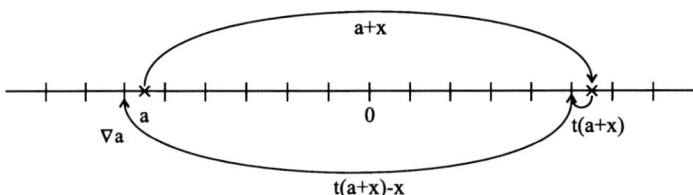

Fig. 6. Execution of the rounding ∇a in case of b-complement representation of negative numbers $a < 0$.

Example: We assume that the decimal number system is used, and that the mantissa has three decimal digits. Then we obtain for the positive real number $a = 0.354672 \cdot 10^3 \in \mathbb{R}$:

$$\nabla a = t(a) = 0.354 \cdot 10^3.$$

For the negative real number $a = -0.354672 \cdot 10^3$ we obtain obviously

$$\nabla a = -0.355 \cdot 10^3.$$

This value is obtained by application of (43) with $x = 1.00....0 \cdot 10^3$:

$$
\begin{aligned}
a + x &= 0.645328 \cdot 10^3, \\
t(a + x) &= 0.645 \cdot 10^3, \\
\nabla a &= t(a + x) - x = 0.355 \cdot 10^3.
\end{aligned}
$$

Here the easily executable b-complement has been taken twice. In between the function $t(a)$ was applied which also is easily executable. These three steps are particularly simple if the binary number system is used.

It is interesting that in case of the $(b - 1)$-complement representation of negative numbers the monotone rounding downwards ∇a cannot be executed by the function $t(a)$. This representation is isomorphic to the sign-magnitude representation.

In the preceding Sections 1 to 8 ideal interval arithmetic for elements of $I\!I\!R$ including division by an interval which contains zero has been developed. In no case did the symbols $-\infty$ and $+\infty$ occur as result of an interval operation. This is not so in this Section where interval arithmetic on the computer is considered. Here $-\infty$ and $+\infty$ can occur as result of the roundings ∇ and $\triangle$, and as result of the operations $a \nabla b$ and $a \triangle b$, $\circ \in \{+, -, *, /\}$, respectively. The interval rounding is defined by $\Diamond A := [\nabla a_1, \triangle a_2]$, and the arithmetic operations for intervals $A, B \in I\!I\!R$ are defined by $A \Diamond\!\!\!\!\diamond B := \Diamond (A \circ B)$, $\circ \in \{+, -, *, /\}$. As a consequence of this the symbols $-\infty$ and $+\infty$ can also occur as bounds of the result of an interval operation.

This happens, for instance, in case of division by an interval which contains zero, see Table 8. The extended Interval Newton Method is an example of this. We have studied this process in detail. Here very large intervals with $-\infty$ and $+\infty$ as bounds only appear intermediately. They disappear again as soon as the intersection with the previous approximation is taken. Finally the diameters of the approximations decrease to small bounds for the solution.

Among the six interval operations addition, subtraction, multiplication, division, intersection, and interval hull, the intersection is the only operation which can reduce an interval which stretches to $-\infty$ or $+\infty$ or both to a finite interval again. This step is advantageously used in the extended Interval Newton Method.

Also certain elementary functions can reduce an interval which stretches to $-\infty$ or $+\infty$ or both to a finite interval again. In such a case continuation of the computation may also be reasonable. The user has to take care that such situations are appropriately treated in his program.

In general, the appearance of $-\infty$ or $+\infty$ in the result of an interval operation indicates that an exponent overflow has occurred or that an operation or an elementary function has been called outside its range of definition. This means that the computation has gotten out of control. In this case continuation of the computation is not really recommendable. An appropriate scaling of the problem may be necessary.

Here the situation is very different from a conventional floating-point computation. In floating-point arithmetic the general directive often is just to "compute" at any price, hoping that at the end of the computation something that is reasonable will be delivered. In this process the non numbers $-\infty$, $+\infty$, and even NaN (not a number) are often treated as numbers and the computation is continued with these entities. Since a floating-point computation often flips out of control anyhow it must be the user's responsibility to control and judge the final result by other means.

In interval mathematics the general philosophy is very different. The user and the computation itself are controlling the computational process at any time. In general, an interval computation is aiming to compute small bounds for the solution of the problem. If during a computation the intervals grow overly large or even an interval appears which stretches to $-\infty$ or $+\infty$ or both, this should be taken as a severe warning. It should cause the user to think about and study the computational process again with the aim of obtaining smaller intervals. Blind continuation of the computation even with non numbers as in the case of floating-point arithmetic hoping that something reasonable will come out at the end is in strong contradiction to the philosophy and basic understanding of interval mathematics.

10   Hardware Support for Interval Arithmetic

An interval operation requires the computation of the lower and the upper bound of the resulting interval. For the four basic operations each of these bounds can be expressed by a single floating-point operation with particular bounds of the interval operands. The lower bound of the resulting interval has to be computed with rounding downwards and the upper bound with rounding upwards. While addition and subtraction are straightforward, multiplication and division require a detailed case analysis and in case of multiplication additionally a maximum / minimum computation if both interval operands contain zero. This may slow down these operations considerably in particular if the case analysis is performed in software. Thus in summary an interval operation is slower by a factor of at least two on a conventional sequential processor in comparison with the corresponding floating-point operation.

We show in this section that with dedicated hardware interval arithmetic can be made more or less as fast as simple floating-point arithmetic. The cost increase for the additional hardware is relatively modest and it is close to zero on superscalar processors. Although different in detail we follow in this Section ideas of [70,71].

We assume in this section that one arithmetic operation as well as one comparison cost one unit of time whereas switches controlled by one bit as the sign bit or data transports inside the unit are free of charge. For simplicity we denote the computer operations for intervals in this section by $+, -, *,$ and $/$. The interval operands are denoted by $A = [a_1, a_2]$ and $B = [b_1, b_2]$. The lower bound, upper bound respectively of the result is denoted by lb, ub respectively, i.e. $[lb, ub] := [a_1, a_2] \circ [b_1, b_2], \circ \in \{+, -, *, /\}$.

10.1   Addition $A + B$ and Subtraction $A - B$

The formulas for addition and subtraction

$$[a_1, a_2] + [b_1, b_2] = [a_1 \triangledown b_1, a_2 \triangle b_2],$$
$$[a_1, a_2] - [b_1, b_2] = [a_1 \triangledown b_2, a_2 \triangle b_1]$$

require no conditionals or exceptions. They show that in comparison with floating-point arithmetic a time factor of 2 is achieved with one arithmetic unit. Duplication of this unit yields a factor of 1. In the case of addition we have with one arithmetic unit sequentially:

(A) $lb := a_1 \triangledown b_1$;

(B) $ub := a_2 \triangle b_2$;

and with two arithmetic units in parallel:

(C) $lb := a_1 \triangledown b_1$; $ub := a_2 \triangle b_2$;

10.2   Multiplication $A * B$

A basic method for the multiplication of two intervals is the method of case distinction. Nine cases have been distinguished in Table 6. In eight of the nine cases one multiplication with directed rounding suffices for the computation of each bound of the resulting interval. When both interval operands contain zero as an interior point two multiplications with directed roundings and one comparison have to be performed for each bound of the resulting interval. The case selection depends on the sign bits of the interval operands. It may be performed by hardware multiplexers which select one of two inputs. On a sequential processor with one multiplier and one comparator the following algorithm solves the problem:

Algorithm 1:

The eight cases with only one multiplication for each bound can be obtained by:

(A) lb := if $(b_1 \geq 0 \vee (a_2 \leq 0 \wedge b_2 > 0))$ then a_1 else a_2

$\triangledown$ if $(a_1 \geq 0 \vee (a_2 > 0 \wedge b_2 \leq 0))$ then b_1 else b_2;

(B) ub := if $(b_1 \geq 0 \vee (a_1 \geq 0 \wedge b_2 > 0))$ then a_2 else a_1

$\triangle$ if $(a_1 \geq 0 \vee (a_2 > 0 \wedge b_1 \geq 0))$ then b_2 else b_1;

and the final case where two multiplications have to be performed for each bound by:

(C) $p := a_1 \triangledown b_2$;

(D) $q := a_2 \triangledown b_1$;

(E) $lb := \min(p, q)$; $r := a_1 \triangle b_1$;

(F) $s := a_2 \triangle b_2$;

(G) $ub := \max(r, s)$;

Taking all parts together we have:

if $(a_1 < 0 \wedge a_2 > 0 \wedge b_1 < 0 \wedge b_2 > 0)$ **then**

$\qquad \{(C),(D),(E),(F),(G)\}$

else

$\qquad \{(A),(B)\}$;

The correctness of the algorithm can be checked against the case distinctions of Table 6. The algorithm needs 5 time steps in the worst case. In all the other cases the product can be computed in two time steps.

If two multipliers and one comparator are provided the same algorithm reduces the execution time to one time step for (A), (B) and three time steps for (C), (D), (E), (F), (G). Two multiplications and a comparison can then be performed in parallel:

Algorithm 2:

(A) $lb := $ **if** $(b_1 \geq 0 \vee (a_2 \leq 0 \wedge b_2 > 0))$ **then** a_1 **else** a_2

$\qquad \triangledown$ **if** $(a_1 \geq 0 \vee (a_2 > 0 \wedge b_2 \leq 0))$ **then** b_1 **else** b_2;

$\quad ub := $ **if** $(b_1 \geq 0 \vee (a_1 \geq 0 \wedge b_2 > 0))$ **then** a_2 **else** a_1

$\qquad \triangle$ **if** $(a_1 \geq 0 \vee (a_2 > 0 \wedge b_1 \geq 0))$ **then** b_2 **else** b_1;

$\quad$ and

(B) $p := a_1 \triangledown b_2$; $q := a_2 \triangledown b_1$;

(C) $lb := \min(p, q)$; $\qquad\qquad\qquad r := a_1 \triangle b_1$; $s := a_2 \triangle b_2$;

(D) $\qquad\qquad\qquad\qquad\qquad\qquad ub := \max(r, s)$;

if $(a_1 < 0 \wedge a_2 > 0 \wedge b_1 < 0 \wedge b_2 > 0)$ **then** $\{(B),(C),(D)\}$ **else** (A);

The resulting interval is delivered either in step (A) or in step (C) (minimum) and step (D) (maximum). In step (A) one multiplication for each bound suffices while in the steps (B), (C), (D) a second multiplication and a comparison are necessasry for each bound. This case where both operands contain zero occurs rather rarely. So the algorithm shows that on a processor with two multipliers and one comparator an interval multiplication can in general be performed in the same time as a floating-point multiplication.

There are applications where a large number of interval products have to be computed consecutively. This is the case, for instance, if the scalar product of two interval vectors or a matrix vector product with interval components is to be computed. In such a case it is desirable to perform the computation in a pipeline. Algorithms like 1 and 2 can, of course, be performed in a pipeline. But in these algorithms the time needed to compute an interval product heavily depends on the data. In algorithm 1 the computation of an interval product requires 2 or 5 time steps and in algorithm 2, 1 or 3 time steps. So the pipeline would have to provide 5 time steps in case of algorithm 1 and 3 in case of algorithm 2 for each interval product. I.e. the worst case rules the pipeline. The pipeline can not easily draw advantage out of the fact that in the majority of cases the data would allow to compute the product in 2 or 1 time step, respectively.

There are other methods for computing an interval product which, although they look more complicated at first glance, lead to a more regular pipeline. These methods compute an interval product in the same number of time steps as algorithms 1 and 2. The following two algorithms display such possibilities.

Algorithm 3:

By (9.4) the interval product can be computed by the following formula:

$$A * B := [\nabla \min(a_1 * b_1, a_1 * b_2, a_2 * b_1, a_2 * b_2),$$
$$\triangle \max(a_1 * b_1, a_1 * b_2, a_2 * b_1, a_2 * b_2)].$$

This leads to the following 5 time steps for the computation of $A * B$ using 1 multiplier, 2 comparators and 2 assignments:

(A) $p := a_1 * b_1$;
(B) $q := a_1 * b_2$;
(C) $r := a_2 * b_1$; $\text{MIN} := \min(p, q)$; $\text{MAX} := \max(p, q)$;
(D) $s := a_2 * b_2$; $\text{MIN} := \min(\text{MIN}, r)$; $\text{MAX} := \max(\text{MAX}, r)$;
(E) $lb := \nabla \min(\text{MIN}, s)$; $ub := \triangle \max(\text{MAX}, s)$;

Note that here the minimum and maximum are taken from the unrounded products of double length. The algorithm always needs 5 time steps. In algorithm 1 this is the worst case.

Algorithm 4:

Using the same formula but 2 multipliers, 2 comparators and 2 assignments leads to:

(A) $p := a_1 * b_1$; $q := a_1 * b_2$;
(B) $r := a_2 * b_1$; $s := a_2 * b_2$; $\text{MIN} := \min(p, q)$; $\text{MAX} := \max(p, q)$;
(C) $\text{MIN} := \min(\text{MIN}, r)$; $\text{MAX} := \max(\text{MAX}, r)$;
(D) $lb := \nabla \min(\text{MIN}, s)$; $ub := \triangle \max(\text{MAX}, s)$;

Again the minimum and maximum are taken from the unrounded products. The algorithm needs 4 time steps. This is one time step more than the corresponding algorithm 2 using case distinction with two multipliers.

10.3   Interval Scalar Product Computation

Let us denote the components of the two interval vectors $A = (A_k)$ and $B = (B_k)$ by $A_k = [a_{k1}, a_{k2}]$ and $B_k = [b_{k1}, b_{k2}]$, $k = 1(1)n$. Then the product $A_k * B_k$ is to be computed by

$$A_k * B_k = [\min_{i,j=1,2} (a_{ki} * b_{kj}), \max_{i,j=1,2} (a_{ki} * b_{kj})], \ k = 1(1)n.$$

The formula for the scalar product now reads:

$$[lb, ub] \ := \ A \Diamond B := \Diamond (A * B) := \Diamond \sum_{k=1}^{n} A_k * B_k$$

$$:= \ [\nabla \sum_{k=1}^{n} \min_{i,j=1,2} (a_{ki} * b_{kj}), \Delta \sum_{k=1}^{n} \max_{i,j=1,2} (a_{ki} * b_{kj})].$$

This leads to the following pipeline using 1 multiplier, 2 comparators, and 2 long fixed-point accumulators (see [40]):

Algorithm 5:

(A) $p := a_{k1} * b_{k1}$;
(B) $q := a_{k1} * b_{k2}$;
(C) $r := a_{k2} * b_{k1}$;   MIN := $\min(p, q)$; MAX := $\max(p, q)$;
(D) $s := a_{k2} * b_{k2}$;   MIN := $\min(\text{MIN}, r)$; MAX := $\max(\text{MAX}, r)$;
(E) $p := a_{k+1,1} * b_{k+1,1}$; MIN := $\min(\text{MIN}, s)$; MAX := $\max(\text{MAX}, s)$;
(F) $q := a_{k+1,1} * b_{k+1,2}$; $lb := lb + \text{MIN}$; $ub := ub + \text{MAX}$;
(G) $r := a_{k+1,2} * b_{k+1,1}$; MIN := $\min(p, q)$; MAX := $\max(p, q)$;
(H) $s := a_{k+1,2} * b_{k+1,2}$; MIN := $\min(\text{MIN}, r)$; MAX := $\max(\text{MAX}, r)$;
                                   MIN := $\min(\text{MIN}, s)$; MAX := $\max(\text{MAX}, s)$;
                                   $lb := lb + \text{MIN}$; $ub := ub + \text{MAX}$;

$$lb := \nabla (lb + \text{MIN}); \ ub := \Delta (ub + \text{MAX});$$

This algorithm shows that in each sequence of 4 time steps one interval product can be accumulated. Again the minimum and maximum are taken from the unrounded products. Only at the very end of the accumulation of the bounds is a rounding applied. Then lb and ub are floating-point numbers which optimally enclose the product $A * B$ of the two interval vectors A and B.

In the algorithms 3, 4, and 5 the unrounded, double length products were compared and used for the computation of their minimum and maximum corresponding to (41). This requires comparators of double length. This can be avoided if formula (42) is used instead:

$$A * B := [\min(a_1 \nabla b_1, a_1 \nabla b_2, a_2 \nabla b_1, a_2 \nabla b_2),$$
$$\max(a_1 \triangle b_1, a_1 \triangle b_2, a_2 \triangle b_1, a_2 \triangle b_2)].$$

Computation of the 8 products $a_i \nabla b_j, a_i \triangle b_j, i, j = 1, 2$, can be avoided if the exact flag of the IEEE arithmetic is used. In general $a \nabla b$ and $a \triangle b$ differ only by one unit in the last place and we have

$$a \nabla b \leq a * b \leq a \triangle b.$$

If the computation of the product $a * b$ leads already to a floating-point number which needs no rounding, then the product is called exact and we have:

$$a \nabla b = a * b = a \triangle b.$$

If the product $a * b$ is not a floating-point number, then it is "not exact" and the product with rounding upwards can be obtained by taking the successor $a \triangle b := succ(a \nabla b)$. This changes algorithm 4, for instance, into

Algorithm 6:

(A) $p := a_1 \nabla b_1; q := a_1 \nabla b_2;$
(B) $r := a_2 \nabla b_1; s := a_2 \nabla b_2;$ $\text{MIN} := \min(p, q); \text{MAX} := \max(p, q);$
(C) $\text{MIN} := \min(\text{MIN}, r); \text{MAX} := \max(\text{MAX}, r);$
(D) $lb := \min(\text{MIN}, s); \text{MAX} := \max(\text{MAX}, s);$
(E) if $\text{MAX} = $ "exact" then $ub := \text{MAX}$ else $ub := succ(\text{MAX});$

The algorithm requires one additional step in comparison with algorithm 4 where products of double length have been compared.

10.4   Division A / B

If $0 \notin B$, 6 different cases have been distinguished as listed in Table 7. If $0 \in B$, 8 cases have to be considered. These are listed in Table 8. This leads directly to the following.

Algorithm 7:

if $b_2 < 0 \lor b_1 > 0$ then
{
$\quad lb := ($ if $b_1 > 0$ then a_1 else $a_2) \ \nabla$
$\qquad\qquad ($ if $a_1 \geq 0 \lor (a_2 > 0 \land b_2 < 0)$ then b_2 else $b_1);$
$\quad ub := ($ if $b_1 > 0$ then a_2 else $a_1) \ \triangle$
$\qquad\qquad ($ if $a_1 \geq 0 \lor (a_2 > 0 \land b_1 < 0)$ then b_1 else $b_2);$

} else {
$\quad$ if $(a_1 \leq 0 \land 0 \leq a_2 \land b_1 \leq 0 \land 0 \leq b_2)$ then $[lb, ub] := [-\infty, +\infty];$
$\quad$ if $((a_2 < 0 \lor a_1 > 0) \land b_1 = 0 \land b_2 = 0)$ then $[lb, ub] := [\ ];$
$\quad$ if $(a_2 < 0 \land b_2 = 0)$ then $[lb, ub] := [a_2 \nabla b_1, +\infty];$
$\quad$ if $(a_2 < 0 \land b_1 = 0)$ then $[lb, ub] := [-\infty, a_2 \triangle b_2];$
$\quad$ if $(a_1 > 0 \land b_2 = 0)$ then $[lb, ub] := [-\infty, a_1 \triangle b_1];$
$\quad$ if $(a_1 > 0 \land b_1 = 0)$ then $[lb, ub] := [a_1 \nabla b_2, +\infty];$
$\quad$ if $(a_2 < 0 \land b_1 < 0 \land b_2 > 0)$ then $\{\ [lb_1, ub_1] := [-\infty, a_2 \triangle b_2];$
$\qquad\qquad\qquad\qquad\qquad\qquad\qquad\quad [lb_2, ub_2] := [a_2 \nabla b_1, +\infty];\ \}$
$\quad$ if $(a_1 > 0 \land b_1 < 0 \land b_2 > 0)$ then $\{\ [lb_1, ub_1] := [-\infty, a_1 \triangle b_1];$
$\qquad\qquad\qquad\qquad\qquad\qquad\qquad\quad [lb_2, ub_2] := [a_1 \nabla b_2, +\infty];\ \}$
}

The algorithm is organized in such a way that the most complicated cases, where the result consists of two separate intervals, appear at the end. It would be possible also in these cases to write the result as a single interval which then would overlap the point infinity. In such an interval the lower bound would then be greater than the upper bound. This could cause difficulties with the order relation. So we prefer the notation with the two separate intervals. On the other hand, the representation of the result as an interval which overlaps the point infinity has advantages as well. The result of an interval division then always consists of just two bounds. In the Newton step the separation into two intervals then would have to be done by the intersection.

In practice, division by an interval that contains zero occurs infrequently. So algorithm 7 shows again that on a processor with two dividers and some multiplexer equipment an interval division can in general be performed in the same time as a floating-point division.

Variants of the algorithms discussed in this Section can, of course, also be used. In algorithm 7, for instance, the sequence of the if-statements after the *else* could be interchanged. If between these if-statements all semicolons are replaced by an *else* the result may be obtained faster.

10.5   Instruction Set for Interval Arithmetic

Convenient high level programming languages with particular data types and operators for intervals, the XSC-languages for instance [68,69,11,12,37,38,26–29,76], have been in use for more than thirty years now. Due to the lack of hardware and instruction set support for interval arithmetic, subroutine calls have to be used by the compiler to map the interval operators and comparisons to appropriate floating-point instructions. This slows down interval arithmetic by a factor close to ten compared to the corresponding floating-point arithmetic.

It has been shown in the last Section that with appropriate hardware support interval operations can be made as fast as floating-point operations. Three additional measures are necessary to let an interval calculation on the computer run at a speed comparable to the corresponding floating-point calculation:

1. Interval arithmetic hardware must be supported by the instruction set of the processor.

2. The high level programming language should provide operators for floating-point operations with directed roundings. The language must provide data types for intervals, and operators for interval operations and comparisons. It must allow overloading of names of elementary functions for interval data types.

3. The compiler must directly map the interval operators, comparisons and elementary functions of the high level programming language onto the instruction set of the processor. This mapping must not be done by slow function or subroutine calls.

From the mathematical point of view the following instructions for interval operations are desirable ($A = [a_1, a_2]$, $B = [b_1, b_2]$):

Algebraic operators:

addition	$C := A + B$	$C := [a_1 \triangledown b_1, a_2 \triangle b_2]$,
subtraction	$C := A - B$	$C := [a_1 \triangledown b_2, a_2 \triangle b_1]$,
negation	$C := -A$	$C := [-a_2, -a_1]$,
multiplication	$C := A * B$	Table 6,
division	$C := A/B, 0 \notin B$	Table 7,
	$C := A/B, 0 \in B$	Table 8,
scalar product	$C := \Diamond(A * B)$	for interval vectors $A = (A_k)$, $B = (B_k)$, see [40].

Comparisons and lattice operations:

equality	$A = B$	$a_1 = b_1, a_2 = b_2,$
less than or equal	$A \leq B$	$a_1 \leq b_1, a_2 \leq b_2,$
greatest lower bound	$C := glb\,(A, B)$	$C := [\min(a_1, b_1), \min(a_2, b_2)],$
least upper bound	$C := lub\,(A, B)$	$C := [\max(a_1, b_1), \max(a_2, b_2)],$
inclusion	$A \subseteq B$	$b_1 \leq a_1, a_2 \leq b_2,$
element of	$a \in A$	$a_1 \leq a \leq a_2,$
interval hull	$C := A \underline{\cup} B$	$C := [\min(a_1, b_1), \max(a_2, b_2)],$
intersection	$C := A \cap B$	$C := $ if $\max(a_1, b_1) \leq \min(a_2, b_2)$

$$\text{then}$$
$$[\max(a_1, b_1), \min(a_2, b_2)]$$
$$\text{else}$$
$$[\,].$$

Other comparisons can directly be obtained by comparison of bounds of the intervals A and B. With two comparators all comparisons and lattice operations can be performed in parallel in one time step.

Fast multiple precision arithmetic and fast multiple precision interval arithmetic can easily be obtained by means of the exact scalar product [40].

10.6   Final Remarks

We have seen that relatively modest hardware equipment consisting of two operation units, a few multiplexers and comparators could make interval arithmetic as fast as floating-point arithmetic. For multiplication various approaches have been compared. The case selection clearly is the most favorable one. Several of the multiplication algorithms use and compare double length products. These are not easily obtainable on most existing processors. Since the double length product is a fundamental operation for other applications as well, like complex arithmetic, the accurate dot product, vector and matrix arithmetic, we require that multipliers should have a fifth rounding mode, namely "unrounded", which enables them to provide the full double length product.

A similar situation appears in interval arithmetic, if division by an interval which contains zero is permitted. The result of an interval division then may consist of two disjoint intervals. In applications of interval arithmetic, division by an interval which contains zero has been used for 30 years now. This forces an extension of the real numbers $I\!R$ by $-\infty$ and $+\infty$ and a consideration of the complete lattice $I\!R^* := I\!R \cup \{-\infty\} \cup \{+\infty\}$ and its computer representable subset $R^* := R \cup \{-\infty\} \cup \{+\infty\}$. In the early days of interval arithmetic attempts were made to define interval arithmetic immediately in $I I\!R^*$ instead of $I I\!R$. See [23–25] and others. This leads to deep mathematical considerations. We did not follow such lines in this study. The Extended Interval Newton Method is practically the only frequent application where $-\infty$ and $+\infty$ are needed. But there they appear only in an intermediate step as auxiliary values and they disappear immediately in the next step when the intersection with the former approximation is taken.

In conventional numerical analysis Newton's method is the key method for nonlinear problems. The method converges quadratically to the solution if the initial value of the iteration is already close enough. However, it may fail in finite as well as in infinite precision arithmetic even in the case of only a single solution in a given interval. In contrast to this the interval version of Newton's method is globally convergent. It never fails, not even in rounded arithmetic. Newton's method reaches its final elegance and strength in the Extended Interval Newton Method. It encloses all (single) zeros in a given domain. It is locally quadratically convergent. The key operation to achieve these fascinating properties is division by an interval which contains zero. It separates different solutions from each other. A Method which provides for computation of all zeros of a system of nonlinear equations in a given domain is much more frequently applied than the conventional Newton method. This justifies taking division by an interval which contains zero into the basic set of interval operations, and supporting it within the instruction set of the computer.

References

1. Adams, E.; Kulisch, U.(eds.): **Scientific Computing with Automatic Result Verification.** I. Language and Programming Support for Verified Scientific Computation, II'. Enclosure Methods and Algorithms with Automatic Result Verification, III'. Applications in the Engineering Sciences. Academic Press, San Diego, 1993 (ISBN 0-12-044210-8).
2. Albrecht, R.; Alefeld, G.; Stetter, H.J. (Eds.): *Validation Numerics – Theory and Applications.* Computing Supplementum **9**, Springer-Verlag, Wien / New York, 1993.
3. Alefeld, G.: *Intervallrechnung über den komplexen Zahlen und einige Anwendungen.* Dissertation, Universität Karlsruhe, 1968.
4. Alefeld, G.: *Über die aus monoton zerlegbaren Operatoren gebildeten Iterationsverfahren.* Computing **6**, pp. 161-172, 1970.
5. Alefeld, G.; Herzberger, J.: **Einführung in die Intervallrechnung.** Bibliographisches Institut (Reihe Informatik, Nr. 12), Mannheim / Wien / Zürich, 1974 (ISBN 3-411-01466-0).
6. Alefeld, G.; Herzberger, J.: **An Introduction to Interval Computations.** Academic Press, New York, 1983 (ISBN 0-12-049820-0).
7. Alefeld, G.; Mayer, G.: *Einschließungsverfahren.* In [22, pp. 155-186], 1995.
8. Alefeld, G.; Frommer, A.; Lang, B. (eds.): *Scientific Computing and Validated Numerics.* Proceedings of SCAN-95. Akademie Verlag, Berlin, 1996. ISBN 3-05-501737-4
9. Baumhof, Ch.: *Ein Vektorarithmetik-Koprozessor in VLSI-Technik zur Unterstützung des Wissenschaftlichen Rechnens.* Dissertation, Universität Karlsruhe, 1996.

10. Blomquist, F.: *PASCAL-XSC, BCD-Version 1.0, Benutzerhandbuch für das dezimale Laufzeitsystem.* Institut für Angewandte Mathematik, Universität Karlsruhe, 1997.

11. Bohlender, G.; Rall, L. B.; Ullrich, Ch.; Wolff v. Gudenberg, J.: *PASCAL-SC: Wirkungsvoll programmieren, kontrolliert rechnen.* Bibliographisches Institut, Mannheim / Wien / Zürich, 1986 (ISBN 3-411-03113-1).

12. Bohlender, G.; Rall, L. B.; Ullrich, Ch.; Wolff v. Gudenberg, J.: *PASCAL-SC: A Computer Language for Scientific Computation.* Perspectives in Computing, Vol. 17, Academic Press, Orlando, 1987 (ISBN 0-12-111155-5).

13. Bohlender, G.: *Literature on Enclosure Methods and Related Topics.* Institut für Angewandte Mathematik, Universität Karlsruhe, pp. 1-68, 2000.

14. Collatz, L.: *Funktionalanalysis und numerische Mathematik.* Springer–Verlag, Berlin / Heidelberg / New York, 1968.

15. Fischer, H.-C.: *Schnelle automatische Differentiation, Einschließungsmethoden und Anwendungen.* Dissertation, Universität Karlsruhe, 1990.

16. Fischer, H.: *Automatisches Differenzieren.* In [22, pp. 53-104], 1995.

17. Hammer, R.; Hocks, M.; Kulisch, U.; Ratz, D.: **Numerical Toolbox for Verified Computing I: Basic Numerical Problems.** (Vol. II see [31], version in C++ see [18]) Springer–Verlag, Berlin / Heidelberg / New York, 1993.

18. Hammer, R.; Hocks, M.; Kulisch, U.; Ratz, D.: **C++ Toolbox for Verified Computing: Basic Numerical Problems.** Springer–Verlag, Berlin / Heidelberg / New York, 1995.

19. Hansen, E.: *Topics in Interval Analysis.* Clarendon Press, Oxford, 1969.

20. Hansen, E.: *Global Optimization Using Interval Analysis.* Marcel Dekker Inc., New York/Basel/Hong Kong, 1992.

21. Herzberger, J. (ed.): *Topics in Validated Computations.* Proceedings of IMACS–GAMM International Workshop on Validated Numerics, Oldenburg, 1993. North Holland, 1994.

22. Herzberger, J.: *Wissenschaftliches Rechnen, Eine Einführung in das Scientific Computing.* Akademie Verlag, 1995.

23. Kaucher, E.: *Über metrische und algebraische Eigenschaften einiger beim numerischen Rechnen auftretender Räume.* Dissertation, Universität Karlsruhe, 1973.

24. Kaucher, E.: *Algebraische Erweiterungen der Intervallrechnung unter Erhaltung der Ordnungs- und Verbandsstrukturen.* In: Albrecht, R.; Kulisch, U. (Eds.): *Grundlagen der Computerarithmetik.* Computing Supplementum 1. Springer-Verlag, Wien / New York, pp. 65-79, 1977.

25. Kaucher, E.: *Über Eigenschaften und Anwendungsmöglichkeiten der erweiterten Intervallrechnung und des hyperbolischen Fastkörpers über **R**.* In: Albrecht, R.; Kulisch, U. (Eds.): *Grundlagen der Computerarithmetik.* Computing Supplementum 1. Springer-Verlag, Wien / New York, pp. 81-94, 1977.

26. Klatte, R.; Kulisch, U.; Neaga, M.; Ratz, D.; Ullrich, Ch.: **PASCAL–XSC — Sprachbeschreibung mit Beispielen.** Springer-Verlag, Berlin/Heidelberg/New York, 1991 (ISBN 3-540-53714-7, 0-387-53714-7).

27. Klatte, R.; Kulisch, U.; Neaga, M.; Ratz, D.; Ullrich, Ch.: **PASCAL–XSC — Language Reference with Examples.** Springer-Verlag, Berlin/Heidelberg/New York, 1992.

28. Klatte, R.; Kulisch, U.; Lawo, C.; Rauch, M.; Wiethoff, A.: **C–XSC, A C++ Class Library for Extended Scientific Computing.** Springer-Verlag, Berlin/Heidelberg/New York, 1993.

29. Klatte, R.; Kulisch, U.; Neaga, M.; Ratz, D.; Ullrich, Ch.: **PASCAL–XSC — Language Reference with Examples (In Russian)**. Moscow, 1994, second edition 2000.

30. Knöfel, A.: *Hardwareentwurf eines Rechenwerks für semimorphe Skalar- und Vektoroperationen unter Berücksichtigung der Anforderungen verifizierender Algorithmen*. Dissertation, Universität Karlsruhe, 1991.

31. Krämer, W.; Kulisch, U.; Lohner, R.: *Numerical Toolbox for Verified Computing II: Theory, Algorithms and Pascal-XSC Programs*. (Vol. I see [17,18]) Springer–Verlag, Berlin / Heidelberg / New York, to appear 2001.

32. Krawczyk, R.; Neumaier, A.: *Interval Slopes for Rational Functions and Associated Centered Forms*. SIAM Journal on Numerical Analysis **22**, pp. 604-616, 1985.

33. Kulisch, U.: **Grundlagen des Numerischen Rechnens — Mathematische Begründung der Rechnerarithmetik.** Reihe Informatik, Band 19, Bibliographisches Institut, Mannheim/Wien/Zürich, 1976 (ISBN 3-411-01517-9).

34. Kulisch, U.; Miranker, W. L.: **Computer Arithmetic in Theory and Practice.** Academic Press, New York, 1981 (ISBN 0-12-428650-x).

35. Kulisch, U.; Ullrich, Ch. (Eds.): **Wissenschaftliches Rechnen und Programmiersprachen.** Proceedings of Seminar held in Karlsruhe, April 2–3, 1982. Berichte des German Chapter of the ACM, Band 10, B. G. Teubner Verlag, Stuttgart, 1982 (ISBN 3-519-02429-2).

36. Kulisch, U.; Miranker, W. L. (Eds.): **A New Approach to Scientific Computation.** Proceedings of Symposium held at IBM Research Center, Yorktown Heights, N. Y., 1982. Academic Press, New York, 1983 (ISBN 0-12-428660-7).

37. Kulisch, U. (Ed.): **PASCAL–SC: A PASCAL extension for scientific computation**, Information Manual and Floppy Disks, Version IBM PC/AT; Operating System DOS'. B. G. Teubner Verlag (Wiley-Teubner series in computer science), Stuttgart, 1987 (ISBN 3-519-02106-4 / 0-471-91514-9).

38. Kulisch, U. (Ed.): **PASCAL–SC: A PASCAL extension for scientific computation**, Information Manual and Floppy Disks, Version ATARI ST'. B. G. Teubner Verlag, Stuttgart, 1987 (ISBN 3-519-02108-0).

39. Kulisch, U. (Ed.): **Wissenschaftliches Rechnen mit Ergebnisverifikation — Eine Einführung.** Ausgearbeitet von S. Geörg, R. Hammer und D. Ratz. Vol. 58. Akademie Verlag, Berlin, und Vieweg Verlagsgesellschaft, Wiesbaden, 1989.

40. Kulisch, U.: *Advanced Aithmetic for the Digital Computer — Design of Arithmetic Units*. Electronic Notes of Theoretical Computer Science, http://www.elsevier.nl/locate/entcs/volume24.html pp. 1-72, 1999.

41. Lohner, R.: *Einschließung der Lösung gewöhnlicher Anfangs- und Randwertaufgaben und Anwendungen*. Dissertation, Universität Karlsruhe, 1988.

42. Lohner, R.: *Computation of Guaranteed Enclosures for the Solutions of Ordinary Initial and Boundary Value Problems*. pp. 425–435 in: Cash, J. R.; Gladwell, I. (Eds.): *Computational Ordinary Differential Equations*. Clarendon Press, Oxford, 1992.

43. Mayer, G.: *Grundbegriffe der Intervallrechnung*. In [39, pp. 101-117], 1989.

44. Moore, R. E.: *Interval Analysis*. Prentice Hall Inc., Englewood Cliffs, N. J.; 1966.

45. Moore, R. E.: *Methods and Applications of Interval Analysis*. SIAM, Philadelphia, Pennsylvania, 1979.

46. Moore, R. E. (Ed.): *Reliability in Computing: The Role of Interval Methods in Scientific Computing*. Proceedings of the Conference at Columbus, Ohio, September 8–11, 1987; Perspectives in Computing **19**, Academic Press, San Diego, 1988 (ISBN 0-12-505630-3).

47. Neumaier, A.: *Interval Methods for Systems of Equations*. Cambridge University Press, Cambridge, 1990.

48. Neumann, J. von; Goldstine, H. H.: *Numerical Inverting of Matrices of High Order*. Bulletin of the American Mathematical Society, 53, 11, pp. 1021-1099, 1947.

49. Rall, L. B.: *Automatic Differentiation: Techniques and Applications*. Lecture Notes in Computer Science, No. 120, Springer-Verlag, Berlin, 1981.

50. Ratschek, H.; Rokne, J.: *Computer Methods for the Range of Functions*. Ellis Horwood Limited, Chichester, 1984.

51. Ratz, D.: *Programmierpraktikum mit PASCAL-SC*. In: Höhler, G.; Staudenmaier, H. M. (Hrsg.): *Computer Theoretikum und Praktikum für Physiker*. Band **5**, Fachinformationszentrum Karlsruhe, 1990.

52. Ratz, D.: *Globale Optimierung mit automatischer Ergebnisverifikation*. Dissertation, Universität Karlsruhe, 1992.

53. Ratz, D.: *Automatic Slope Computation and its Application in Nonsmooth Global Optimization*. Shaker Verlag, Aachen, 1998.

54. Ratz, D.: *On Extended Interval Arithmetic and Inclusion Isotony*. Preprint, Institut für Angewandte Mathematik, Universität Karlsruhe, 1999.

55. Rump, S. M.: *Kleine Fehlerschranken bei Matrixproblemen*. Dissertation, Universität Karlsruhe, 1980.

56. Rump, S. M.: *How Reliable are Results of Computers? / Wie zuverlässig sind die Ergebnisse unserer Rechenanlagen?* In: *Jahrbuch Überblicke Mathematik*, Bibliographisches Institut, Mannheim, 1983.

57. Rump, S.M.: *Validated Solution of Large Linear Systems*. In [2, pp. 191-212], 1993.

58. Rump, S.M.: *Verification Methods for Dense and Sparse Systems of Equations*. In [21, pp. 63-135], 1994.

59. Schmidt, L.: *Semimorphe Arithmetik zur automatischen Ergebnisverifikation auf Vektorrechnern*. Dissertation, Universität Karlsruhe, 1992.

60. Shiriaev, D. V.: *Fast Automatic Differentiation for Vector Processors and Reduction of the Spatial Complexity in a Source Translation Environment*. Dissertation, Universität Karlsruhe, 1994.

61. Sunaga, T.: *Theory of an interval algebra and its application to numerical analysis*. RAAG Memoires 2, pp. 547-564, 1958.

62. Teufel, T.: *Ein optimaler Gleitkommaprozessor*. Dissertation, Universität Karlsruhe, 1984.

63. Ullrich, Ch. (Ed.): **Computer Arithmetic and Self-Validating Numerical Methods.** (Proceedings of SCAN 89, held in Basel, Oct. 2-6, 1989, invited papers). Academic Press, San Diego, 1990.

64. Walter, W. V.: *FORTRAN-SC, A FORTRAN Extension for Engineering / Scientific Computation with Access to ACRITH: Language Description with Examples*. In [46, pp. 43-62], 1988.

65. Walter, W. V.: *Einführung in die wissenschaftlich-technische Programmiersprache FORTRAN-SC*. ZAMM **69**, 4, T52-T54, 1989.

66. Walter, W. V.: *FORTRAN-SC: A FORTRAN Extension for Engineering / Scientific Computation with Access to ACRITH, Language Reference and User's Guide.* 2nd ed., pp. 1-396, IBM Deutschland GmbH, Stuttgart, Jan. 1989.

67. Walter, W. V.: *Flexible Precision Control and Dynamic Data Structures for Programming Mathematical and Numerical Algorithms.* Dissertation, Universität Karlsruhe, 1990.

68. Wippermann, H.-W.: *Realisierung einer Intervallarithmetik in einem ALGOL-60 System.* Elektronische Rechenanlagen **9**, pp. 224-233, 1967.

69. Wippermann, H.-W.: *Implementierung eines ALGOL-60 Systems mit Schrankenzahlen.* Elektronische Datenverarbeitung **10**, pp. 189-194, 1968.

70. Wolff v. Gudenberg, J.: *Hardware Support for Interval Arithmetic, Extended Version.* Report No. 125, Institut für Informatik, Universität Würzburg, 1995.

71. Wolff v. Gudenberg, J.: *Hardware Support for Interval Arithmetic.* In [8, pp. 32-38], 1996.

72. Wolff v. Gudenberg, J.: *Proceedings of Interval'96.* International Conference on Interval Methods and Computer Aided Proofs in Science and Engineering, Würzburg, Germany, Sep. 30 - Oct. 2, 1996. Special issue 3/97 of the journal Reliable Computing, 1997.

73. Yohe, J.M.: *Roundings in Floating-Point Arithmetic.* IEEE Trans. on Computers, Vol. C-22, No. 6, June 1973, pp. 577-586.

74. IBM: *IBM System/370 RPQ'. High Accuracy Arithmetic.* SA 22-7093-0, IBM Deutschland GmbH (Department 3282, Schönaicher Strasse 220, D-71032 Böblingen), 1984.

75. IBM: **IBM High-Accuracy Arithmetic Subroutine Library (ACRITH).** IBM Deutschland GmbH (Department 3282, Schönaicher Strasse 220, D-71032 Böblingen), 3rd edition, 1986.
    1. General Information Manual. GC 33-6163-02.
    2. Program Description and User's Guide. SC 33-6164-02.
    3. Reference Summary. GX 33-9009-02.

76. IBM: **ACRITH-XSC: IBM High Accuracy Arithmetic — Extended Scientific Computation. Version 1, Release 1.** IBM Deutschland GmbH (Schönaicher Strasse 220, D-71032 Böblingen), 1990.
    1. General Information, GC33-6461-01.
    2. Reference, SC33-6462-00.
    3. Sample Programs, SC33-6463-00.
    4. How To Use, SC33-6464-00.
    5. Syntax Diagrams, SC33-6466-00.

77. American National Standards Institute / Institute of Electrical and Electronics Engineers: *A Standard for Binary Floating-Point Arithmetic.* ANSI/IEEE Std. 754-1985, New York, 1985 (reprinted in SIGPLAN **22**, 2, pp. 9-25, 1987). Also adopted as IEC Standard 559:1989.

78. American National Standards Institute / Institute of Electrical and Electronics Engineers: *A Standard for Radix-Independent Floating-Point Arithmetic.* ANSI/IEEE Std. 854-1987, New York, 1987.

79. IMACS; GAMM: *IMACS-GAMM Resolution on Computer Arithmetic.* In Mathematics and Computers in Simulation **31**, pp. 297-298, 1989. In Zeitschrift für Angewandte Mathematik und Mechanik **70**, no. 4, p. T5, 1990.

80. IMACS; GAMM: *GAMM-IMACS Proposal for Accurate Floating-Point Vector Arithmetic.* GAMM, Rundbrief 2, pp. 9-16, 1993. Mathematics and Computers

in Simulation, Vol. **35**, IMACS, North Holland, 1993. News of IMACS, Vol. 35, No. 4, pp. 375-382, Oct. 1993.

81. SIEMENS: **ARITHMOS (BS 2000) Unterprogrammbibliothek für Hochpräzisionsarithmetik. Kurzbeschreibung, Tabellenheft, Benutzerhandbuch.** SIEMENS AG, Bereich Datentechnik, Postfach 83 09 51, D-8000 München 83. Bestellnummer U2900-J-Z87-1, Sept. 1986.

82. Sun Microsystems: **Interval Arithmetic Programming Reference, Fortran 95.** Sun Microsystems, Inc., 901 San Antonio Road, Palo Alto, CA 94303, USA, 2000.

Highly Accurate Verified Error Bounds for Krylov Type Linear System Solvers

Axel Facius

Institut für Angewandte Mathematik, Universität Karlsruhe (TH),
D-76128 Karlsruhe, Germany.
axel.facius@math.uni-karlsruhe.de

Abstract. Preconditioned Krylov subspace solvers are an important and frequently used technique for solving large sparse linear systems. There are many advantageous properties concerning convergence rates and error estimates. However, implementing such a solver on a computer, we often observe an unexpected and even contrary behavior.

The purpose of this paper is to show that this gap between the theoretical and practical behavior can be narrowed by using a problem-oriented arithmetic. In addition we give rigorous error bounds to our computed results.

1   Introduction

In Section 2 we illustrate the effect of finite precision arithmetic on preconditioned iterative solvers. Recalling some basic properties of Krylov subspace methods, Section 3 gives a deeper insight to the importance of the underlying arithmetic. Consequently, in Section 4 we discuss some arithmetic improvements which aim to narrow the gap between the theoretical and computational behavior. Since these improvements are more or less heuristic, we have to prove their effect on the quality of the computed solutions. Therefore we give a short introduction to *a-posteriori* verification techniques for linear systems in Section 5. Finally we present some numerical results in Section 6.

2   Iterative Solvers and Finite Precision

Preconditioned Krylov subspace solvers are frequently used for solving large sparse linear systems. There are many advantageous properties concerning convergence rates and error estimates but unfortunately, if we implement such a solver on a computer, we often observe an unexpected and even contrary behavior (see e.g. [10,22]).

2.1   Preconditioning

Preconditioning is very important in solving linear systems because well conditioned systems are much easier and particularly faster to solve. However, preconditioning in finite precision can cause a drastically perturbed solution. Thus an important question is: *How does preconditioning affect the solution of a linear system?*

Since in modern Krylov subspace solvers preconditioning is no separate step but an inherent part to the solver itself (see Algorithm 1), it can hardly be distinguished which part of the deviation between the solutions of a preconditioned and a non-preconditioned system is caused by the preconditioning itself and which part is induced by various other error sources.

To give an idea of the magnitude of the perturbation that can be caused by preconditioning we consider the following example. We use a Jacobi preconditioner, which is so easy to apply to an entire linear system, that it is often used to transform A and b in advance and afterwards a non-preconditioned solver is applied. This means we scale A to get a unit diagonal, i.e., we just perform one division for each element of A and b. This operation is done in IEEE double-precision. In order to identify the error caused by this scaling operation, we apply a verifying solver to the scaled system. Here we solve systems Beam(n) which result from an n-point discretization of a fourth order ODE describing the bending line of a weighted beam. For some n the resulting verified solutions of the scaled systems are shown in Figure 1. For comparison we also plotted the solutions of the original systems and the solution of the underlying continuous problem. The latter two differ so little that they appear as a single line.

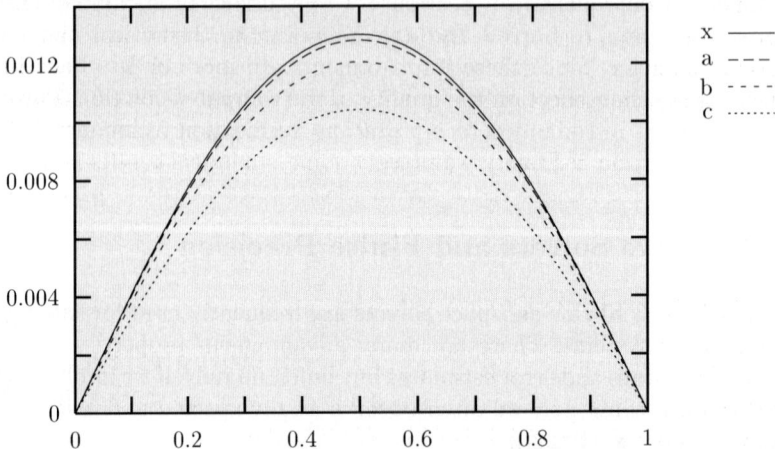

Fig. 1. The effect of preconditioning on ill-conditioned linear systems. The curves represent the solutions of the preconditioned systems a: Beam(8191), b: Beam(12287), and c: Beam(16383) with cond $\approx 10^{14}$, 10^{15}, and 10^{16}, resp. The exact solutions of the non-preconditioned systems differ so little, that they appear as a single line (x).

This example demonstrates that preconditioning in this traditional way may introduce unacceptably large errors which can be significantly larger than the discretization error ($\approx 10^{-5}$ in these cases) of the underlying continuous problem. It is possible to avoid these problems by doing verified preconditioning but this leads to an interval matrix and therefore to an often unacceptable computational effort to solve these systems. Another, maybe more practical, way might be to use a more accurate arithmetic to avoid the significant propagation of the various errors.

Applying the preconditioner at each step of an iterative solver (see Algorithm 1, right) rather than precondition the matrix and the right hand side vector in advance then we don't need any additional storage and we have the preconditioning done with the same accuracy as the iterated solution.

2.2   Convergence

One well known Krylov subspace method is the conjugate gradient algorithm (CG) of Hestenes and Stiefel [11]. It can be interpreted as a Lanczos procedure which reduces the system matrix A to a tridiagonal matrix T and a subsequent LDL^T-factorization to solve this tridiagonal system $T\tilde{x} = \tilde{b}$ [8]. In Figure 2 the Euclidean norms of the residual and the error in each step are plotted during solving the Beam(1023) system. Additionally we show the level of orthogonality of the new residual-vector r_{m+1} to the previous ones: $\max_{k=1}^{m}\{\langle r_k \mid r_{m+1}\rangle/(\|r_k\|_2\|r_{m+1}\|_2)\}$. As we can see there is no convergence at all up to step $m = 1.5n$ and particularly no convergence at the theoretically guaranteed step $m = n$. One reason is easy to identify: the residuals and thus the basis vectors of the Krylov subspace lose their orthogonality completely at $m \approx 400$ and the *basis*-vectors may even become linearly dependent. So CG can't minimize the residual in the entire $\mathbb{R}^m$ but only in a smaller subspace ($\text{span}\{r_1,\ldots,r_m\}$). Further we can observe that the error norm runs into saturation at a level of approximately 10^{-6}. This matches with the well known rule of thumb saying that we may lose up to $\log(\text{cond}(A))$ (≈ 10 in this case) digits from the 16 digits we have in IEEE double [3].

3   Krylov Subspace Methods

Let us briefly recall the basic ideas of these iterative methods and give some important properties, compare [8]. Applying Krylov methods subdivides the process of solving $Ax = b$ into two subproblems

(*i*) Compute an orthonormal basis $V_m = \{v_1,\ldots,v_m\}$ of the Krylov subspace $\mathcal{K}_m := \text{span}\{b, Ab, \ldots, A^{m-1}b\}$.

(*ii*) Orthogonal projection of the linear system from $\mathbb{R}^n$ into $\mathbb{R}^m$, i.e., find an $x_m \in \mathcal{K}_m$ that optimally solves $Ax_m = b$ (according to some additional condition).

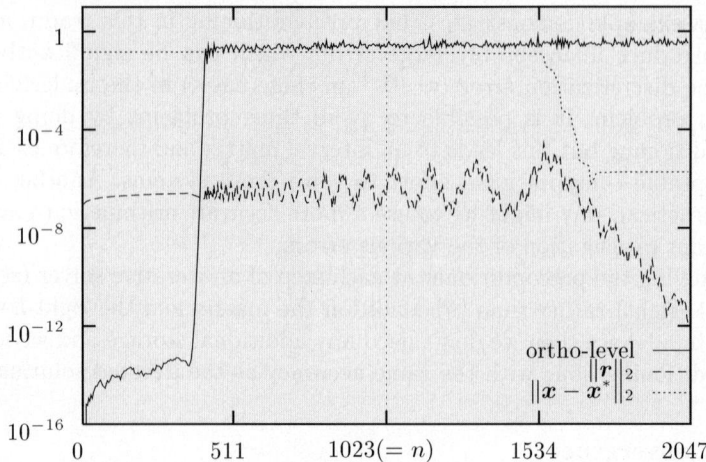

Fig. 2. The Euclidean norms of the residual and the error, and the level of orthogonality during solving the Beam(1023) system. (ortho-level $= \max_{k=1}^{m}\{\langle r_k \mid r_{m+1}\rangle/(\|r_k\|_2\|r_{m+1}\|_2)\}$)

The parts (i) and (ii) have to be repeated for $m = 1, 2, \ldots$ iteratively. In each iteration there are two possibilities. First, $A^{m-1}b \in \mathcal{K}_{m-1}$, i.e., $\dim(\mathcal{K}_m) < m$ and thus there exists no m-dimensional basis or second, $\dim(\mathcal{K}_m) = m$ and everything works fine. The good news is, that in the first case we have already computed the solution in part (ii) of the previous iteration (if we have started with $v_1 = b - Ax_0$). However, at least after n iterations[1] this situation occurs and thus we need at most n steps to solve $Ax = b$, at least in exact arithmetic.

The two subproblems (i) and (ii) can also be interpreted as first finding a matrix H_m which is unitarily similar to A and which is particularly structured, and afterwards solving the transformed system $V_m^T A V_m V_m^T x = V_m^T b \Leftrightarrow H_m \tilde{x} = \tilde{b}$ (Galerkin condition).

With the *Arnoldi-algorithm* one iteration step of (i) is performed as follows

$$\tilde{v}_{m+1} = A v_m \qquad \text{(compute a prototype for } v_{m+1})$$

$$\tilde{v}_{m+1} = \tilde{v}_{m+1} - \sum_{i=1}^{m} \underbrace{\langle v_i \mid \tilde{v}_{m+1}\rangle}_{=:h_{im}} v_i \qquad \text{(orthogonalize it against } V_m) \quad (1)$$

$$v_{m+1} = \underbrace{\frac{\tilde{v}_{m+1}}{\|\tilde{v}_{m+1}\|}}_{=:h_{m+1,m}} . \qquad \text{(and finally normalize it)}$$

---

[1] Throughout this article the letter 'n' always denotes the system dimension.

This algorithm can be written in matrix form as

$$AV_m = V_m H_m + v_{m+1} h_{m+1,m} e_m^T, \quad (V_m^T V_m = I), \tag{2}$$

where H_m is a upper Hessenberg matrix with the elements h_{ij}.

For symmetric A the situation in the Arnoldi-algorithm becomes much more favorable. By multiplying (2) with V_m^T from left we see

$$H_m = V_m^T A V_m = V_m^T A^T V_m = H_m^T.$$

So H_m turns out to be symmetric upper Hessenberg, i.e., a symmetric tridiagonal matrix T_m. The *Lanczos-algorithm* (see Algorithm 1) [15] computes $T_m = \text{tridiag}(\beta_{j-1}, \alpha_j, \beta_j)$ directly by its two 'new' components per row.

Given $\tilde{v}_0$, e.g. $\tilde{v}_0 = b - Ax_0$	Given $\tilde{v}_0$, e.g. $\tilde{v}_0 = b - Ax_0$
for $m = 1, 2, \ldots$	for $m = 1, 2, \ldots$
	Solve $M \tilde{w}_{m-1} = \tilde{v}_{m-1}$
$\beta_{m-1} = \langle \tilde{v}_{m-1} \mid \tilde{v}_{m-1} \rangle^{1/2}$	$\beta_{m-1} = \langle \tilde{v}_{m-1} \mid \tilde{w}_{m-1} \rangle^{1/2}$
$v_{m-1} = \beta_{m-1}^{-1} \tilde{v}_{m-1}$	$v_{m-1} = \beta_{m-1}^{-1} \tilde{v}_{m-1}, \quad w_{m-1} = \beta_{m-1}^{-1} \tilde{w}_{m-1}$
if $m = 1$	if $m = 1$
$\tilde{v}_m = A v_{m-1}$	$\tilde{v}_m = A w_{m-1}$
else	else
$\tilde{v}_m = A v_{m-1} - \beta_{m-1} v_{m-2}$	$\tilde{v}_m = A w_{m-1} - \beta_{m-1} v_{m-2}$
$\alpha_m = \langle \tilde{v}_{m-1} \mid \tilde{v}_{m-1} \rangle$	$\alpha_m = \langle \tilde{v}_{m-1} \mid \tilde{w}_{m-1} \rangle$
$\tilde{v}_m = \tilde{v}_m - \alpha_m v_{m-1}$	$\tilde{v}_m = \tilde{v}_m - \alpha_m v_{m-1}$

Algorithm 1: Lanczos-algorithm without (left) and with (right) preconditioning (preconditioner M).

In this case the recurrence of depth m in (1) reduces to one which is of depth 2. Again in matrix form we get

$$AV_m = V_m T_m + \beta_m v_{m+1} e_m^T, \quad (V_m^T V_m = I). \tag{3}$$

For a non-symmetric A we get similar results iterating a pair of *bi-orthogonal* bases V_m and W_m:

$$AV_m = V_m \hat{T}_m + \gamma_m v_{m+1} e_m^T \quad \text{and} \quad A^T W_m = W_m \hat{T}_m^T + \beta_m w_{m+1} e_m^T,$$
$$\text{with} \quad V_m^T W_m = I.$$

In this case $\hat{T}_m$ is not symmetric anymore but remains to be tridiagonal ($\hat{T}_m = \text{tridiag}(\gamma_{j-1}, \alpha_j, \beta_j)$) [10].

An error analysis of the Lanczos procedure [17,21] shows that in finite precision, formula (3) has to be replaced by

$$AV_m = V_m T_m + v_{m+1} \beta_m e_m^T + F_m, \tag{4}$$

where F_m contains the rounding errors and $\|F_m\|$ is approximately of size $\epsilon\|A\|$ [19]. Here ϵ denotes the machine precision, e.g. $\approx 10^{-16}$ in IEEE double-precision. To be correct, we actually had to use different symbols for the exact and the finitely precise computed values V_m, T_m and β_m but we omitted these extra symbols for readability reasons. Now the question is: *What happens to the orthogonality condition if V_m is computed in finite precision?* The answer is given in

Theorem 1. *Let (Y_m, Λ_m) with $Y_m = (y_1|\cdots|y_m)$ and $\Lambda_m := \mathrm{diag}(\lambda_1,\ldots, \lambda_m)$ be the exact spectral factorization of the finitely precise computed T_m, i.e., the (z_j, λ_j) are the computed Ritz-pairs of A if we define $z_j := V_m y_j$. Then we have*

$$\angle(z_j, v_{m+1}) \approx \mathrm{Arccos}\left(\frac{\epsilon\|A\|}{|\,\|Az_j - \lambda_j z_j\| - \epsilon\|A\|\,|}\right).$$

Notice that v_{m+1} should stay orthogonal to $\mathcal{K}_m = \mathrm{span}(v_1,\ldots,v_m)$ and since z_j is an element of $\mathcal{K}_m$, v_{m+1} should stay orthogonal to z_j, too.

However, if (z_j, λ_j) is a good approximation to an eigenpair of A, i.e., $\|Az_j - \lambda_j z_j\|$ is small, then $\angle(z_j, v_{m+1}) \approx \mathrm{Arccos}(1) = 0$. This means, ironically, if a Ritz-pair is converging, and that is exactly what we want in eigenvalue computations, we completely lose orthogonality in V_m. Nevertheless we will show in Section 6 that we are able to delay this loss of orthogonality using a multiple precision arithmetic, i.e., a small machine precision ϵ.

See [6] for a proof.

4   Improved Arithmetic

In many numerical algorithms there is a large gap between the theoretical, i.e., mathematical, behavior on the one hand and the finite precision behavior on the other hand. In cases where the accuracy of a result is insufficient or no results can be obtained at all due to poorly conditioned problems, it is desirable to have a *better* arithmetic.

To close the gap between exact and finite precision arithmetic, often some minor arithmetic improvements suffice to get the desired results. One is simply to use only a more precise arithmetic instead of an exact one (see Section 4.1). A second possibility is to leave the data type but control the rounding errors introduced by arithmetic operations performed on these numbers as described in Section 4.2. The last approach we describe, is to pick out some important operations or functions and improve their accuracy. In particular, we focus on the scalar product (Section 4.3) which is a fundamental operation in numerical linear algebra.

4.1   Multiple Precision

There are various tools and libraries providing numbers with a higher precision than the *build-in* data types (usually IEEE double precision). We can

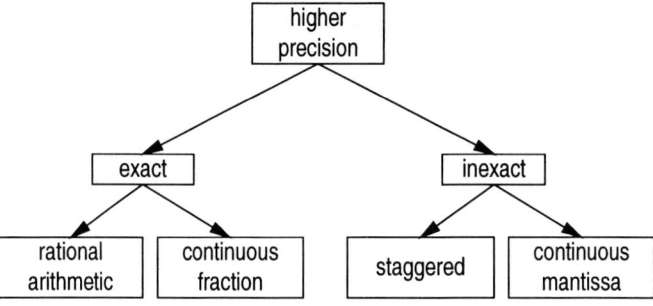

Fig. 3. Higher precision data types.

subdivide them in two fundamental types. The first implements exact numbers, i.e., numbers with infinite precision while the latter offers higher but finite precision. Because of storage and computing time requirements, we only focus on these multiple precision numbers. They are subdivided according to different implementation techniques in so called *staggered* numbers which are basicly a sum of ordinary floating point numbers and numbers with a continuous mantissa, i.e., big floating point numbers. The latter are mostly implemented using an integer field for the mantissa and some additional memory for the exponent and sign (see e.g. [9]).

The staggered technique is implemented for example in the XSC languages [13] and is massively based on the availability of the exact scalar product. The special case where the number of ordinary floating point numbers used to define a staggered number is fixed to 2 can also be coded by some arithmetic tricks without the exact scalar product. These tricks a mainly based on ideas by Dekker and Kahan [4].

4.2   Interval Arithmetic

In order to get reliable results, e.g. for error bounds of linear systems, it is not sufficient only to reduce the rounding errors. We have to get rid off this arithmetic uncertainty completely. Based on floating point arithmetic, we cannot avoid rounding errors, but we can select the rounding direction to obtain guaranteed bounds for the exact, but generally not representable result of an arithmetic operation. Thus, if we are interested for example in an upper bound of a complex computation, we have to adjust the rounding mode of each single operation. Unfortunately, that doesn't imply that each single operation has to be performed with an upward directed rounding and it is a tedious work to adjust the direction of every rounding by hand.

A very useful tool to get reliable bounds is the interval arithmetic [2]. There, always the upper and lower bound of an operation is computed simultaneously and we can finally select, which one we need.

4.3   Exact Scalar Product

Since scalar products occur very frequently and are important basic operations in numerical linear algebra, it would be advantageous if we could perform this operation with the same precision as the other basic operations like addition or subtraction. Usually, scalar products are implemented by repeated use of ordinary multiplication and addition and we have to beware of a lot of roundoff errors which might be dramatically enlarged by cancellation. This is not necessary as has been shown by Kulisch [14].

The basic idea is first to multiply the floating-point vector elements exactly, that means we have to store the result in a floating-point format with double mantissa length and doubled exponent range[2], and second we have to accumulate these products without any roundoff error, see Figure 4. One

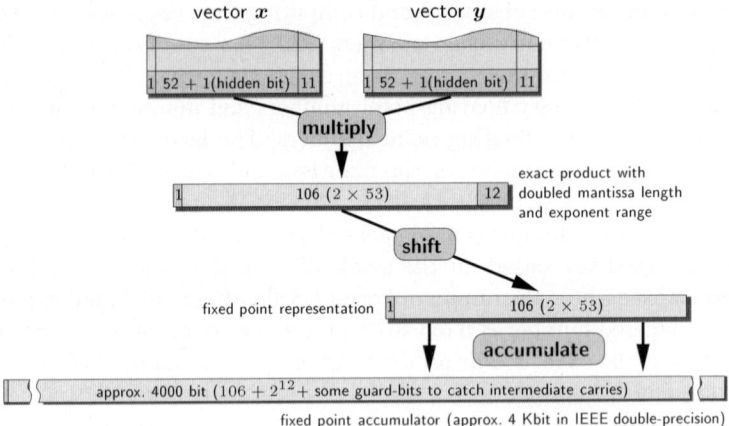

Fig. 4.   The basic idea of the exact scalar product, implemented by means of a long accumulator.

possibility to achieve this is by using a fixed-point accumulator that covers the whole floating-point number range plus some extra bits for intermediate overflows. At a first glance one might think that this accumulator must be very large, but in fact for the IEEE double-precision format, a little more than half a kilobyte is sufficient: 106 mantissa bits (for a zero exponent) plus 2^{11} binary digits for all possible left shifts and the same for right shifts plus one sign bit and some guard bits. With this technique the entire scalar product operation can be computed with just one rounding at all and therefore

---

[2] That is a sign bit, 2×53 mantissa bits and $11 + 1$ bits for the exponent, i.e., a total of 119 bits for the IEEE double-precision format.

we have the much sharper bound

$$\left| \sum_{i=1}^{n} x_i y_i - \square \left(\sum_{i=1}^{n} x_i y_i \right) \right| \le \epsilon \left| \sum_{i=1}^{n} x_i y_i \right|$$

for the relative error[3] than we usually have for scalar products if we use an ordinary floating point arithmetic with a rounding after each multiplication and accumulation (see [12])

$$\left| \sum_{i=1}^{n} x_i y_i - \boxed{\sum_{i=1}^{n}} x_i \boxdot y_i \right| \le \frac{n\epsilon}{1 - n\epsilon} \sum_{i=1}^{n} |x_i y_i|,$$

which can be arbitrary bad if $\sum_i |x_i y_i| \gg |\sum_i x_i y_i|$.

5   Verified Error Bounds

Usually, stopping criteria for iterative solvers of linear systems are based on the norm of the residual $r = b - A\tilde{x}$ of the computed approximate solution $\tilde{x}$. Since $\|\tilde{x} - x^*\|_2 \le \mathrm{cond}_2(A) \cdot \|r\|_2$, this gives a rough idea about the distance to the the the exact solution $x^* = A^{-1}b$ if we have some information about the condition of A. Sometimes 'cheap' condition estimators are used to estimate $\mathrm{cond}_2(A)$ but this approach gives only error estimates instead of bounds. Instead of using uncertain estimates we compute a verified upper bound of $\|\tilde{x} - x^*\|_2$.

This section is essentially based on ideas of S. Rump [20]. A well known method to compute the smallest singular value of a matrix A is the inverse power method (with shift 0). Therefore it is necessary to have a factorization, say (L, U) of A that enables us to compute $(LU)^{-1}z$ for arbitrary z easily. Mostly, $LU = A$ doesn't hold exactly but $LU = \tilde{A} \approx A$ is sufficient. Often it is possible to get $\|\Delta A\| = \|\tilde{A} - A\|$ fairly small. The next theorem clarifies how $\|\tilde{x} - x^*\|$ depends on the smallest singular value $\sigma_{\min}(A)$, ΔA and $\|r\|$.

Theorem 2. *Let $A \in \mathbb{R}^{n \times n}$, $b \in \mathbb{R}^n$ be given as well as a non-singular $\tilde{A} \in \mathbb{R}^{n \times n}$ and $\tilde{x} \in \mathbb{R}^n$. Define $\Delta A := \tilde{A} - A$, $r := b - A\tilde{x}$ and suppose $\sigma_{min}(\tilde{A}) > n^{1/2} \cdot \|\Delta A\|_\infty$.*

Then A is non-singular and for $x^* := A^{-1}b$ holds

$$\|x^* - \tilde{x}\|_\infty \le \frac{n^{1/2} \cdot \|r\|_\infty}{\sigma_{min}(\tilde{A}) - n^{1/2} \cdot \|\Delta A\|_\infty}.$$

See [20] for a proof.

---

[3] The boxed operators denote rounded operators and $\square(\cdot)$ is the rounding operator itself.

Of course for sparse matrices this 2-norm bound is a rough overestimation to the ∞-norm. Suppose B to have at most m nonzero elements per row and let $\beta := \max_{i,j}\{|B_{i,j}|\}$ then both $\|B\|_1$ and $\|B\|_\infty$ are bound by $m \cdot \beta$. Using $\|B\|_2^2 \le \|B\|_1 \cdot \|B\|_\infty$ we get

$$\|B\|_2 \le (\|B\|_1 \cdot \|B\|_\infty)^{1/2} \le m \cdot \beta.$$

Theorem 3. *Let $A \in \mathrm{I\!R}^{n\times n}$, $b \in \mathrm{I\!R}^n$ be given as well as a nonsingular $\tilde{A} \in \mathrm{I\!R}^{n\times n}$ and $\tilde{x} \in \mathrm{I\!R}^n$. Define $\Delta A := \tilde{A} - A$, $r := b - A\tilde{x}$ and suppose $\sigma_{min}(\tilde{A}) > (\|\Delta A\|_1 \cdot \|\Delta A\|_\infty)^{1/2}$.*
Then A is non-singular and for $x^* := A^{-1}b$ holds

$$\|x^* - \tilde{x}\|_\infty \le \|x^* - \tilde{x}\|_2 \le \frac{\|r\|_2}{\sigma_{min}(\tilde{A}) - (\|\Delta A\|_1 \cdot \|\Delta A\|_\infty)^{1/2}}.$$

Again see [20] for a proof.

In practical computations it is a difficult task to get the smallest singular value of an arbitrary matrix A or at least a reliable lower bound to $\sigma_{min}(A)$. But if we have an approximate decomposition, say (L, U) with $\tilde{A} = LU$ we can apply inverse power iteration to LU to compute $\sigma_{min}(\tilde{A})$. Of course, if we can compute a lower bound to $\sigma_{min}(A)$ directly then by setting $\tilde{A} := A$ we get $\Delta A = 0$ and thus

$$\|x^* - \tilde{x}\|_\infty \le \|x^* - \tilde{x}\|_2 \le \sigma_{min}(A)^{-1} \cdot \|r\|_2. \tag{5}$$

It is possible to sharpen this error bound by computing a smaller coefficient than $\sigma_{min}(A)^{-1}$ in (5). Usually, the best one can get is a error bound which is reduced by a factor of 20 compared to (5) (see [7]). However, using a higher precision arithmetic to reduce $\|r\|_2$ as demonstrated in section 6 is generally much more effective.

5.1   Verified Computation

Of course, the results of the preceeding section assumes that all computations are *exact* or at least valid bounds of the exact values. Since we cannot guarantee this by using floating-point arithmetic, we have to bring the tools from Section 4 into action.

In all subsequent algorithms, the variable *accu* represents a long accumulator in the sense of section 4.3. In particular, expressions of the form $accu = accu \pm x \cdot y$ denote exact accumulation of $x \cdot y$ in *accu*.

Decomposition Error $\|LU^T - A\|_2$ Suppose L, U to be a nonsingular lower triangular matrices. Then Algorithm 2 computes a rigorous upper bound for $\|LU^T - A\|_2$.

$$e_1^{\text{row}} := e_2^{\text{row}} := \ldots := e_n^{\text{row}} := 0$$
$$e_1^{\text{col}} := e_2^{\text{col}} := \ldots := e_n^{\text{col}} := 0$$
for $i = 1, \ldots, n$
   for $j = 1, \ldots, i - 1$
      $accu := -a_{i,j}$
      for $k = 1, \ldots, j$
         $accu = accu + l_{i,k} * u_{j,k}$
         $e_i^{\text{row}} = e_i^{\text{row}} \triangle\triangle(|accu|)$
         $e_j^{\text{col}} = e_j^{\text{col}} \triangle\triangle(|accu|)$
   $accu := -a_{i,i}$
   for $k = 1, \ldots, i$
      $accu = accu + l_{i,k} * u_{i,k}$
   $e_i^{\text{row}} = e_i^{\text{row}} \triangle\triangle(|accu|)$
   $e_i^{\text{col}} = e_i^{\text{col}} \triangle\triangle(|accu|)$
$$e_{\max}^{\text{row}} := \max\{e_1^{\text{row}}, e_2^{\text{row}}, \ldots, e_n^{\text{row}}\}$$
$$e_{\max}^{\text{col}} := \max\{e_1^{\text{col}}, e_2^{\text{col}}, \ldots, e_n^{\text{col}}\}$$
return $\overset{\triangle}{\sqrt{e_{\max}^{\text{row}} \triangle e_{\max}^{\text{col}}}}$

Algorithm 2: Compute a verified upper bound for $\|LU^T - A\|_2$ via the inequality $\|B\|_2 \le \sqrt{\|B\|_1 \cdot \|B\|_\infty}$.

Smallest Singular Value Due to ideas of Rump [20], we compute the smallest singular value of a matrix A in two steps. First we factorize A approximately in a product of two triangular matrices, say T_1 and T_2 (e.g. an LU or Cholesky factorization) and then we compute a lower bound of $\sigma_{\min}(T_1 T_2^T)$ via[4] $\sigma_{\min}(T_1 T_2^T) \ge \sigma_{\min}(T_1)\sigma_{\min}(T_2)$.

Now we have to compute the smallest singular values of these triangular matrices. The basic idea is first to compute an approximation $\tilde{\sigma} \approx \sigma_{\min}(T)$ and then proving that $TT^T - \kappa \tilde{\sigma}^2 I$ is positive semidefinite, where κ is slightly less than one. In case of success, $\sqrt{\kappa}\tilde{\sigma}$ is a lower bound of $\sigma_{\min}(T)$. To decide whether the shifted TT^T remains positive semidefinite, we try to compute its Cholesky factorization LL^T. Since this decomposition is usually not exact, we have to apply the following theorem from Wilkinson to guarantee that LL^T, if it exists, is not too far from $TT^T - \kappa \tilde{\sigma}^2 I$ so that the positive definiteness of LL^T is sufficient for the smallest eigenvalue of $TT^T - \kappa \tilde{\sigma}^2 I$ to be nonnegative.

Theorem 4. *Let B, $\tilde{B} \in \mathbb{R}^{n \times n}$ be symmetric and $\lambda_i(B)_{i=1}^n$, respectively $\lambda_i(\tilde{B})_{i=1}^n$ be the eigenvalues ordered by magnitude.*
   Then from $\|B - \tilde{B}\|_\infty \le d$ it follows that $|\lambda_i(B) - \lambda_i(\tilde{B})| \le d$.

See [23] for a proof.

---

[4] In this computation of the smallest singular value hides the $\mathrm{O}(n^3)$ effort which seems to be necessary to compute error bounds [5].

That is, if $\|LL^T - (TT^T - \kappa\tilde{\sigma}^2 I)\|_\infty \leq d$ then $\sigma_{\min}(T)^2 = \lambda_{\min}(TT^T) \geq \kappa\tilde{\sigma}^2 - d$. Thus, if d is a verified upper bound for $\|B - \tilde{B}\|_\infty$ and $\kappa\tilde{\sigma}^2 \geq d$ we have $\sigma_{\min}(T) \geq \sqrt{\kappa\tilde{\sigma}^2 - d}$.

Verifying Positive Definiteness of $TT^T - \sigma^2 I$ Suppose T to be a nonsingular lower triangular matrix. Then Algorithm 3 computes a rigorous lower bound for its smallest singular value.

$\sigma = 0.9 * approx\_smallest\_singular\_value(T)$
start:
$e_1 := e_2 := \ldots := e_n := 0$
for $i = 1, \ldots, n$
  for $j = 1, \ldots, i - 1$
    $accu := 0$
    for $k = 1, \ldots, j$
      $accu = accu + t_{i,k} * t_{j,k}$
    for $k = 1, \ldots, j - 1$
      $accu = accu - l_{i,k} * l_{j,k}$
    $l_{i,j} = \Box(accu)/l_{j,j}$
    $e_i = e_i \triangle\triangle(|accu - l_{i,j} * l_{j,j}|)$
    $e_j = e_j \triangle\triangle(|accu - l_{i,j} * l_{j,j}|)$
  $accu := -\sigma^2$
  for $k = 1, \ldots, i$
    $accu = accu + t_{i,k} * t_{i,k}$
  for $k = 1, \ldots, i$
    $accu = accu - l_{i,k} * l_{i,k}$
  if $accu < 0$
    if $iter < max\_iter$
      $\sigma = 0.9 * \sigma$
      $iter = iter + 1$
      goto start
    else
      return failed
  $l_{i,i} = \sqrt{\Box(accu)}$
  $e_i = e_i \triangle\triangle(|accu - l_{i,i}^2|)$
$e_{\max} := \max\{e_1, e_2, \ldots, e_n\}$
if $\sigma \triangledown \sigma \geq e_{\max}$
  return $\sqrt[\triangledown]{\sigma \triangledown \sigma \triangledown e_{\max}}$
else
  return failed

Algorithm 3: Compute a verified lower bound $\sigma \leq \sigma_{\min}(T)$ and a lower triangular matrix L with $LL^T - (TT^T - \sigma^2 I) \leq e_{\max}$.

6 Computational Results

6.1 Level of Orthogonality

As a first example of the effectiveness of higher precision arithmetic, we solved again the system Beam(1023) (compare Section 2.2). The residual norms and error norms, achieved by using a staggered precision arithmetic, are plotted in Figure 5. The letter l denotes the *staggered length*, i.e., the number of floating-point numbers defining a staggered number (see Section 4.1). The case $l = 1$ corresponds with the IEEE double arithmetic used in Figure 2. Since the staggered arithmetic is simulated in software, it cannot compete with the built-in double arithmetic ($l = 1$) in computing time. However, despite getting more accurate solutions, (which might be unnecessary for practical problems) we observe a significant saving in the number of iterations.

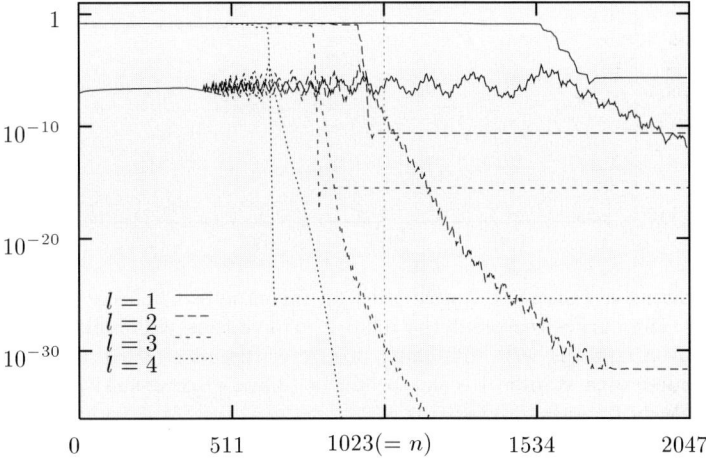

Fig. 5. The Euclidean norms of the residuals (oscillating) and errors (more or less piecewise constant) during solving the Beam(1023) system with staggered length l from 1 to 4.

6.2 High Precision and Exact Scalar Products

In this example, we demonstrate the effect of high precision in conjunction with exact scalar products. Our example linear system is called `fidap009` and is taken fro the MatrixMarket [16]. We solved this linear system with a preconditioned Conjugate Gradient solver. For the preconditioner we used an incomplete Cholesky factorization with drop-tolerance 10^{-10}. For several larger droptolerances, i.e. more sparse preconditioners, we got no convergence, neither with `double`, nor with `extended` arithmetic.

The used arithmetics were IEEE double precision without (double) and with (doubleX) exact scalar products as well as Intel's extended precision format again without (extended) and with (extendedX) exact scalar products.

Figure 6 shows the relative error norms vs. number of iterations. As we can see, the higher precision used for accumulation results in faster convergence and increased accuracy.

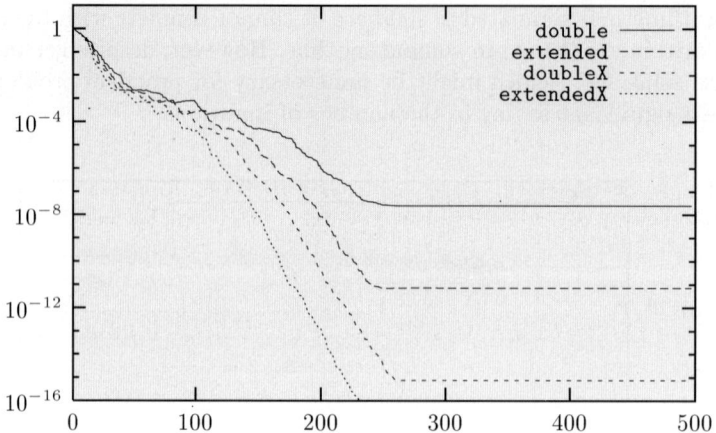

Fig. 6. Solving a fidap009 system with an incomplete Cholesky preconditioned CG solver. The curves represent the relative error norms vs. number of iterations achieved by computing with differently precise arithmetics (IEEE double resp. Intel's extended with standard scalar products (double/extended) and with exact scalar products (doubleX/extendedX.)

Taking into consideration that exact scalar products, sufficiently supported in hardware, need not to be slower than ordinary scalar products, there is simply no reason not to utilize this technique.

Of course, for practical problems, the exact scalar product often provides much more precision than is actually needed. To get an idea on how many precision would suffice, we solved the fidap009 system with basic data type double and different accumulator precisions.

Namely, we used accumulation in double, extended, multiple<2> (128 bit mantissa[5]), and exact accumulation. The results are shown in Figure 7. The curves that corresponds with exact accumulation and multiple<2>-accumulation differ so little that they appear as a single line. That is, in this

---

[5] In general multiple<n> is a multiple precision datatype with a $n \times 64$ bit mantissa.

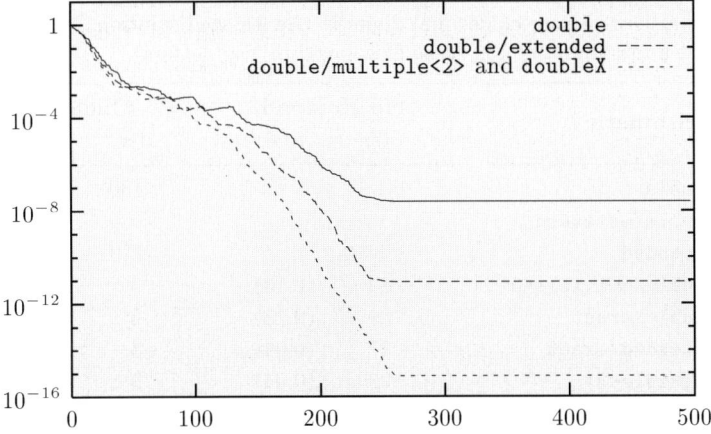

Fig. 7. This figure shows the same experiments as Figure 6 but now we used different scalar products while leaving the basic data type fixed at `double`.

particular case, 128 bit mantissa length was sufficient for almost error free accumulation.

6.3   Beyond Ordinary Floating-Point Arithmetic

While in the latter section we only saved iterations by using a more precise arithmetic, we now show that there are examples that are not solvable in standard floating-point arithmetics at all.

For this purpose, we solved a linear system with the Hilbert matrix of dimension 13. To get a simple stopping criterion, we first computed a verified solution that is guaranteed to have at least 16 correct decimal digits and then stopped each of the following iterations, when the approximated solutions coincide with the verified solution within the first five digits.

Each experiment was carried out twice, once with a (complete) Cholesky preconditioner and once without preconditioning. The results for various arithmetics are displayed in Table 1. If there are two entries in the 'Arithmetic' column (separated by a slash), then the first denotes the data-type and the second is the accumulation precision. If there is only one arithmetic data type given, then all computations are performed with this type.

As we can see, the *smallest arithmetic* enabling convergence, is IEEE double with 128 bit accumulation (`double/mutiple<2>`). Using `extended` precision (with `mutiple<2>` accumulation), significantly speeds up the computation in the non-preconditioned case and further increasing the precision saves up to 85% computing time. With 320 bit mantissa length (`multiple<5>`), we match the *exact precision* property of convergence after at most n steps.

Table 1. This table shows the number of iterations and the computing time, needed to obtain at least 5 correct decimal digits in the iterated solution. We stopped the process at a maximum of 130 steps (displayed in gray letters).

Arithmetic	No Precond.		Cholesky	
	iter	*time*	*iter*	*time*
double	>130	(—)	>130	(—)
double/extended	>130	(—)	>130	(—)
extended	>130	(—)	>130	(—)
double/multiple<2>	89	(0.13)	3	(<0.01)
double/exact	89	(0.13)	3	(<0.01)
extended/exact	37	(0.04)	3	(<0.01)
multiple<2>	23	(0.04)	3	(0.01)
multiple<4>	16	(0.02)	3	(0.01)
multiple<5>	13	(0.02)	3	(0.01)

Though we used a (complete) Cholesky preconditioner in the right column, i.e. a direct solver in each iteration, we also got convergence only with at least 128 bit accumulation. Further increasing the precision does not save more iterations and consequently does not save computing time. However, even with this direct solver in each step, we need 3 iterations to get 5 correct digits.

6.4   Does Higher Precision Increase the Computational Effort?

Inspecting, for example, the MATLAB implementations of Krylov subspace solvers, we can see that the residual norm which is used for evaluating the stopping criterion, is always computed as `normr = norm(b - A * x)`; instead of using the norm of the updated residual. This effort is often necessary because in finite precision these two theoretically equal values tend to differ significantly after sometimes only a few iterations (see Figure 8).

However, there is one extra matrix-vector multiplication at each step and this information is solely used to decide whether the iteration should be stopped or not. Fortunately, we can do much better. Assume we have the internal precision adjusted to staggered with length 2. For generating the Krylov spaces, we need

$$Ar = A(r^{(1)} + r^{(2)}) = Ar^{(1)} + Ar^{(2)}.$$

That means, *increasing the precision by one* is comparable with doing one extra matrix-vector product. However, with this approach, we do not only improve the stopping criterion, but also significantly improve the accuracy of the iterated solution, see Figure 8. This effect is demonstrated with the `mcca` matrix [16].

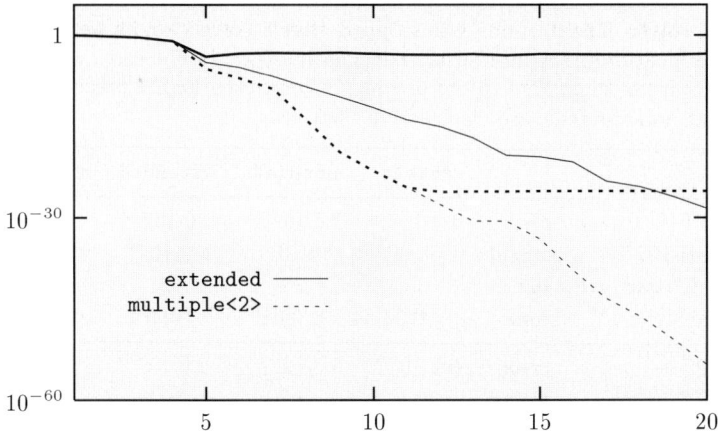

Fig. 8. This figure compares the norm of the iteratively updated residual (*thin lines*) and the exact residual $\|b - Ax\|_2$ (*thick lines*) computed with differently precise arithmetics (solid: `extended`, dashed: `multiple<2>`).

6.5 Solving Ill-Conditioned Test-Matrices

Here we solved some `Beam`(n) systems with a preconditioned `CG` algorithm with Cholesky preconditioner (see Table 2).

In this and all following examples, we stopped the iteration as soon as five correct digits of the solution could be guaranteed or after stagnation of the residual norm (marked by an *). With *error* we denote the actual relative error of the approximate solution. Usually, one has no exact solution available and therefore cannot compute the *error*. However, for these examples we use a very tight enclosure of the exact solution, computed with a high precision arithmetic. With *bound* we denote the computed upper bound of the error and *iter* and *time* are the number of iterations and the time needed for these iterations (in sec, measured on a PentiumII/400). The quantity *vtime* denotes the time needed to compute a rigorous bound for the smallest singular value of A.

As we can see, verification with the techniques described in Section 5, is only possible in conjunction with higher precision arithmetic. Even if we are only interested in a non-verified solution, we cannot trust in standard floating-point (`double`) arithmetic. However, simply replacing all floating-point scalar products by exact scalar products suffices to deliver always enough correct digits in the approximate solution, although non-verified. Note that the very slow convergence speed for `doubleX` is caused by the slow software simulation of the exact scalar product. Sufficiently supported in hardware, `doubleX` should need the same time as `double`. Using a 128 bit arithmetic we always achieved fast convergence and highly accurate verified solutions.

Table 2. Solving some Beam(n) systems with an incomplete Cholesky preconditioned CG solver. The iteration was stopped after 5 correct digits were guaranteed (by the verification procedure) or after stagnation (gray). The cases where we have 5 digits accuracy, compared to the previously computed highly precise verified solution (but non-verified), are displayed in dark gray.

Matrix		double	doubleX	extended	multiple<2>
$n = 100$	error	$1.3 \cdot 10^{-9}$	$1.6 \cdot 10^{-15}$	$2.0 \cdot 10^{-12}$	$1.4 \cdot 10^{-23}$
$\sigma_{min} = 8.4 \cdot 10^{-7}$	bound	$7.5 \cdot 10^{-3}$	$2.0 \cdot 10^{-3}$	$4.2 \cdot 10^{-6}$	$4.9 \cdot 10^{-15}$
vtime < 0.01sec	iter	3^*	3^*	2	2
	time	< 0.01	< 0.01	< 0.01	0.01
$n = 1000$	error	$9.6 \cdot 10^{-5}$	$5.3 \cdot 10^{-15}$	$9.0 \cdot 10^{-8}$	$6.9 \cdot 10^{-25}$
$\sigma_{min} = 9.7 \cdot 10^{-11}$	bound	> 1	> 1	> 1	$1.0 \cdot 10^{-11}$
vtime $= 0.04$sec	iter	3^*	3^*	3^*	3
	time	< 0.01	0.13^a	0.02	0.15
$n = 10\,000$	error	> 1	$3.2 \cdot 10^{-14}$	$2.7 \cdot 10^{-3}$	$4.8 \cdot 10^{-25}$
$\sigma_{min} = 9.7 \cdot 10^{-15}$	bound	> 1	> 1	> 1	$2.5 \cdot 10^{-6}$
vtime $= 0.47$sec	iter	3^*	5^*	3^*	5
	time	0.08	2.33^a	0.10	3.06
$n = 50\,000$	error	> 1	$2.5 \cdot 10^{-13}$	> 1	$1.9 \cdot 10^{-14}$
$\sigma_{min} = 2.2 \cdot 10^{-17}$	bound	> 1	> 1	> 1	$3.3 \cdot 10^{-7}$
vtime $= 3.01$sec	iter	3^*	12^*	13^*	9
	time	0.45	28.66^a	2.83	28.73
$n = 100\,000$	error	> 1	$2.8 \cdot 10^{-13}$	> 1	$4.7 \cdot 10^{-14}$
$\sigma_{min} = 4.1 \cdot 10^{-18}$	bound	> 1	> 1	> 1	$7.0 \cdot 10^{-7}$
vtime $= 4.83$sec	iter	3^*	11^*	7^*	12
	time	0.93	52.55	2.96	77.31

In the next example we solve some Hilbert systems, again with a Cholesky preconditioned CG solver, see Table 3. With a standard double or extended arithmetic for the solver, we can only handle very small dimensions, while a mantissa length of 128 bit always suffices to get fast convergence and good approximations for the solution. At dimension 14, the Cholesky decomposition (computed in double) fails. Increasing the precision used to compute the Cholesky decomposition enables us to handle larger Hilbert matrices. In Table 3, rows 8-13 we used the data type double whereas in rows 15-21 we used the data type multiple<2> for computing the preconditioner. Now the Cholesky decomposition is sufficient to solve the linear system in one step.

Table 3. Here we aimed to verify 5 correct digits in the solution of various Hilbert systems. For dimensions up to 13, the Cholesky preconditioner was computed in `double` while the higher dimensional Hilbert matrices were factorized with a `multiple<2>` arithmetic.

dim	σ_{min}	double iter	time	extended iter	time	multiple<2> iter	time	Chol. prec.
8	$3.60 \cdot 10^{-05}$	3	< 0.01	1	< 0.01	1	< 0.01	
10	$2.29 \cdot 10^{-05}$	> 10	—	2	< 0.01	2	< 0.01	*double*
12	$5.11 \cdot 10^{-07}$	> 12	—	> 12	—	3	< 0.01	
13	$3.05 \cdot 10^{-08}$	> 13	—	> 13	—	4	< 0.01	
15	$4.92 \cdot 10^{-09}$	> 15	—	> 15	—	1	0.01	
17	$2.83 \cdot 10^{-10}$	> 17	—	> 17	—	1	0.02	*multiple<2>*
19	$9.62 \cdot 10^{-12}$	> 19	—	> 19	—	1	0.02	
21	$3.60 \cdot 10^{-13}$	> 21	—	> 21	—	1	0.02	

This high precision preconditioning also allows us to handle larger dimensions for the **Beam** matrices. Table 4 shows the results of the **Beam**(2 000 000) system.

Table 4. This table shows the same experiment as Table 2 but now with dimension 2 000 000 (this was the largest possible dimension solvable on my PC due to storage limitations).

Matrix		doubleX	extended	multiple<2>	multiple<3>
$n = 2\,000\,000$	error	—	—	—	—
$\sigma_{min} = 4.2 \cdot 10^{-24}$	bound	> 1	> 1	$1.8 \cdot 10^{-2}$	$6.7 \cdot 10^{-7}$
$vtime = 123.27$sec	iter	8*	2*	2*	2
	time	903.34	18.15	261.42	306.54

In the Hilbert example above we stoped at dimension 21, because it is not possible to store higher dimensional Hilbert matrices in IEEE double exactly. Just to see how far we can go with Hilbert matrices, we stored them in a multiple precision format in order not to have conversion errors in the input data. With this extension we are even able to solve Hilbert systems of dimension 42 [1] and higher[6].

---

[6] In fact, we solved the Hilbert 42 system in less than 8 seconds with 113 guaranteed decimals (using a data type with 2560 bits mantissa length, i.e. approximately 770 decimal digits on a PentiumII/400).

6.6   Verified Solutions for 'Real-Life' Problems

In this section we investigate some example systems taken from various application areas such as fluid dynamics, structural engineering, computer component design, and chemical engineering [16].

Symmetric positive definite systems we always solved with a Cholesky preconditioned Conjugate Gradient solver. In the nonsymmetric case we show only the results of the fastest ILU preconditioned Krylov solver.

Particularly we utilized various solvers (BiCG, CGS, and BiCGStab) and incomplete preconditioners. In this context of high precision arithmetics, the GMRES algorithm couldn't compete with the *short recurrence solvers*. Since GMRES needs most arithmetic operations and storage anyway, this lack is even reinforced by the increased requirements in memory and computing time for the high precision arithmetic operations.

In Table 6.6 we compare the results achieved by using several arithmetics. Table 6 gives a quick overview over some systems we solved.

Table 5. Solving some 'real-life' problems with different arithmetics. See Table 2 for explanation of the used notations.

Matrix		double	doubleX	extended	multiple<2>
fidap009	*error*	$1.9 \cdot 10^{-4}$	$1.1 \cdot 10^{-13}$	$1.6 \cdot 10^{-6}$	$6.2 \cdot 10^{-11}$
$n = 4683$	*bound*	0.33	$6.6 \cdot 10^{-2}$	$2.0 \cdot 10^{-4}$	$1.1 \cdot 10^{-8}$
$\sigma_{min} = 2.9 \cdot 10^{-4}$	*iter*	3*	4*	3*	2
vtime = 22.55sec	*time*	0.53	6.72^a	0.68	2.82
s3rmt3m1	*error*	$3.4 \cdot 10^{-5}$	$8.6 \cdot 10^{-14}$	$2.5 \cdot 10^{-8}$	$4.3 \cdot 10^{-11}$
$n = 5489$	*bound*	>1	>1	$9.7 \cdot 10^{-2}$	$5.2 \cdot 10^{-7}$
$\sigma_{min} = 3.5 \cdot 10^{-7}$	*iter*	3*	3*	3*	2
vtime = 507.6sec	*time*	1.14	18.40^a	1.37	12.42
e30r5000	*error*	$1.1 \cdot 10^{-11}$	$1.5 \cdot 10^{-14}$	$5.4 \cdot 10^{-12}$	$1.2 \cdot 10^{-38}$
$n = 9661$	*bound*	>1	>1	0.7	$1.4 \cdot 10^{-32}$
$\sigma_{min} = 3.7 \cdot 10^{-12}$	*iter*	3*	3*	3*	2
vtime = 2h09.47min	*time*	3.86	69.4^a	5.45	70.41
e40r5000	*error*	$2.8 \cdot 10^{-11}$	$1.9 \cdot 10^{-14}$	$2.3 \cdot 10^{-14}$	$1.6 \cdot 10^{-32}$
$n = 17281$	*bound*	>1	>1	>1	$1.3 \cdot 10^{-29}$
$\sigma_{min} = 1.5 \cdot 10^{-13}$	*iter*	3*	3*	3*	2
vtime = 6h48.09min	*time*	9.04	232.3^a	13.00	848.0

Again, we often have enough correct digits in the approximate solution but usually we are not aware of this fact. Particularly, we are only able to prove this by using a higher precision arithmetic.

Table 6. A quick overview over some systems we solved. The objective was to get 5 correct digits in the iterated solution. Since convergence sometimes was very fast, we overshot at times. We only display the results of the smalles arithmetic that delivers these 5 digits (almost always `multiple<2>`). Note that 'time' denotes the overall time for solving and verification.

Name	dim	nnz	cond	bound	time
fs 680 1	680	2646	$2.1 \cdot 10^4$	$3.56 \cdot 10^{-38}$	21.75 sec
west2021	2021	7353	$7.5 \cdot 10^{12}$	$9.43 \cdot 10^{-25}$	705.93 sec
mvmtls4000	4000	8784	$2.7 \cdot 10^7$	$3.62 \cdot 10^{-30}$	1h03.55 min
pores2	1224	9613	$3.31 \cdot 10^8$	$4.87 \cdot 10^{-17}$	292.68 sec
bcsstk08	1074	12960	$4.7 \cdot 10^7$	$9.49 \cdot 10^{-28}$	82.72 sec
pde2961	2961	14585	$9.49 \cdot 10^2$	$1.77 \cdot 10^{-14}$	7.22 sec
add32	4960	23884	$2.14 \cdot 10^2$	$6.89 \cdot 10^{-15}$	1h08.28 min
fidap009	4683	95053	$1.04 \cdot 10^7$	$1.1 \cdot 10^{-8}$	25.37 sec
s3rmt3m1	5489	112505	$1.33 \cdot 10^{10}$	$5.2 \cdot 10^{-7}$	520.02 sec
e30r5000	9661	306356	$1.27 \cdot 10^{11}$	$1.2 \cdot 10^{-38}$	2h10.57 min
e40r5000	17281	553956	$1.4 \cdot 10^{16}$	$1.3 \cdot 10^{-29}$	7h02.09 min

6.7   Verification via Normal Equations

In the nonsymmetric case, we have to meet the assumption

$$\sigma_{\min}(LU) > \|LU - A\|_2 \tag{6}$$

which is sometimes a problem if either $\sigma_{\min}(LU)$ is very small or A is ill-conditioned and its elements are large. In such cases it is often advantageous to switch to the normal equations. We stress that we actually do not have to compute $A^T A$ and $A^T b$ [18]. Although we have squared the smallest singular value (which possibly makes it more difficult to find a verified lower bound if $\sigma_{\min}(LU) < 1$), we now do not have to fulfill (6) anymore.

Using this technique we solved, e.g., the `mcca` system. For this matrix, we got a lower bound for the smallest singular value as $2.4 \cdot 10^{-2}$ (this seems to be roughly underestimated due to the very high condition number) and an upper bound for $\|LU - A\|_2$ as $2.31 \cdot 10^9$. That is we cannot apply Theorem 3 directly. Switching to the normal equations, we found $7.67 \cdot 10^3$ as a lower bound for $\sigma_{\min}(A^T A)$. Applying a Cholesky preconditioned `CG` algorithm, we are able to verify five decimal digits in less than three seconds.

References

1. D. Adams. *Life, the Universe and Everything.* Ballantine Books, 1995.
2. G. Alefeld and J. Herzberger. *Introduction to Interval Computations.* Academic Press, 1983.

3. American National Standards Institute / Institute of Electrical and Electronic Engineers, New York. *A Standard for Binary Floating-Point Arithmetic*, 1985. ANSI/IEEE Std. 754-1985.

4. T. J. Dekker. A floating-point technique for extending the available precision. *Numer. Math.*, 18:224–242, 1971.

5. J. Demmel, B. Diament, and G. Malajovich. On the complexity of computing error bounds. `www.cs.berkeley.edu/~demmel/`, 1999.

6. A. Facius. The need for higher precision in solving ill-conditioned linear systems. to be published in Iterative methods in Scientific Computation II, 2000.

7. A. Frommer and A. Weinberg. Verified error bounds for linear systems through the Lanczos process. *Reliable Computing*, 5(3):255–267, 1999.

8. G. Golub and C. van Loan. *Matrix Computations*. Johns Hopkins, third edition, 1996.

9. T. Granlund. The GNU Multiple Precision Arithmetic Library. `www.csd.uu.se/documentation/~programming/gmp/`

10. A. Greenbaum. *Iterative Methods for Solving Linear Systems*. SIAM, Philadelphia, 1997.

11. M. Hestenes and E. Stiefel. Methods of conjugate gradients for solving linear systems. *Journal of Research of the National Bureau of Standards*, 49:409–436, 1952.

12. N. J. Higham. *Accuracy and Stability of Numerical Algorithms*. SIAM, Philadelphia, 1996.

13. R. Klatte, U. Kulisch, A. Wiethoff, C. Lawo, and M. Rauch. *C-XSC*. Springer Verlag, 1992.

14. U. Kulisch. *Advanced Aithmetic for the Digital Computer — Design of Arithmetic Units*. Electronic Notes of Theoretical Computer Science, `www.elsevier.nl/locate/entcs/volume24.html` pp. 1-72, 1999.

15. C. Lanczos. An iteration method for the solution of the eigenvalue problem of linear differential and integral operators. *Journal of Research of the National Bureau of Standards*, 45:255–282, 1950.

16. The matrix market: a visual web database for numerical matrix data. `math.nist.gov/MatrixMarket/`

17. C. Paige. *The Computation of Eigenvalues and Eigenvectors of Very Large Sparse Matrices*. PhD thesis, University of London, 1971.

18. C. Paige and M. Saunders. LSQR: An algorithm for sparse linear equations and sparse least squares. *ACM Trans. Math. Soft.*, 8:43–71, 1982.

19. B. N. Parlett and D. S. Scott. The Lanczos algorithm with selective orthogonalization. *Math. Comput.*, 33:217–238, 1979.

20. S. M. Rump. Validated solution of large linear systems. In R. Albrecht, G. Alefeld, and H. J. Stetter, editors, *Validation Numerics*, number 9 in Computing Supplementum, pages 191–212. Springer Verlag, 1993.

21. H. D. Simon. *The Lanczos Algorithm for Solving Symmetric Linear Systems*. PhD thesis, University of California, Berkeley, 1982.

22. Z. Strakoš. On the real convergence rate of the conjugate gradient method. *Linear Algebra Appl.*, 154/156:535–549, 1991.

23. J. H. Wilkinson. *The Algebraic Eigenvalue Problem*. Clarendon Press, Oxford, 1988.

Elements of Scientific Computing

J. Hartmut Bleher

KommunikationsSysteme,
Bismarckstr. 31, D-72622 Nürtingen, Germany.
kommsys@t-online.de

Abstract. In the past decades, computer performance has increased dramatically and is still doing so, putting tremendous power at the fingertips of every computer user. Many of these capabilities are used for new and complex functions and for a (hopefully) better user interface, but also sometimes wasted for questionable gadgets. Unfortunately, in the same time frame little has been done to increase the *confidence in the answers computers produce* in the scientific field, because, so far, computer designers are much more motivated by mass market needs than by scientific requirements. With the circuit densities achievable on computer chips today and with 'clean' architectures already in place, only very little additional effort would have to be spent to provide the means to *allow computers to deliver results with guarantees*, i.e. with verified accuracy.

1 Hardware Requirements

Scientific computing is done mostly with *Floating Point representation* of numbers. A 'clean' Floating Point architecture is a recognized requirement for the survival of today's single-chip or multi-chip computers. This means that for basic arithmetic functions ($+$, $-$, $\times$, $/$) and even for a set of standard functions (sqrt, sin, tan, exp, etc.) *the presented accuracy is one half of the Unit in the Last Place* (1/2 ULP).

It is in principle immaterial which Floating Point format is used, like Hexadecimal as in IBM /370, /390 architecture or Binary as in IEEE 754 architecture, although a Decimal format would sometimes alleviate conversion to 'human-readable' results.

It is in principle also relatively unimportant which precision of the numerical data presented is chosen, i.e. how many bits are used in the mantissa and in the exponent of the Floating Point number. The precision, being Single (Short), Double (Long) or Extended (Quadrupel), is mainly influenced by the data paths of the computer core and the associated memory (32, 64, 128 or 256 bits). It may be freely selected by the programmer as long as it allows computation without loss of accuracy in combined and iterative arithmetic functions, and representation of the results with the necessary number of accurate digits. However, the *danger of losing accuracy in solving (ill-conditioned) problems* which span a very large area of the computer representable number space is still immanent, even with the highest practical precision. Therefore, at least one more encompassing basic arithmetic function has to be provided which completely avoids loss of accuracy, especially *through cancellation* in the 'Multiply-and-Add' function sequence applied to an arbitrarily large number set.

It has been shown by U. Kulisch et al. that an *architected inner product* (Dot Product •) can satisfy this requirement in an advantageous way, and will thus become a key element of scientific computing. In essence, it requires the entire architectured computer number space to be in Fixed Point representation for the 'Multiply-and-Add' function. This can be easily accomplished by a moderately enhanced addressing and storing mechanism for the L1-cache which is not difficult at today's circuit count of computer chips.

Computation with absolute accuracy requires *Directed Rounding* (up, down, to 0, to nearest) to avoid accumulation of rounding errors and to allow bounding of results using iterative algorithms. All rounding modes must be intrinsically executable with the basic arithmetic instructions, requiring individual OP-codes to not lose any cycles for setting a rounding mode. This *architected Interval Arithmetic capability* is another key element of today's scientific computing.

It would be quite helpful to have, besides the *REAL Floating Point number space*, an extension of the basic computer data types to the *COMPLEX Floating Point number space* in hardware, with all the intricacies of errorfree handling of boundaries and exceptions being removed from the hands of the not so experienced programmer.

An extended set of Floating Point registers may simultaneously contain upper and lower bounds of COMPLEX Floating Point numbers fed by multiple arithmetic and logic units *executing in parallel COMPLEX interval arithmetic instructions.*

Hardware Performance, usually measured in Floating Point Operations Per Second (FLOPS), is determined by the number of machine cycles required to execute a Floating Point instruction in a particular precision. Again, the high circuit count of today's processor chips makes the objective of single cycle Floating Point instructions a realistic goal, even for the highest precision implemented. Additional circuitry may be used for instruction and data pipelining and for providing numbers of identical execution units working in parallel on larger data sets. Thus, several Floating Point instructions may be executed per machine cycle in a multi-way fashion on a multi-million circuit computer chip in the very near future.

2   Software Requirements

Operating Systems supporting scientific computing must be able to *effectively move large amounts of data* in a deep storage hierarchy. They also have to be able to *handle exception cases* on the system level without necessarily impeding the flow of data, i.e. *in a benign predictive manner*. For large problems and large data sets appropriate *partitioning mechanisms* have to be put into place to effectively use the available storage and execution hardware resources.

Basic Algorithms and Procedures have to be provided to handle REAL and COMPLEX Floating Point interval number manipulation for basic and also for more complicated arithmetic functions in a *basic logic and arithmetic system* (blas). Special care is required to *minimize the path length* of individual procedures and to pay attention to the *efficiency of algorithms* used to achieve a high thruput, which is a measure for overall system performance.

The blas will become the base for Scientific Computing Languages, supporting in userfriendly dialects comprehensive *errorfree Operators on multiple* REAL and COMPLEX Floating Point (interval) *data types*.

Problem Solvers are needed to solve a large variety of mathematical and real life numerical problems in the fields of biology, chemistry, physics, and mechanical and electrical engineering with strong bounds of the results presented. They may be built using interval arithmetic and the software facilities described above. *Algorithmic proof of solutions, and verification and guarantee of result accuracy* can be provided using the comprehensive work of U. Kulisch and his associates.

3   Modelling Requirements

Having available the Hardware and Software Elements of Scientific Computing, one must not forget the fact that the *results delivered by the computer can only be as good as the theory, the model and the resulting input data* used to describe the real life circumstances and processes. Therefore it is imperative that the enhanced arithmetic capabilities of the computer described above are already considered when the models are described by equations, as well as during the process of representing the equations numerically. The scientist solving a particular problem has to decide whether to use straight forward traditional arithmetic or to embark on interval methods in order to verify the results of his/her computations even in ill-conditioned cases.

4   Conclusion

In today's world, with the rather mature hardware and software computer technologies in place and with the Very Large Scale Integration capabilities of the semiconductor foundries, there is no excuse whatsoever for not providing the key elements of scientific computing with each and every general purpose or scientific computer core delivered for main frames, servers and personal computers. *Interval Arithmetic for the REAL and COMPLEX Floating Point number space with single cycle execution* of basic instructions, including the Multiply-and-Add function for the architected Inner Product, shall be supported to allow guaranteed accuracy of computed results.

Hardware
Floating Point Architecture → 'clean'
IBM /370, /390   Format   IEEE 754 Hex   (Decimal)   Binary
Precision (# of Bits used in Mantissa, Exponent) Single (Short)   Double (Long)   Extended (Quadrupel)
Basic Arithmetic   Standard Functions + − × / •   (sin, tan, exp, etc.)
Accuracy →   1/2 ULP (Unit in the Last Place)
Directed Rounding →   up   down   to 0   to nearest
Interval Arithmetic

Performance (FLOPS)
Cycles per Instruction Single Cycle Instructions   Pipelining Instructions per Cycle ←   Parallelism

Performance (Thruput)
Pathlength Algorithm Efficiency Problem- and Data-Partitioning

Operating Systems   Exception Handling
Algorithms and Procedures
Languages
Problem Solvers
Algorithmic Proof of Solutions
Verification and Guarantee of Result Accuracy
Software

Fig. 1. Elements of Scientific Computing

References

1. Kulisch U. W., Miranker W. L. (1981) Computer Arithmetic in Theory and Practice, Academic Press, New York.
2. Kulisch U. W., Miranker W. L., eds. (1983) A New Approach to Scientific Computation, Academic Press, New York.
3. IBM (1983) High-Accuracy Arithmetic Subroutine Library, Program No. 5664-185, General Information Manual, GC-33-6163-0, and Program Description and User's Guide, SC-33-6164-1.
4. IBM (1984) IBM S/370 RPQ High-Accuracy Arithmetic, SA-22-7093-0.
5. ANSI/IEEE (1985) A Standard for Binary Floating-Point Arithmetic, ANSI/IEEE Std. 754-1985, New York, printed in SIGPLAN 22, 2 (1987), pp. 9–25.
6. IBM (1986) High-Accuracy Arithmetic Subroutine Library, Program No. 5665-337/5666-320 General Information Manual, GC-33-6163-02, and Program Description and User's Guide, SC-33-6164-02.
7. Bleher J. H., Rump S. M., Kulisch U. W., Metzger M., Ullrich Ch., Walter W. (1987, 1988) FORTRAN-SC, A Study of a FORTRAN Extension for Engineering/Scientific Computation with Access to ACRITH, Computing 39 and Computing, Suppl. 6.
8. Adams E., Kulisch U. W., eds. (1993) Scientific Computing with Automatic Result Verification, Academic Press, New York.
9. Kulisch U. W., Teufel T., Höfflinger B. (1994) Genauer und trotzdem schneller, ein neuer Coprozessor für hochgenaue Matrix- und Vektoroperationen, Electronic 26, Franzis, Poing.
10. Gustafson J. (1998) Computational Verifiability and Feasibility of the ASCI Program, IEEE Computational Science and Engineering, January-March 1998.

Biography

J. Hartmut Bleher is a Consultant for Information and Communication Systems. He has been involved in Solid State and Low Temperature Physics and Integrated Circuit Technology at the Universities of Stuttgart and Aachen (Germany), and at the IBM Laboratory in East Fishkill, New York (U.S.A.). At the IBM Laboratory in Böblingen (Germany) he has managed Hardware and Software System Development for Computer Graphics and for Scientific Computing. He has been Product Manager for the IBM/9370 Processor Family and for IBM/370 and IBM/390 Integrated Parallel Systems. He has managed IBM's cooperations with the University of Karlsruhe (Germany) in the field of Scientific Computing and with the CERN High Energy Physics Laboratory in Geneva (Switzerland) in the field of Parallel Processing. J. Hartmut Bleher received his M.S. and his Ph.D. from the Technical University of Stuttgart (Germany).

The Mainstreaming of Interval Arithmetic

John L. Gustafson

Sun Microsystems, Inc.
901 San Antonio Road, UMPK24-201, Palo Alto, CA 94303-4900.
John.Gustafson@eng.sun.com

Abstract. Interval arithmetic and validated arithmetic methods are almost unknown in the United States, and are absent in the federally funded High Performance Computing efforts of the last twenty years. The focus has been on floating-point operations per second (FLOPS) to the exclusion of any concern for the correctness of the result. However, the treaty-mandated need to validate nuclear weapons without physical experiments (the ASCI program) may prove to be the key to changing this. Radiation transport provides an example where bounded intervals can provide much more useful answers than existing point methods, whether they are used for modeling nuclear reactions or for computer-generated graphics. This example, and others, can be used to illustrate a general strategy that will allow us to move interval arithmetic into the mainstream of high-speed computing.

1 Introduction

Why hasn't interval arithmetic become a widely adopted technique?

Much of the effort of the last several decades of interval arithmetic research has focused on the making interval bounds tighter for mathematical problems. It has not yet addressed what is required to replace *physical simulation* point algorithms in common use with interval algorithms. In the area of technical computing, as compared to business use of computers, the results of point algorithms are generally regarded as advice to an engineer or scientist and not the proof of a result.

While only a very small percentage of technical computer users have knowledge of numerical analysis, most understand that rounding, discretization, and coding errors are ever-present hazards. Because of the other vertices of the triangle, experiment and theory, the error-prone nature of computational science has not prevented it from being a useful approach.

The ASCI (Accelerated Strategic Computing Initiative) program initially seemed like it might provide a watershed for this attitude. In 1996, international treaties demanded cessation of all experimental testing of nuclear weapons [6], so the ASCI program was begun to provide a way to completely replace experiments with computations. This decision potentially represented as great a change in the business of science as was the rise of the experimental method during the Renaissance.

Unfortunately, instead of making use of the well-established body of knowledge of interval arithmetic and verifiable computing, the ASCI program has

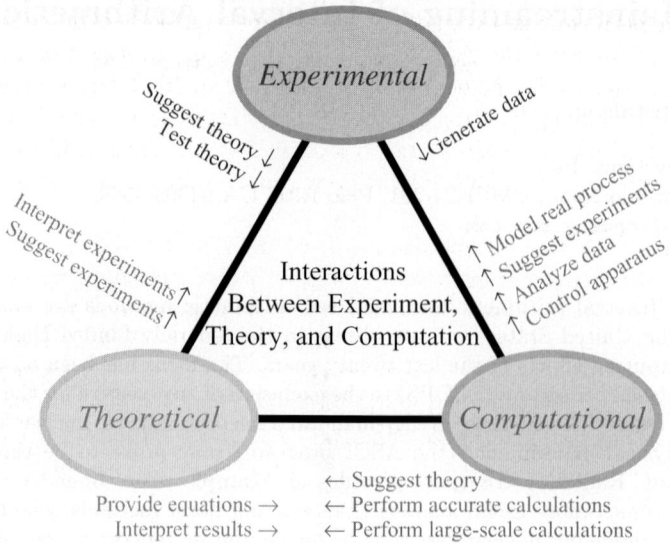

Fig. 1. Computational Science as a Third Branch of Science

retreated to comparing conventional point algorithms to historical experiments and some near-critical nuclear experiments, instead of directly addressing the issue of how to obtain certainty from computer simulations alone.

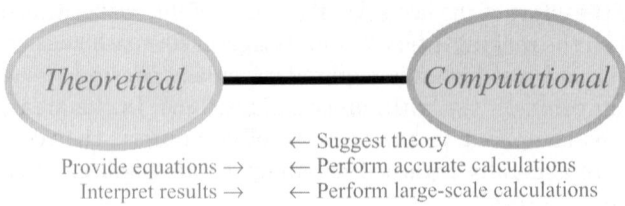

Fig. 2. "Experimentless Science"

Perhaps verifiable computing techniques will be invoked later in the ASCI program, or perhaps the independent software vendors (ISVs) of technical software will lead the process of adoption; here we discuss some of the fundamental obstacles that must be overcome in either case.

2   Moore's Law and Precision

Pressure to confront issues of computing veracity come from another source: Moore's law. While not stated explicitly until 1967, the doubling of part

density and performance every 18 to 22 months can be backdated at least
to the late 1930s with the first binary computers of Atanasoff or Zuse [4].
The Zuse Z1 used 22-bit floating-point arithmetic in 1939, and used six bits
to address 64 memory locations. In the year 2000, a large server like a Sun
Starfire uses 38 bits to address 128 billion bytes in a linear address space.
This represents $38 - 6 = 32$ doublings per Moore's law, over a period of 61
years.

Does precision also follow Moore's law? In fits and starts, the size of
floating-point numbers has increased along with the number of address bits.
The 1977 Cray-1 made 64-bit floating-point arithmetic available with full
hardware support, and at the time, it seemed like the 47-bit mantissa (14-
decimal) precision would make worries about numerical precision unnecessary. The IEEE 754 floating-point standard extended the mantissa to 53 bits
(almost 16 decimals), and both Intel and Motorola have offered processors
that have floating-point registers with extra precision for register-to-register
scratch calculations.

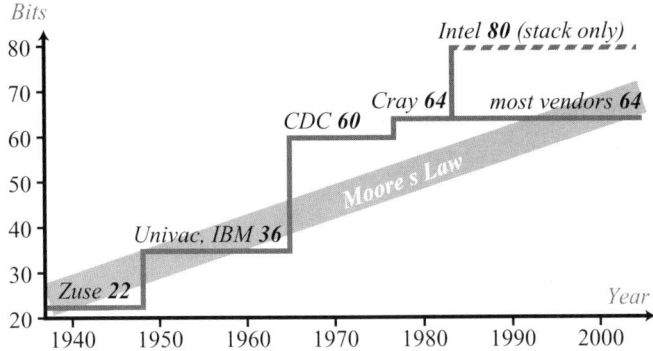

Fig. 3. Moore's Law and Floating-Point Precision

With computers that can perform 10^{12} operations per second becoming
common, it would appear that we are now due for another step up in precision.
A biased loss of only 0.5 ULP per operation means an entire 53-bit mantissa
can lose all meaning in only half an hour's worth of arithmetic. However,
to change to 128-bit precision is a daunting prospect. Besides halving the
number of floating-point data we can hold in a given amount of memory, the
use of 128-bit data doubles the burden on the memory bus and caches, which
are already the limiting parts of modern computer architectures. This creates
some very practical pressures on computer manufacturers to revisit the issue
of veracity in technical computing and explore alternatives.

For example, it may be that intervals with 32-bit precision numbers are
much more powerful than point values with 64-bit precision, eliminating the

need to increase the number of bits per datum. In trying to persuade users that interval arithmetic might actually be faster than conventional arithmetic, the following points serve as useful starting points for discussions:

- Intervals form a closed algebra that eliminates the need for complicated error trapping; this can save both time and chip area.
- Strassen matrix multiplication becomes numerically safe, allowing substantial speedups for certain applications.
- Fast Fourier Transforms (FFTs) do not require extra precision for the "twiddle factors."
- Sums and dot products can be performed without need for sorting to guard against rounding errors.
- Vendors such as Sun are now supporting interval arithmetic in compilers, and to a growing extent, in processor designs.
- Argument reduction for functions like $\sin(x)$, $\cos(x)$, $\exp(x)$, and $\log(x)$ is simplified via intervals.
- Compilers can avoid unnecessary precision effort by recognizing imprecise inputs.

These are offered not as theorems, but more as points for discussion. In practice, they may or may not hold. These issues must be explored if interval arithmetic is to enter the mainstream of scientific computing.

3   Interval Physics

Consider the usual sequence by which physical problems are converted to computer programs:

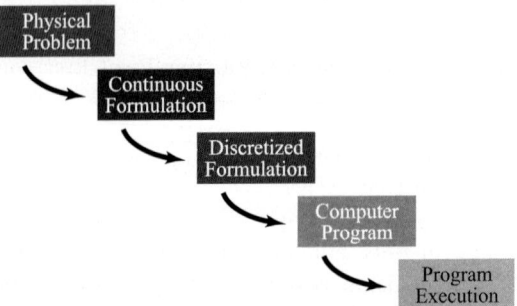

Fig. 4. Where to Apply Interval Concepts

The last step shown in Figure 4, between the Computer Program and the Program Execution, is where most of the effort to improve veracity in computing has been. Either by using extra guard bits in temporary calculations or using high-precision mantissas, vendors have sought to improve answer quality without requiring any effort on the part of programmer or user.

Within the Interval Arithmetic community, most of the effort has been one or two levels up from this. For example, one might explore how to perform Cholesky factorization using interval arithmetic, and under what conditions the bounds remain reasonable. An ordinary differential equation with bounded coefficients might be shown to have a bounded solution.

When such accomplishments are shown to technical ISVs, the reaction is predictably unenthusiastic. The ISV developer might have questions like the following:

- Does a particular car design protect the passengers in a 20 m/s crash?
- Will a jet engine vibrate enough to destroy itself at some operating RPM?
- Does the steady-state heat flow in a circuit design permit it to operate correctly?
- What will the weather be, three days from now?

There doesn't seem to be any obvious way to proceed from questions like these to the solutions provided by interval arithmetic. What is needed is not so much interval arithmetic, but what I call *interval physics*. This is an attack on the first two steps of the sequence shown in Figure 4.

We are taught, early in our technical careers, that physical problems are continuum problems; they therefore map to partial differential equations (PDEs), and a solution of those PDEs will predict behavior in a manner that is as accurate as our understanding of the physics. Since closed-form solutions of PDEs are rare, the differential equations are approximated by finite differences that are discrete and amenable to calculation. However, this is exactly the point at which verifiability is lost in the model. Any use of interval arithmetic after this is useless if the desired result is to provably contain the true answer for a physical simulation. No matter how clever we are about controlling the expansion of intervals in an algorithm, we will never convince the mainstream scientific computing community that interval arithmetic is a good idea until we can bound the *physical* behavior of a system.

3.1   Measuring Physical Answer Quality

Suppose the result of a physical simulation is a scalar quantity F that is a function of space and time variables. Suppose further that one can establish by physical and mathematical reasoning that it is impossible for F to be larger than F^+ or smaller than F^-. We can then define the *total error*, E, as

$$E = \iiiint_{\mathcal{D}} (F^+ - F^-)\, dx\, dy\, dz\, dt \tag{1}$$

and the answer quality, Q, as

$$Q = 1/E, \tag{2}$$

where $\mathcal{D}$ represents the domain of interest in space-time.

These two definitions permit scientific comparison of algorithms and a way to assess performance rigorously. Two very different computers using very different approaches to solving a problem can be compared based on the answer quality achieved in a given amount of time. The trick now is to establish $F^+ - F^-$ for practical situations.

3.2   A Radiation Example

Some physical problems can be expressed as integral equations instead of differential equations. Integral equations of the second kind are particularly amenable to interval techniques:

$$f(x) - \int_{\mathcal{D}} K(x,s)f(s)\,ds = g(x). \tag{3}$$

The $g(x)$ term makes the integral equation inhomogeneous. This equation can be restated in a form that suggests an iterative technique based on initial values for f:

$$f(x) = g(x) + \int_{\mathcal{D}} K(x,s)f(s)ds. \tag{4}$$

Suppose we can place physical bounds $f^- < f < f^+$ based on physical reasoning (such as causality, conservation of energy, etc.) The K and g functions must also be bounded above and below. These all can be very wide intervals initially, corresponding to very low answer quality (as defined in the previous section) if the action of the right-hand side operator above is a contractive mapping. With the bounds initialized, iterate the above equation until the answer quality Q stops improving.

The domain $\mathcal{D}$ is now subdivided. It could be subdivided in space or in time; what matters is that once $\mathcal{D}$ is split, bounds can be improved on K and g. With the subdivided geometry, the interval representing f can be again refined successively until Q stops improving. Each part of the subdivided domain has a measurable contribution to the error, E. What seems most efficient is to repeatedly subdivide the part of the domain that makes the largest contribution to E, maintaining a sorted queue of subdomains to be split.

Radiosity provides a specific example of this technique [5]. Radiosity problems arise both in radiative heat transfer and in computer graphics where surfaces are assumed diffuse reflectors. The equation, using conventional variable names, is

$$b(r) = e(r) + \rho(r) \int_{S} F(r,r')b(r')\,dr', \tag{5}$$

where b is the radiation given off at a point, r and r' are points in the set of all reflecting surfaces S, e is the radiation that is emitted (as a heat or light source, not simply a reflector), ρ is the reflectivity at a point, and $F(r,r')$ is the form factor that determines how much radiation from point r' is received at point r.

For a given geometry, e is stated as an exact value and not an interval. The reflectivity r is between 0.05 and 0.95 in practice. Empirically, a perfectly "white" surface reflects no more than 95% of light that strikes it, and a perfectly "black" surface always reflects at least 5% of the light that strikes it. The form factors are between 0 (completely occluded) and 1 (completely visible) if integrated over r'. If we solve

$$b_{max} = e_{max} + \rho_{max}(1.0 \times b_{max}), \qquad (6)$$

we obtain $b_{max} = e_{max}/(1 - 0.95)$ as a physical upper bound. In other words, the brightest that the radiation can be is if every surface has perfect visibility of the emitting surfaces and the maximum 0.95 reflectivity, and the multiple reflections create the infinite series $e_{max} \times (1 + 0.95 + 0.95^2 + 0.95^3 + \ldots) = e_{max}/(1 - 0.95)$. Similarly, the darkest that the radiation can be is zero, since it is possible for F to be exactly zero. This treats the entire geometry as a single point, so the integral equation becomes a simple scalar equation. Now the geometry can be subdivided, and this improves the bounds on all the quantities in the radiosity integral equation. A simple test geometry in two dimensions is shown in Figure 5.

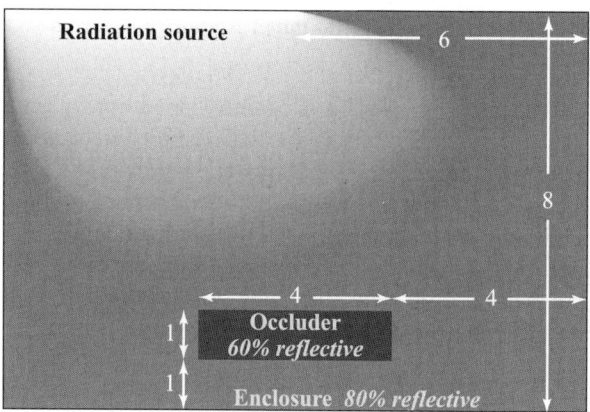

Fig. 5. Simple Radiosity Test Geometry

Results after several iterations are shown in Figure 6 that produce a progressively better set of intervals that bound the radiation, as the z axis of Figure 5.

Interestingly, this research revealed that the so-called "hierarchical radiosity" methods [3] for radiosity are very inefficient at obtaining high answer quality compared to the method described here. While the work done on a

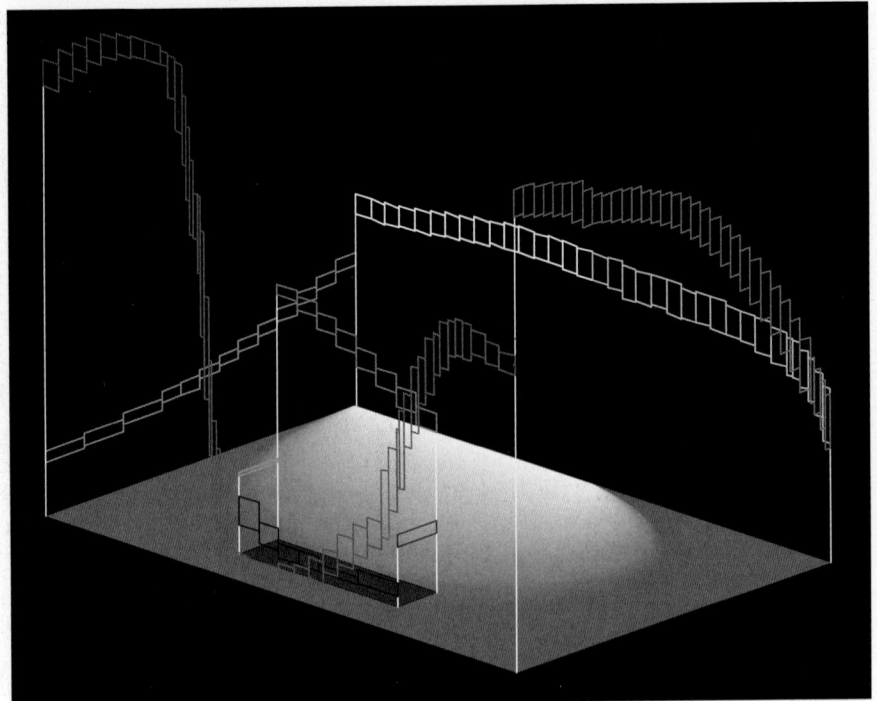

Fig. 6. Rigorously Bounded Radiosity Solution

system of N patches is order N, this is accomplished not by minimizing work but by proliferating patches. The $Q = 1/E$ definition of answer quality is quite useful for resolving arguments about the superiority of any particular approach; the traditional $O(N)$ arguments can be misleading because they equate answer quality with the number of discrete variables, which may have nothing to do with the confidence in their computed value. Figure 7 shows the error bounds that result from the "hierarchical radiosity" method, using the same number of operations as was used to obtain Figure 6.

3.3   Bypassing PDEs

Consider the limiting process used in deriving a PDE and then a difference equation. The physics is assumed continuous, the rules of calculus are applied to the continuous functions that represent the variables and their derivatives, and then the difference equations are derived as approximations to the differential equations.

It may be better practice to bypass the PDE formulation and *discretize the physics directly*. Mass distributions can be regarded as collections of point masses with potential forces between them ("large atoms"). These can be made smaller as more computing power is available, moving the limiting

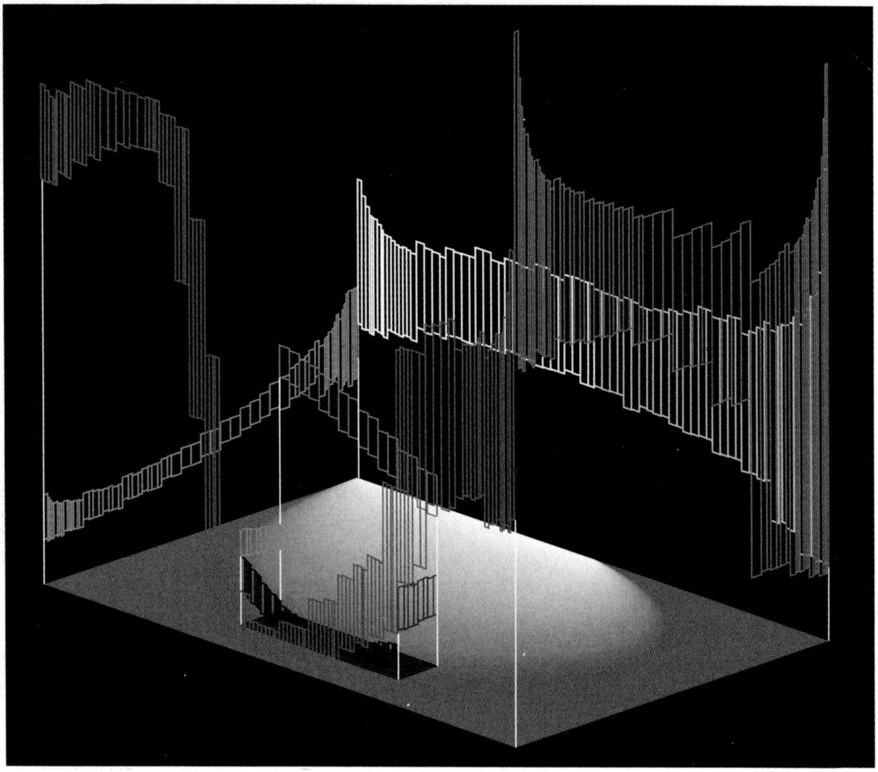

Fig. 7. Poorer Bound via "Hierarchical Radiosity"

process to the physics and not to the difference approximation of the PDE. Similarly, electromagnetic fields might be modeled as quanta, or fluid dynamics as interacting particles. In a way, this is what lattice-gauge theories do.

The advantage of this approach is that it permits provable bounds on physical simulations via interval physics. Interval arithmetic can then be used on the interval physics, which insures that rounding errors cannot spoil the containment of the answer. In creating an example to illustrate this approach of "bypassing PDEs," I stumbled on a surprisingly simple case where the two limiting approaches do not yield the same answer, and for which the traditional PDE approach is the one making the more dubious approximations: A one-dimensional wave equation with an initial velocity given by a step function. That is, a stretched string set in motion by being struck with a hammer.

In the traditional solution [1], the PDE is exactly solved by functions of the form $y = f(x + ct) + g(x - ct)$. This gives rise to the following familiar-looking time sequence:

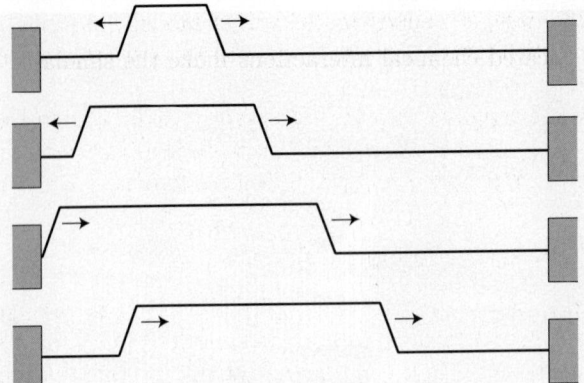

Fig. 8. One-Dimensional Wave Equation by Traditional PDE

Suppose instead that we model the string as a set of point masses on springs. By bounding the behavior based on conservation of energy (spring potentials plus the kinetic energies of the masses), and momentum (components in x and y), the string behaves in a strikingly different manner:

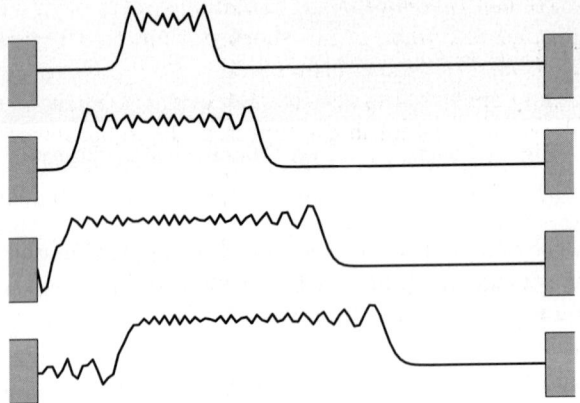

Fig. 9. Wave Equation as the Limit of Point Mass Behavior

One might recognize this sort of "ringing" behavior as one resembling the Gibbs Overshoot Phenomenon, the high-frequency oscillations that never disappear when approximating a step function with a Fourier series. How can this be valid when the PDE shows something so different?

By taking the limiting process of turning the string physics into a PDE, one masks the error of using a forcing function that has discontinuities. Yet, it is reasonable that one could accelerate a string with a hammer that touches the string completely at one point but not at all at the adjacent point. Doing so produces "ringing" that does not disappear as the number of mass points increases. Of course, a realistic model would model the string as a cylinder of mass points with intermolecular forces that give rise to stiffness and other complications.

This example suggests that in addition to providing rigorous bounds for physical simulations, the interval physics approach can sometimes yield insights missed by approximations to PDE formulations.

3.4   The N-Body Problem

Interval physics can be applied to the N-body problem, though the method of getting an initial bound on the particle motion is not obvious. The difficulty arises in the fact that the potential function for inverse-square attractive forces has a singularity if any two particles contact each other. If that happens, their velocity goes to infinity. For any time step size, the possibility of infinite velocity exists, so there are no bounds. While there is ample literature for the bounding of ordinary differential equations (ODEs) by interval methods, it is usually assumed that the coefficients of the ODE can be bounded.

The key is to use *space steps* instead of time steps. Let r be the minimum separation between any pair of particles in the ensemble. Before any particles can move into contact with each other, they would first have to move a distance $r/3$. We can therefore put a bounding sphere of radius $r/3$ around every particle, and ask what is the shortest time t that would allow any particle to move that far. Since the forces on particles can be bounded within the range of those spheres, the acceleration of any particle is also bounded, and this can be used to establish the timestep. Time becomes the dependent variable, not the independent variable, if the calculations are to be rigorous. Because the minimum time must be greater than zero, the simulation will make progress. The amount of temporal progress will vary from one space step to the next.

The bounding spheres can be reduced in size iteratively for that timestep, since each particle's range of motion can be made progressively more precise which in turn allows the forces to be bounded more precisely for all the other particles. The error is the sum of the volumes of all the bounding spheres for all the time steps, so once again we can precisely define the answer quality of the calculation. It is worth noting that doubling the answer quality of the N-body problem seems to require quadrupling the amount of work. The arguments of Greengard, Barnes-Hut, Appel *et al.* regarding clustering methods that reduce the force computations from $O(N^2)$ to $O(NlogN)$ or even $O(N)$ [2] do not take the time stepping error into account; they only state an error bound for a static force calculation. Again, we see that the use of interval physics and a rigorous definition of simulation quality shed insight, on the relative merits of algorithms that may be contrary to commonly held beliefs.

The solution of the N-body problem by rigorous methods points the way to interval physics methods for computational fluid dynamics, structural anal-

ysis, and material property calculations. The missing piece is a definition of
the potential function. Complicated chemical interactions make the simulation, say, of an ensemble of water molecules a very challenging proposition. If
firm bounds can be found on the potential functions, it might be possible to
do accurate fluid simulations without the need the need for the Navier-Stokes
equations. The uncertainty of the potential would make clear what the limits
of predictability in the fluid simulation are.

3.5   "Conservation Laws"

In trying to find interval physics formulations, every situation is different.
Here are some examples of "conservation laws" that can be useful in bounding
physical behavior:

- Energy, momentum, and mass are conserved (non-relativistic).
- Surfaces cannot reflect more radiant energy than they receive.
- Before a particle can move a distance x, it must first move distance $x/2$.
- Probabilities are in the range $[0, 1]$.
- The time to do a task cannot be negative.
- Group wave velocity cannot be greater than c.
- Local minima or maxima cannot occur in a Laplacian interior.
- Effect cannot precede cause.
- The energy at one point cannot exceed the energy of a closed system.
- The Mean Value Theorem holds for continuous physical variables.
- Rigid bodies cannot occupy the same region of space.
- Coefficients of friction cannot be negative.

...and so forth. These indicate the style of thinking that is needed to
formulate simulation problems in a manner that will allow the application
of interval physics, and thus interval arithmetic. The use of interval physics
controls the discretization errors and other uncertainties, while the interval
arithmetic controls the rounding error to preserve containment of the answer
as the computation progresses.

4   Summary

The main hurdle to the mainstream use of interval arithmetic is the adoption
of the methods by commercial vendors of scientific simulation software. While
Moore's law and the market for experimentless computing are applying pressure to find verifiable ways of computing, and vendors like Sun have supplied
the tools in language compilers, there has not been a way to replace conventional PDEs with interval methods, and thus interval computing methods
have remained confined to a small but ardent group of followers.

If we elevate the point at which interval-type thinking is introduced when
a physical problem is being converted into a computer algorithm, we may find

that the PDE formulation can be avoided altogether and "interval physics" used instead. The success of the small-scale programs discussed here suggests that it is time to attempt a simulation code that is on the scale of a NAS-TRAN or an LS-DYNA or a FLUENT, and see if the problems addressed by these programs can instead be done by rigorous means that provably contain the answer. Accomplishing this would be as significant as the introduction of the experimental method, since it bring computational science to the same standing as experimental and theoretical science as a means of obtaining scientific truths.

Acknowledgments

While at Ames Laboratory, the patient work of post-doctoral researchers Quinn Snell and Mary Oman, and research assistant Cassandra Biggerstaff, made possible many of the conceptual breakthroughs for the examples presented here. The constant encouragement of Sun colleague William Walster to develop and present these ideas is what led to the current effort, and for his unwavering zealotry, I am grateful.

References

1. Boyce, W.E., and DePrima, R.C., *Elementary Differential Equations and Boundary Value Problems*, Second Edition. New York: John Wiley & Sons, 1969.
2. Greengard, L. and Rokhlin, V., A fast algorithm for particle simulations. Journal of Computational Physics, 73, 1988.
3. Hanrahan, P., Salzman, D., and Aupperle, L. A rapid hierarchical radiosity algorithm. Computer Graphics 25, 1991; 4:197-206.
4. Rojas, R., and Hashagen, U., editors, *The First Computers: History and Architectures*. Cambridge: MIT Press, 2000.
5. Sillion, F. and Puech, C. *Radiosity and Global Illumination*. San Francisco: Morgan Kaufmann Publishers, Inc., 1994.
6. U.S. Department of Senate, *Daily Press Briefing*, DOT #147, Sept. 11, 1999.

Bounds for Eigenvalues with the Use of Finite Elements

Henning Behnke[1] and Ulrich Mertins[2]

Institut für Mathematik, TU Clausthal,
Erzstr. 1, D-38678 Clausthal-Zellerfeld, Germany.
[1] behnke@math.tu-clausthal.de
[2] mertins@math.tu-clausthal.de

Abstract. Verified upper and lower bounds for the smallest eigenvalues of eigenvalue problems with self-adjoint partial differential equations are computed. Upper bounds are obtained by the Rayleigh-Ritz method, a suitable Goerisch method provides lower bounds. The trial functions are constructed with the use of finite elements. All computations are carried out with interval arithmetic thus the results are protected against rounding errors. Numerical results are given for the L-shaped membrane.

1   Introduction

Usually the finite element method for eigenvalue problems with self-adjoint partial differential equations is understood as a procedure for approximating eigenvalues and eigenfunctions. Since it is based on the well known Rayleigh-Ritz method (which on its part is based on Poincaré's principle) it does not only yield approximations for eigenvalues but *upper bounds*. With the use of interval arithmetic it's more or less straight forward to turn the method into a procedure for the computation of *verified* upper bounds for the smallest eigenvalues.

The aim of our paper is to show that it is possible to compute *verified lower bounds* for the smallest eigenvalues with the use of finite elements as well. The theory for this aim is the Goerisch method. It is based on a variational characterization for the eigenvalues as well. We explain this in the first two sections. After that we discuss some computational aspects.

Besides numerical examples the rest of our paper is dedicated to a setup for the Goerisch method which is suitable for finite element computations. Since the Goerisch method permits us to choose certain quantities in many different ways, this is one of the critical questions.

2   Setting for the Problem

Let $\mathbf{H}_a$ and $\mathbf{H}_b$ be two separable, complex Hilbert spaces with inner products $a(\,.\,,\,.\,)$ and $b(\,.\,,\,.\,)$, respectively. Suppose $\mathbf{H}_a$ is a dense subspace of $\mathbf{H}_b$

continuously embedded in $\mathbf{H}_b$ such that for $\kappa > 0$

$$\kappa\, b(u, u) \leq a(u, u) \text{ for all } u \in \mathbf{H}_a$$

holds true. We then consider the following variationally posed eigenvalue problem:

$$\left. \begin{array}{l} \text{Find eigenpairs } (\lambda, u) \in \mathbb{R} \times \mathbf{H}_a \,, \ u \neq 0 \,, \text{ such} \\ \text{that } a(u, v) = \lambda b(u, v) \text{ holds for all } v \in \mathbf{H}_a. \end{array} \right\} \tag{1}$$

Denote by $B \in \mathcal{L}(\mathbf{H}_a)$ the bounded self-adjoint operator that satisfies

$$a(Bu, v) = b(u, v) \text{ for all } u, v \in \mathbf{H}_a \,.$$

By assumption B possesses a self-adjoint inverse $A = B^{-1} : \mathbf{H}_a \supset \mathcal{D}(A) \longrightarrow \mathbf{H}_a$ and (1) is equivalent to the eigenvalue problem for A. Hence, $\sigma(A)$ and $\sigma_e(A)$ represent the spectrum σ and the essential spectrum σ_e of (1), respectively.

We suppose that for some $N \in \mathbb{N}$ the lower part of σ consists of at least $N + 1$ isolated eigenvalues of finite multiplicity

$$0 < \kappa \leq \lambda_1 \leq \lambda_2 \leq \cdots \leq \lambda_{N+1} < \inf \sigma_e \,.$$

These eigenvalues are characterized by the variational principle

$$\lambda_j = \inf_{\substack{V \subset \mathbf{H}_a \\ \dim V = j}} \ \max_{0 \neq v \in V} \ \frac{a(v, v)}{b(v, v)} \,. \tag{2}$$

This formula can be obtained (see [17, Chapter 3]) from Poincaré's principle for the eigenvalues $\mu_j = 1/\lambda_j$ of the operator $B \in \mathcal{L}(\mathbf{H}_a)$.

A discretization of (2) gives the famous Rayleigh-Ritz method for an efficient and straightforward computation of upper bounds to the eigenvalues below the essential spectrum.

In order to compute lower bounds we have to establish an other variationally posed characterization of the eigenvalues as it is given in [18]. Since, the method acts on the Hilbert space $\mathbf{H}_a$ and the inner product $a(.\,,.)$ stands on the left-hand side of (1) this procedure is called the left definite case.

Assume that $\mathbf{X}$ is a further complex Hilbert space with inner product $s(.\,,.)$ and equipped with an isometric embedding $\mathbf{T} : \mathbf{H}_a \longrightarrow \mathbf{X}$ such that

$$s(\mathbf{T}u, \mathbf{T}v) = a(u, v) \text{ for all } u, v \in \mathbf{H}_a \,. \tag{3}$$

Additionally, the method makes use of a separating parameter $\rho \in \mathbb{R}$ with

$$\lambda_N < \rho < \lambda_{N+1} \,,$$

where this lower bound for the $(N + 1)$th eigenvalue is known a priori.

Now, if the eigenvalues λ_j are represented in the form

$$\lambda_j = \rho + \frac{\rho}{\tau_j - 1} \quad \text{with} \quad \tau_j = \frac{\lambda_j}{\lambda_j - \rho}, \quad j = 1, \ldots, N, \tag{4}$$

we have (see [18, Corollary 2.1]) for $j = 1, \ldots, N$ the variational characterization

$$\tau_j = \inf_{\substack{V \subset \mathbf{H}_a \\ \dim V = j}} \max_{0 \neq v \in V} \min_w \frac{a(v, v) - \rho b(v, v)}{a(v, v) - 2\rho b(v, v) + \rho^2 s(w, w)}, \tag{5}$$

where the minimum is taken over $w \in \mathbf{X}$ such that

$$s(w, \mathbf{T}u) = b(v, u) \text{ for all } u \in \mathbf{H}_a. \tag{6}$$

Note, that by the spectral mapping theorem the numbers $\tau_N \leq \cdots \leq \tau_1 < 0$ are the eigenvalues of the bounded and self-adjoint operator-function $T = A(A - \rho I)^{-1} \in \mathcal{L}(\mathbf{H}_a)$ giving the lower part of its spectrum.

A discretization of (5) using a Rayleigh-Ritz procedure to the operator $T \in \mathcal{L}(\mathbf{H}_a)$ combined with a complementary variational principle gives upper bounds to τ_j and hence, by the transformation $\tau \longmapsto \rho + \frac{\rho}{\tau - 1}$ lower bounds to the eigenvalues λ_j, $j = 1, \ldots, N$, of (1).

3   Calculation of Bounds

For a discretization of (2) and (5) let $n \in \mathbb{N}$, $m \in \mathbb{N}_0$ and suppose the following:

L1. $v_1, \ldots, v_n \in \mathbf{H}_a$ are linearly independent trial functions.
L2. $w_1^*, \ldots, w_n^* \in \mathbf{X}$ satisfy $s(w_i^*, \mathbf{T}u) = b(v_i, u)$ for all $u \in \mathbf{H}_a$, $i = 1, \ldots, n$.
L3. $w_0^\circ, \ldots, w_m^\circ \in \mathbf{X}^\circ = \{w \in \mathbf{X} : s(w, \mathbf{T}u) = 0 \text{ for all } u \in \mathbf{H}_a\}$ where $w_0^\circ = 0$ and $w_1^\circ, \ldots, w_m^\circ$ are linearly independent.

Then, we construct matrices $A_1 = (a_{ik}^{(1)})$, $A_2 = (a_{ik}^{(2)})$ by

$$\begin{aligned} a_{ik}^{(1)} &= a(v_k, v_i) \\ a_{ik}^{(2)} &= b(v_k, v_i) \end{aligned} \quad \text{for } i, k = 1, \ldots, n$$

and matrices $C_{11} = (c_{ik}^{(11)})$, $C_{12} = (c_{ik}^{(12)})$, $C_{22} = (c_{ik}^{(22)})$ by

$$c_{ik}^{(11)} = s(w_k^*, w_i^*) \quad \text{for } i, k = 1, \ldots, n,$$

$$c_{ik}^{(12)} = \begin{cases} s(w_k^\circ, w_i^*) & \text{if } m > 0 \\ 0 & \text{if } m = 0 \end{cases} \quad \text{for } i = 1, \ldots, n; \ k = 1, \ldots, \max\{1, m\},$$

$$c_{ik}^{(22)} = \begin{cases} s(w_k^\circ, w_i^\circ) & \text{if } m > 0 \\ 1 & \text{if } m = 0 \end{cases} \quad \text{for } i, k = 1, \ldots, \max\{1, m\}.$$

Now, we consider the following matrix eigenvalue problems

$(A^{[n]}, x) \in \mathbb{R} \times \mathbb{C}^n$:

$$A_1 x = A^{[n]} A_2 x \,, \tag{7}$$

$(\tau^{\rho[n,m]}, x) \in \mathbb{R} \times \mathbb{C}^n$:

$$(A_1 - \rho A_2)x = \tau^{\rho[n,m]}(A_1 - 2\rho A_2 + \rho^2(C_{11} - C_{12}C_{22}^{-1}C_{12}^H))x \tag{8}$$

and arrange the N lowest eigenvalues of both problems in the order

$$\tau_N^{\rho[n,m]} \leq \cdots \leq \tau_2^{\rho[n,m]} \leq \tau_1^{\rho[n,m]} < 0 < A_1^{[n]} \leq A_2^{[n]} \leq \cdots \leq A_N^{[n]} . \tag{9}$$

Such an arrangement is always possible if the trial functions fulfill the assumption

$$\bigcup_{n \in \mathbb{N}} \operatorname{span}\{v_1, \ldots, v_n\} \text{ is dense in } \left(\mathbf{H}_a, a(\,.\,,\,.\,)\right).$$

Then, for sufficiently large $n \in \mathbb{N}$ and arbitrary $m \in \mathbb{N}_0$ the problem (8) gives exactly N negative eigenvalues (see [12, Theorem 3.4]).

Obviously, the values $A_j^{[n]}$ are upper Rayleigh-Ritz bounds for the eigenvalues λ_j of (1). Whereas the values $\tau_j^{\rho[n,0]}$ are upper Rayleigh-Ritz bounds for the eigenvalues τ_j of the operator $T \in \mathcal{L}(\mathbf{H}_a)$ with respect to the trial functions $(I - \rho B)v_1, \ldots, (I - \rho B)v_n \in \mathbf{H}_a$ (see [12, Theorem 2.6]). A complementary variational principle provides $\tau_j \leq \tau_j^{\rho[n,0]} \leq \tau_j^{\rho[n,m]}$ for $m \in \mathbb{N}$ (see [12, Lemma 3.3]). Hence, setting

$$A_j^{\rho[n,m]} = \rho + \frac{\rho}{\tau_j^{\rho[n,m]} - 1} \,, \quad j = 1, \ldots, N \,,$$

we arrive at

Theorem *Lower and upper bounds to the eigenvalues of (1) are given by*

$$A_j^{\rho[n,m]} \leq \lambda_j \leq A_j^{[n]} \,, \quad j = 1, \ldots, N \,.$$

For convergence results see [10–12].

Remark The choice $m = 0$ yields the procedure of Lehmann [8] and Maehly [9] (cf. also [18, Remark 2.2]) given by

$$\mathbf{X} = \mathbf{H}_a, \quad s(\,.\,,\,.\,) = a(\,.\,,\,.\,), \quad \mathbf{T} = I \quad \text{and} \quad \mathbf{X}^\circ = \{0\} \,.$$

Generalizing the procedure of Lehmann and Maehly the discretization (8) was originally given by Goerisch [2,5,6]. Therefore, we refer to Goerisch method and to Goerisch bounds $A_j^{\rho[n,m]}$.

4   Verified Computation

In this section we explain computational aspects of a procedure for obtaining verified bounds for eigenvalues of variationally posed eigenvalue problems (1).

Firstly, it is clear, that the upper and lower bounds result from two different matrix eigenvalue problems and both of them have to be treated with interval arithmetic.

Secondly, we intend to construct our trial functions v_i with the use of finite elements, thus all the matrices A_1, A_2, C_{11}, C_{12} and C_{22} are large. We will demonstrate examples with dimensions of more than $10\,000$, so the naive approach to compute bounds for the eigenvalues of the corresponding large sparse interval matrix eigenvalue problems can not be recommended.

Thirdly, the right hand side matrix in the Goerisch matrix eigenvalue problem (8) is

$$A_1 - 2\rho A_2 + \rho^2(C_{11} - C_{12}C_{22}^{-1}C_{12}^H).$$

Here the verified inverse of C_{22} or the verified solution of the linear systems $C_{22}X = C_{12}^H$ should be avoided; in our finite element context the dimension of C_{22} is even larger than the dimension of A_1 and A_2.

Now, we explain the tasks in more detail:

- We compute interval enclosures for the matrices A_1, A_2, C_{11}, C_{12} and C_{22}. This can be done with Pascal-XSC or C-XSC by Kulisch or Profil/BIAS by Rump and Knüppel [14,13,7]. The enclosure matrices are denoted $[A_1]$, $[A_2]$, $[C_{11}]$, $[C_{12}]$ and $[C_{22}]$.
- Approximations for the p smallest eigenvalues and corresponding eigenvectors $x_1, \ldots, x_p$ of the problem

$$\mathrm{mid}([A_1])\, x = \tilde{\Lambda}\, \mathrm{mid}([A_2])\, x$$

  are computed with a suitable algorithm. Here $\mathrm{mid}([A_i])$ denotes the real midpoint matrix and $p \in \mathbb{N}$ is small (we usually use $p = 10$).
- In the next step we apply the Rayleigh-Ritz procedure to the interval matrix eigenvalue problem

$$[A_1]\, x = \Lambda\, [A_2]\, x$$

  with $x_1, \ldots, x_p$ as "trial vectors". This results in a $p \times p$ interval matrix eigenvalue problem which has to be solved with guaranteed bounds (see for example [1]).
- From the enclosures (what we really need are upper bounds) for Λ we obtain guaranteed upper bounds for the $\Lambda_j^{[n]}$.
- Instead of the matrix eigenvalue problem (8) we can consider the problem

$$(\tau, x) \in \mathbb{R} \times \mathbb{C}^{n+m} :$$
$$\begin{pmatrix} A_1 - \rho A_2 & 0 \\ 0 & 0 \end{pmatrix} x = \tau \begin{pmatrix} A_1 - 2\rho A_2 + \rho^2 C_{11} & \rho^2 C_{12} \\ \rho^2 C_{12}^H & \rho^2 C_{22} \end{pmatrix} x \qquad (10)$$

for the computation of lower bounds. Both matrix eigenvalue problems provide the same *negative* eigenvalues. Therefore, we can avoid the *exact* solution of the linear systems

$$C_{22}\,X = C_{12}^H$$

at the expense of a matrix eigenvalue problem of dimension $m+n$ instead of n. Since the lower bound computation is a Rayleigh-Ritz procedure (for the eigenvalues τ_j of the operator $T \in \mathcal{L}(\mathbf{H}_a)$) we can apply the Rayleigh-Ritz procedure once again to the matrix eigenvalue problem (10) and the assertion of our inclusion theorem remains valid. We do this with some subspace of $\mathbb{C}^{n+m}$ whose basis is given by the column vectors of the matrix

$$\begin{pmatrix} I \\ X \end{pmatrix}.$$

Here $I \in \mathbb{C}^{(n\times n)}$ is the identity matrix and $X \in \mathbb{C}^{(m\times n)}$ can be chosen arbitrarily. This results in the following matrix eigenvalue problem for τ:

$$(\tau, x) \in \mathbb{R} \times \mathbb{C}^n \ :$$

$$(A_1 - \rho A_2)\,x = \tag{11}$$
$$\tau(A_1 - 2\rho A_2 + \rho^2(C_{11} + C_{12}X + X^H C_{12}^H + X^H C_{22}X))\,x$$

For the special choice $X = -C_{22}^{-1}C_{12}^H$ problem (11) reduces to (8); for practical computations we choose a floating point approximation

$$X \approx -\mathrm{mid}([C_{22}])^{-1}\mathrm{mid}([C_{12}])^H\,.$$

- In the next step we apply the Rayleigh-Ritz procedure to the interval matrix eigenvalue problem

$$([A_1] - \rho[A_2])x =$$
$$\tau([A_1] - 2\rho[A_2] + \rho^2([C_{11}] + [C_{12}]X + X^H[C_{12}]^H + X^H[C_{22}]X))x$$

with $x_1, \ldots, x_p$ (from the upper bound computation) as "trial vectors". This results in a $p \times p$ interval matrix eigenvalue problem which has to be solved with guaranteed bounds.
- From the enclosures (what we really need are upper bounds) for τ we obtain guaranteed lower bounds for the $\Lambda_j^{\rho[n,m]}$.

Besides the computation of interval enclosures for the matrices $[A_1]$, $[A_2]$, $[C_{11}]$, $[C_{12}]$ and $[C_{22}]$ the essential computational tasks are the approximate solution of one big $n \times n$ matrix eigenvalue problem, the approximate solution of linear systems with a $m \times m$ coefficient matrix, some verified matrix - matrix multiplications and the verified solution of two small $p \times p$ interval matrix eigenvalue problems.

5   Application: the Membrane Problem

In this section we will explain how to apply our theory to the model problem:

$$-\Delta u + \alpha\, u \;=\; \lambda\, u \quad \text{in }\; \Omega$$

$$u \;=\; 0 \qquad \text{on }\; \Gamma \subset \partial\Omega \tag{12}$$

$$\tfrac{\partial u}{\partial n} \;=\; 0 \qquad \text{on }\; \partial\Omega\backslash\Gamma,$$

where $\alpha \in L^\infty(\Omega), 0 < \alpha_0 \le \alpha(x,y) \le \alpha_1$ and $\Omega \subset \mathbb{R}^2$ is a polygonal domain.
We define $\mathbf{H}_a := \{u \in H^1(\Omega) \;:\; u = 0 \;\text{ on }\; \Gamma\}$ and $\mathbf{H}_b := L_2(\Omega)$ and the inner products in $\mathbf{H}_a$ and $\mathbf{H}_b$ by

$$a(u,v) := \int_\Omega \big((\operatorname{grad} u)(\operatorname{grad} v) + \alpha\, u\, v\big) d\Omega$$

and

$$b(u,v) := \int_\Omega u\, v\; d\Omega\,.$$

With these definitions the variationally posed eigenvalue problem (1) is a weak formulation of the membrane problem. The main emphasis is a construction of the quantities $\mathbf{X}$, $s(.,.)$ and $\mathbf{T}$ which is suitable for finite element applications. We will discuss three possibilities:

a) The first choice of $\mathbf{X}$, $s(.,.)$ and $\mathbf{T}$ yields the procedure of Lehmann and Maehly; we define

$$\mathbf{X} \;=\; \mathbf{H}_a,$$
$$s(u,v) \;=\; a(u,v) \;\text{ for }\; u,v \in \mathbf{H}_a,$$
$$\mathbf{T}\, u \;=\; u \;\text{ for }\; u \in \mathbf{H}_a$$

Now, the requirement **L2** is to find $w_1^*,\dots,w_n^* \in \mathbf{X}$ such that

$$s(w_i^*, \mathbf{T}\, u) = b(v_i, u) \;\text{ for all }\; u \in \mathbf{H}_a \,,\; i = 1,\dots,n\,.$$

This is a weak formulation of the boundary value problem

$$-\Delta w_i^* + \alpha\, w_i^* \;=\; v_i \quad \text{in }\; \Omega$$

$$w_i^* \;=\; 0 \qquad \text{on }\; \Gamma$$

$$\tfrac{\partial w_i^*}{\partial n} \;=\; 0 \qquad \text{on }\; \partial\Omega\backslash\Gamma$$

which has to be solved *exactly*. This is hard if not impossible.

b) To make it easy to find elements $w_i^* \in \mathbf{X}$ such that **L2** is fulfilled, $\mathbf{X}$ has to be larger. This is taken into account with our next choice:

$$\mathbf{X} = \mathbf{H}_a \times L_2(\Omega),$$

$$s\left(\begin{pmatrix} u_I \\ u_{II} \end{pmatrix}, \begin{pmatrix} v_I \\ v_{II} \end{pmatrix}\right) = \int_\Omega \left((\operatorname{grad} u_I)(\operatorname{grad} v_I) + \alpha\, u_{II}\, v_{II}\right) d\Omega$$

$$\text{for } \begin{pmatrix} u_I \\ u_{II} \end{pmatrix}, \begin{pmatrix} v_I \\ v_{II} \end{pmatrix} \in \mathbf{X},$$

$$\mathbf{T}\, u = \begin{pmatrix} u \\ u \end{pmatrix} \text{ for } u \in \mathbf{H}_a$$

Here the requirement **L2** can easily be fulfilled by

$$w_i^* := \begin{pmatrix} w_i \\ \frac{1}{\alpha}(v_i + \Delta w_i) \end{pmatrix}$$

for *arbitrary* functions $w_i \in \mathbf{H}_a \cap H^2(\Omega)$. The advantage of this choice is that the w_i^* can easily be constructed since we have not to fulfill a differential equation, but the *drawback* is that we require higher regularity for w_i^* than for v_i. The aim is a setting for $\mathbf{X}$, s(.,.) and $\mathbf{T}$ which takes this into account.

c) We define

$$\mathbf{X} = \left(L_2(\Omega) \times L_2(\Omega)\right) \times L_2(\Omega),$$

$$s\left(\begin{pmatrix} u_I \\ u_{II} \end{pmatrix}, \begin{pmatrix} v_I \\ v_{II} \end{pmatrix}\right) = \int_\Omega \left(u_I \cdot v_I + \alpha\, u_{II}\, v_{II}\right) d\Omega$$

$$\text{for } \begin{pmatrix} u_I \\ u_{II} \end{pmatrix}, \begin{pmatrix} v_I \\ v_{II} \end{pmatrix} \in \mathbf{X},$$

$$\mathbf{T}\, u = \begin{pmatrix} \operatorname{grad} u \\ u \end{pmatrix} \text{ for } u \in \mathbf{H}_a.$$

Obviously $s(\mathbf{T}\,u, \mathbf{T}\,v) = a(u,v)$ holds true for all $u, v \in \mathbf{H}_a$.
The requirement **L2** is a weak formulation of the following boundary value problem:

$$-\operatorname{div} w_{i,I} + \alpha\, w_{i,II} = v_i \text{ in } \Omega$$
$$w_{i,I} \cdot n = 0 \text{ on } \partial\Omega \backslash \Gamma$$

where n denotes the outer normal to the boundary $\partial\Omega \backslash \Gamma$. This system is easyly solved by $w_i^* := \left(0, \frac{1}{\alpha} v_i\right) \in \left(L_2(\Omega) \times L_2(\Omega)\right) \times L_2(\Omega) = \mathbf{X}$ giving $s(w_i^*, \mathbf{T}\,u) = b(v_i, u)$ for all $u \in \mathbf{H}_a$.
In order to fulfill the homogeneous equation **L3** we choose

$$w \in H^1(\operatorname{div}, \Omega) \quad \text{such that}$$
$$w \cdot n = 0 \quad\quad\quad\quad \text{on } \Omega \backslash \Gamma$$

and define

$$w^\circ := \left(w, \frac{1}{\alpha} \operatorname{div} w\right) \in \mathbf{X}$$

giving $s(w^\circ, \mathbf{T}\,u) = \int_\Omega \left(w \cdot \operatorname{grad} u + (\operatorname{div} w)\,u\right)d\,\Omega = 0$ for all $u \in \mathbf{H}_a$. Hence, we require the same regularity for w as for v in the Rayleigh-Ritz procedure. For our numerical examples we choose

$$w = \begin{pmatrix} \hat{v} \\ \tilde{v} \end{pmatrix} \in H^1(\Omega) \times H^1(\Omega) \subset H^1(\operatorname{div}, \ \Omega)$$

where $\hat{v}$ and $\tilde{v}$ are functions constructed by finite elements on the same grid as for the Rayleigh-Ritz procedure.

6   Numerical Examples

In this section we will demonstrate some numerical examples to illustrate the power of our method. The first example is the L-shaped membrane. The equation is

$$-\Delta u + \alpha\,u = (\lambda + \alpha)\,u \quad \text{in } \Omega$$

$$u = 0 \quad \text{on } \Gamma$$

$$\tfrac{\partial u}{\partial n} = 0 \quad \text{on } \partial\Omega\backslash\Gamma$$

Here α is a positive real constant (we used $\alpha = 1$ for the computation) and $\partial\Omega\backslash\Gamma$ is the line form $(0,0)$ to $(-1,1)$ (see Fig. 1). This reduces the problem to functions of even symmetry with respect to $\partial\Omega\backslash\Gamma$. Due to the term α on the right hand side of the equation, the eigenvalues λ of our problem are the the usual membrane eigenvalues.

Table 1. Bounds for the smallest eigenvalues of the L-shaped membrane

	Goerisch	Rayleigh-Ritz
1	9.6397230_{53}	9.6397230_{92}
2	19.7392088_{11}	19.7392088_{24}
3	31.912633_{1}	31.912636_{2}
4	41.4745_{07}	41.47451_{1}
5	49.3480219_{31}	49.3480220_{18}
6	56.7096_{05}	56.70961_{1}

$$\lambda_8 < \rho = 88 < \lambda_9$$

We used Bell elements for the construction of trial functions and the grid had 596 nodal points and 1022 triangles. The resulting Rayleigh-Ritz dimension is $n = 3227$, the number of homogeneous equations for the Goerisch method is $m = 6987$. The question how to determine the spectral parameter ρ will not be discussed in this paper. This problem is treated in [3] even for complicated regions Ω.

From Table 1 one can see, that our results are quite accurate. The bounds are comparable to those of [4] but their method is restricted to this special problem (and takes care of the singularity in the corner) while our method is a general one.

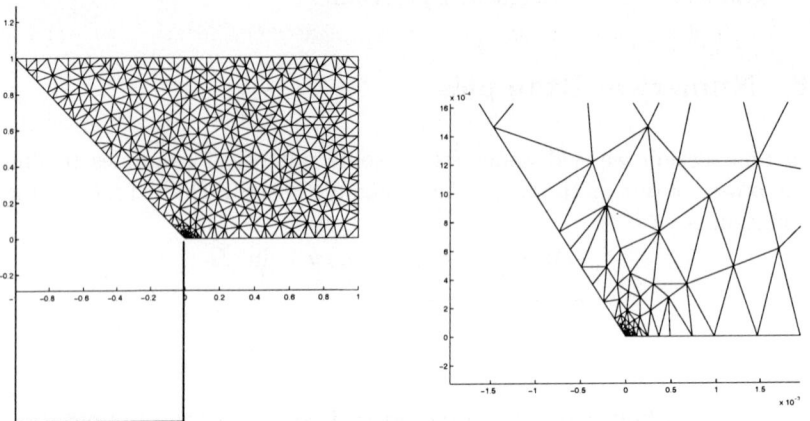

Fig. 1. Triangulation for the L-shaped membrane and a magnification near the corner

The Bell elements result in trial functions $v_i \in H^2(\Omega)$. Of course, this is not necessary for a second order differential equation; we used the Bell elements due to their superior convergence properties. We demonstrate this by a comparison of results with similar dimensions n and different finite elements: Bell elements, Zienkiewicz elements and linear elements. Even though the dimension for the linear element computation is larger the bounds are much worse (see Tables 2, 3 and 4).

Our next example is taken from the finite element book of Schwarz [15,16]. He claims that it is a model for the acoustic eigenfrequencies in the interior of a car. The domain Ω (a two-dimensional cross section) can be seen in Fig. 2; the problem has Neumann boundary conditions.

We used Bell elements, the grid had 1902 nodes, 3285 triangles, the Rayleigh-Ritz dimension is $n = 11412$, the number of homogeneous equations is $m = 21219$.

Table 2. Bell elements: Rayleigh-Ritz dimension $n = 1745$

	Goerisch	Rayleigh-Ritz
1	9.638	9.641
2	19.73920880	19.73920881
3	31.905	31.915
4	41.466	41.476

$$\lambda_6 < \rho = 70 < \lambda_7$$

Table 3. Zienkiewicz elements: Rayleigh-Ritz dimension $n = 1593$

	Goerisch	Rayleigh-Ritz
1	9.633	9.642
2	19.739212	19.739225
3	31.88	31.92
4	41.44	41.48

$$\lambda_6 < \rho = 70 < \lambda_7$$

Table 4. Linear elements: Rayleigh-Ritz dimension $n = 4439$

	Goerisch	Rayleigh-Ritz
1	9.60	9.65
2	19.36	19.76
3	31.01	31.97
4	35.18	41.56

$$\lambda_6 < \rho = 70 < \lambda_7$$

Table 5. Bounds for the smallest eigenvalues of a car-shaped region

	Goerisch	Rayleigh-Ritz
1	-0.00000000_{11}	0.0000000000_{17}
2	0.0125714_{20}	0.0125714_{93}
3	0.043931_{8}	$0.04393\ _{32}$
4	0.056045_{7}	$0.05604\ _{72}$
5	0.11581_{2}	0.11583_{5}
6	0.1362903	0.1362953

$$\lambda_8 < \rho = 0.25 < \lambda_9$$

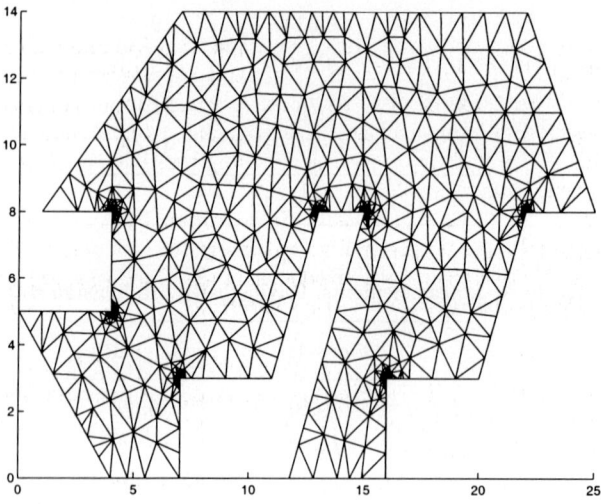

Fig. 2. Triangulation for the car-shaped region

References

1. Behnke H. (1991) The calculation of guaranteed bounds for eigenvalues using complementary variational principles. Computing **47**, 11–27
2. Behnke H., and Goerisch F. (1994) Inclusions for eigenvalues of selfadjoint problems. In: Herzberger, J. (Ed.) Topics in validated computations. Elsevier, Amsterdam, Lausanne, New York, Oxford, Shannon, Tokyo, 277–322
3. Behnke H., Mertins U., Plum M., and Wieners C. (2000) Eigenvalue inclusions via domain decomposition. Proc. R. Soc. Lond. A **456**, 2717–2730
4. Fox L., Henrici P., and Moler C. (1967) Approximations and bounds for eigenvalues of elliptic operators. SIAM J. Numer. Anal. **4**, 89–102
5. Goerisch F. (1980) Eine Verallgemeinerung eines Verfahrens von N.J. Lehmann zur Einschließung von Eigenwerten. Wiss. Z. Tech. Univ. Dresden **29**, 429 – 431
6. Goerisch F., and Haunhorst H. (1985) Eigenwertschranken für Eigenwertaufgaben mit partiellen Differentialgleichungen. Z. Angew. Math. Mech. **65**, 129–135
7. Knüppel O. (1994) Profil / bias — a fast interval library. Computing **53**, 277–287
8. Lehmann N.J. (1949) and (1950) Beiträge zur Lösung linearer Eigenwertprobleme I und II. Z. Angew. Math. Mech. **29**, 341–356 and **30**, 1–16
9. Maehly H.J. (1952) Ein neues Variationsverfahren zur genäherten Berechnung der Eigenwerte hermitescher Operatoren. Helv. Phys. Acta **25**, 547–568
10. Mertins U. (1991) Zur Konvergenz des Rayleigh-Ritz-Verfahrens bei Eigenwertaufgaben. Numer. Math. **59**, 667–682
11. Mertins U. (1992) Asymptotische Fehlerschranken für Rayleigh-Ritz-Approximationen selbstadjungierter Eigenwertaufgaben. Numer. Math. **63**, 227–241
12. Mertins U. (1996) On the convergence of the Goerisch method for self-adjoint eigenvalue problems with arbitrary spectrum. Z. Anal. Anwendungen **15**, 661–686
13. Klatte R., Kulisch U., Law C., and Rauch M. (1993) C-XSC – A C++ Class Library for Extended Scientific Computing. Springer, Heidelberg, New York
14. Klatte R., Kulisch U., Neaga M., Ratz D., and Ullrich Ch. (1991) Pascal-XSC – Language Reference with Examples. Springe, Heidelberg, New York
15. Schwarz H.-R. (1980) Methode der finiten Elemente. Teubner, Stuttgart
16. Schwarz H.-R. (1988) Finite element methods. Academic Press Inc., London
17. Weinberger H.F. (1974) Variational Methods for Eigenvalue Approximation. Regional Conference Series in Applied Mathematics, 15. SIAM, Piladelphia
18. Zimmermann S., and Mertins U. (1995) Variational bounds to eigenvalues of self-adjoint problems with arbitrary spectrum. Z. Anal. Anwend. **14**, 327–345

Algorithmic Differencing

Louis B. Rall[1] and Thomas W. Reps[2] *

University of Wisconsin-Madison,
Madison WI 53706, USA.
[1] rall@math.wisc.edu
[2] reps@cs.wisc.edu

Abstract. An algorithmic representation of a function is a step-by-step specification of its evaluation in terms of known operations and functions, such as a computer program. In addition to function values, the algorithmic representation can be used to compute related quantities such as derivatives of the function. A process similar to automatic (or algorithmic) differentiation will be applied to obtain differences and divided differences of functions. Advantages of this approach are that it often reduces the sometimes catastrophic cancellation errors in computation of differences and divided differences and provides numerical convergence of divided differences to derivatives.

1   Algorithmic Representation of Functions

An *algorithmic representation* R_f of a function f is a step-by-step specification of evaluation of $f(x)$ for given x in terms of previously defined operations and functions. Typically, application of R_f will result in a sequence of values

$$R_f(x) = \{t_1, \ldots, t_n\}, \tag{1}$$

where $t_1 = g_1(x)$, and the value at the kth step

$$t_k = g_k(t_1, \ldots, t_{k-1}), \quad k = 2, \ldots, n, \tag{2}$$

is well-defined in terms of the values at one or more of the previous steps, and $t_n = f(x)$. Possible dependence of $n = n(x)$ on x will be suppressed for simplicity of notation.

   Examples of algorithmic representations are entire or parts of computer programs, which will be referred to simply as *routines* for $f(x)$. Numerical routines consist of arithmetic operations and evaluation of standard (or library) functions, and may include loops and branches. It is assumed that a routine R_f for evaluation of f will give the exact value of $f(x)$ if performed in

---

* Supported in part by the National Science Foundation under grants CCR–9625667, CCR–9619219, and CCR–9986308, by the United States-Israel Binational Science Foundation under grant 96–00337, by the Office of Navel Research under contract N00014–00–1–0607, by the Alexander von Humboldt Foundation, and by the John Simon Guggenheim Memorial Foundation.

exact arithmetic, or at least a sufficiently accurate approximation to $f(x)$ in case exact evaluation requires an infinite number of steps. Here, "sufficiently accurate" is a problem specific and user defined concept which could range from exactness to just the correct sign of the result. With this in mind, it will be convenient to refer to the result of executing the routine R_f as $f(x)$.

2   Transformation of Algorithms

Often, values $\Omega f(x)$ are required in addition to values of $f(x)$, where Ω is some operator. For this purpose, transformation of an algorithmic representation R_f of f into a corresponding algorithmic representation $\Omega R_f = R_{\Omega f}$ is sought. For example, this can be done if a (generalized) *chain rule* holds for the functions in (2),

$$\Omega t_k = \Omega g_k = h_k(t_1, \ldots, t_{k-1}, \Omega t_1, \ldots, \Omega t_{k-1}), \quad k = 1, \ldots, n, \qquad (3)$$

so if both t_1 and $\Omega t_1 = \Omega g_1(x)$ are known, then $\Omega f(x) = t_n$. This *forward mode* transformation generates the algorithmic representation

$$\Omega R_f = \Omega\{t_1, \ldots, t_n\} = \{\Omega t_1, \ldots, \Omega t_n\}, \qquad (4)$$

for Ωf in a step-by-step fashion. There are many other ways to transform algorithmic representations. Suppose $\Omega = \Omega_m \cdots \Omega_2 \Omega_1$, then the sequence

$$\begin{aligned} u_1 &= \Omega_1 t_n = h_1(t_1, \ldots, t_n), \\ u_k &= \Omega_k u_{k-1} = h_k(t_1, \ldots, t_n, u_1, \ldots, u_{k-1}), \quad k = 2, \ldots, m, \end{aligned}$$

provides an algorithmic representation of $u_m = \Omega_m \cdots \Omega_1 t_n = \Omega t_n = \Omega f$. This is an example of a *reverse mode* transformation, since it starts with t_n and uses all or part of the previously computed sequence $\{t_1, \ldots, t_n\}$ at each step. Combinations of forward and reverse modes can also be formulated. However the transformation is performed, it is essential to know how the functions g_k in the algorithm (2) for f behave under action by the operator or operators involved.

Current techniques for transformation of numerical routines include:

1. **Interpretation.** The original routine R_f is treated as a sequence of subroutine calls to the functions $h_k = \Omega g_k$ in (4).
2. **Overloading.** The definitions of the quantities and functions involved in the routine R_f are changed to those appropriate to the result $\Omega f(x)$. For example, a routine for a real function $f(x)$ can be evaluated in interval arithmetic to obtain an interval inclusion of $f(x)$ called its *united extension* [17]. This feature is available in programming languages such as Pascal-SC [2], Pascal-XSC [10], and C-XSC [11].

3. **Code Transformation.** The routine R_f is analyzed and converted into a routine $\Omega R_f = R_{\Omega f}$ for Ωf. While interpretation and overloading are usually limited to forward mode, code transformation can employ reverse or combination modes. Furthermore, appropriate "rewrite rules" can make the transformed algorithm more efficient or suitable for a new data type. An example is the reverse mode of gradient calculation by AD as implemented in ADIFOR [1], see also [8]. Other examples will be given below.

An important application of algorithm transformation is *algorithmic* (or automatic) *differentiation* (AD), in which a routine R_f for $f(x)$ is used to obtain a corresponding routine DR_f for evaluation of the derivative $f'(x)$ [8]. In this case, (3) is the ordinary chain rule for derivatives.

The present purpose is to obtain routines for forward and divided differences,

$$\Delta_h f(x) = f(x+h) - f(x), \tag{5}$$

and for $h \neq 0$,

$$[x]_h f = \frac{1}{h} \Delta_h f(x) = \frac{f(x+h) - f(x)}{h}, \tag{6}$$

respectively, from R_f. These routines are designed to improve accuracy in actual computation over straightforward implementation of (5), (6).

Before the introduction of AD, the divided difference (6) was frequently used as an approximation to the derivative $f'(x)$, in spite of the fact that it suffers from truncation error for large h and roundoff error if h is small. Here, the interest is in accurate calculation of the divided difference itself, not as an approximation to the derivative. The method presented has the advantage of numerical convergence in the sense that h becomes small, the computed value of $[x]_h f$ approaches the value of $f'(x)$ obtained by AD for f differentiable, in which case we define $[x]_0 f = f'(x)$.

In terms of $k = \Delta_h g(x) = h[x]_h g$, the chain rule (3) for forward and divided differences of the composite function $f(g(x))$ can be expressed respectively as

$$\Delta_h f(g(x)) = f(g(x+h)) - f(g(x)) = f(\Delta_h g(x) + g(x)) - f(g(x)),$$

and

$$[x]_h (fg) = [g]_k f \cdot [x]_h g. \tag{7}$$

In the above, an operator notation for divided differences [3] is used instead of the more traditional designations such as $f[x]_h$, $f[x, x+h]$, or $[f(x), f(x+h)]$, see [15, Appendix A] for others.

3   Finite Precision Calculations.

The differences (5) and (6) present little challenge under the assumptions of exact data and exact arithmetic. In actual practice, one often has neither.

Calculations are performed in some finite precision (f.p.) arithmetic such as the generally available floating-point systems. For clarity, f.p. numbers will be denoted in `courier` (typewriter) font where appropriate. Thus, the result of executing the routine R_f in f.p. arithmetic yields an f.p. result `f(x)` with no more significant digits than the datum `x`, usually fewer. This presents a computational problem in general and in particular for differences because of cancellation of significant digits.

In f.p. arithmetic, an f.p. number `b` is said to be *negligible* with respect to an f.p. number `a` if `a ± b = a`. Thus, for example, if `h` is negligible with respect to `x`, then the difference formulas (5),(6) yield the generally incorrect f.p. result `0`. Similarly, for continuous functions and sufficiently small `h` it is to be expected that `f(x + h) - f(x) = 0`, which gives the same results. The treatment of differences below will be based on the assumptions that f is differentiable and the execution of the routines R_f and DR_f in f.p. arithmetic yield sufficiently accurate f.p. values of `f(x)` and `f'(x)`, respectively. Verification of this may be as simple as the use of f.p. interval arithmetic in the routines for the function and its derivative, for example. As in the case of AD, algorithmic differencing (AΔ) is based on replacement of the arithmetic operations and standard functions in the routine R_f by the corresponding formulas for differences. The object is to preserve significant digits as much as possible, and a byproduct is the numerical convergence of the divided difference to the derivative. The possibility of doing this is based on the well-known relationships

$$\Delta_h f(x) = f'(x)h + o(h), \quad f[x]_h = f'(x) + O(h).$$

For small h, the dominant terms depend on the derivative $f'(x)$, a quantity which can be computed with acceptable accuracy by assumption. If the subordinate terms turn out to be negligible in f.p. arithmetic for `h` sufficiently small, then

$$\Delta_\mathbf{h} \mathbf{f(x)} = \mathbf{df(x)} = \mathbf{f'(x)h}, \quad \mathbf{f[x]_h} = \mathbf{f'(x)},$$

so one has numerical convergence of the difference to the differential and the divided difference to the derivative, respectively.

The problem of reliable computation of forward and divided differences can also be formulated in terms of given points x_0, x_1, in which case (5),(6) become

$$\Delta f(x_0) = f(x_1) - f(x_0), \tag{8}$$

and

$$[x_0, x_1]f = \frac{\Delta f(x_0)}{\Delta x_0} = \frac{f(x_1) - f(x_0)}{x_1 - x_0}, \tag{9}$$

respectively. While mathematically to equivalent to (5),(6) with $x = x_0$, $h = \Delta x_0 = x_1 - x_0$, these definitions can lead to distinct computational process due to the fact that they use different input data: x, h in the case of (5), (6), and x_0, x_1 in the case of (8), (9). In the former case, only $R_f(x)$ and

h are needed, while the latter assumes computation of $R_f(x_0), R_f(x_1)$. If h is negligible with respect to x, then x + h = x, so $x_1 = x_0$ even though h $\neq$ 0.

4   First Order Difference Operators

The forward and divided differences (5),(6) result from the application of the operators Δ_h to $f(x)$ and $[x]_h$ to f, respectively. Once the action of these operators on arithmetic operations and standard functions is known, they can be applied to the AR R_f of f to obtain algorithmic representations ΔR_f of (5) or $D\Delta R_f$ of (6). Similarly, (8), (9) can be viewed as corresponding transformations of $R_f(x_0), R_f(x_1)$.

In the following, u, v denote generic elements of $R_f(x)$, with $u(x_j), v(x_j)$ the same for $R(x_j)$, $j = 0, 1$.

4.1   Arithmetic Operations

First, the operators Δ_h and $[x]_h$ will be applied to $R_f(x)$:

$$\Delta_h(c_1 u + c_2 v) = c_1 \Delta_h u + c_2 \Delta_h v,$$
$$\Delta_h(uv) = u\Delta_h v + v\Delta_h u + \Delta_h u \Delta_h v,$$
$$\Delta_h(u/v) = (\Delta_h u - (u/v)\Delta_h v)/(v + \Delta_h v),$$

where c_1, c_2 are constants. Correspondingly,

$$[x]_h(c_1 u + c_2 v) = c_1[x]_h u + c_2[x]_h v,$$
$$[x]_h(uv) = u[x]_h v + v[x]_h u + h[x]_h u[x]_h v,$$
$$[x]_h(u/v) = ([x]_h u - (u/v)[x]_h v)/(v + h[x]_h v).$$

Note that the linear combination $c_1 u + c_2 v$ is *internal* to the calculation of f(x). Since this is assumed to be of sufficient accuracy, excessive loss of significant digits does not occur due to these operations. Furthermore, these formulas for divided differences reduce to the corresponding formulas for derivatives for $h = 0$ given $[x]_0 u = u'(x)$, $[x]_0 v = v'(x)$. Similarly, if $R_f(x_0), R_f(x_1)$ are available, we know that $[x_0, x_1]x = 1$ exactly. The action of the divided difference operator $[x_0, x_1]$ on arithmetic operations is given by

$$[x_0, x_1](c_1 u + c_2 v) = c_1[x_0, x_1]u + c_2[x_0, x_1]v,$$
$$[x_0, x_1](uv) = u(x_0)[x_0, x_1]v + v(x_1)[x_0, x_1]u$$
$$= u(x_1)[x_0, x_1]v + v(x_0)[x_0, x_1]u,$$
$$[x_0, x_1]\left(\frac{u}{v}\right) = \frac{v(x_1)[x_0, x_1]u - u(x_0)[x_0, x_1]v}{v(x_0)v(x_1)}$$
$$= \frac{v(x_0)[x_0, x_1]u - u(x_1)[x_0, x_1]v}{v(x_0)v(x_1)}.$$

The last two equations reflect the identity $[x_0, x_1]f = [x_1, x_0]f$, and again furnish the corresponding formulas for derivatives if $x_0 = x_1$ [20].

4.2   Standard Functions

Action of difference operators on some basic standard functions is also easy to derive. In terms of x, h, these follow the methods used in elementary calculus to obtain formulas for derivatives, see Kahan and Fateman [9]. It will be assumed that $h \neq 0$, since AD formulas will give the corresponding derivatives if $h = 0$. Formulas will be given for differences with $k = \Delta_h u = h[x]_h u$. The corresponding divided differences are obtained by division by h, a stable operation in f.p. arithmetic. For $sqr(u) = u^2$, one has

$$\Delta_h sqr(u) = 2uk + k^2 = (2u + k)k.$$

More generally, let n be a positive integer. Then,

$$\Delta_h u^n = \sum_{i=1}^{n} \binom{n}{i} u^{n-k} k^i. \tag{10}$$

Since

$$\sqrt{u + k} - \sqrt{u} = \frac{k}{\sqrt{u + k} + \sqrt{u}},$$

$$\Delta_h sqrt(u) = \frac{k}{sqrt(u) + sqrt(u + h)},$$

and cancellation of significant digits does not occur. Trigonometric identities for cosine and sine give

$$\Delta_h \cos(u) = -2\sin(k/2)\sin(u + k/2),$$
$$\Delta_h \sin(u) = 2\sin(k/2)\cos(u + k/2),$$

with multiplication instead of subtraction.

Other standard functions, such as $\exp(u)$, $\ln(u)$, $\arctan(u)$, are treated similarly. For the exponential function, one has

$$\Delta_h \exp(u) = \exp(u + k/2)(\exp(k/2) - \exp(-k/2))$$
$$= 2\exp(u + k/2)\sinh(k/2),$$

This requires the standard function $\sinh(z)$, which is usually provided.
Similarly,

$$\Delta_h \ln(u) = \ln(u + k)u - \ln(u) = \ln\left(1 + \frac{k}{u}\right).$$

For the arctangent, let

$$\alpha = \arctan(u + k), \quad \beta = \arctan(u).$$

Then,

$$\tan(\alpha - \beta) = \frac{\tan \alpha - \tan \beta}{1 + \tan \alpha \tan \beta} = \frac{k}{1 + u(u + k)},$$

and thus

$$\Delta_h \arctan(u) = \arctan \left(\frac{k}{1 + u(u + k)} \right), \tag{11}$$

and so on.

The formulas given above are readily adaptable to two-point divided differences $[x_0, x_1]\phi(u)$, where ϕ is one of the standard functions considered. Under the assumption that $h = x_1 - x_0$ is exact, the substitution $k = h[x_0, x_1]u$ followed by division by h gives the desired result. In particular, (10) takes the simple form

$$[x_0, x_1]u^n = \left(\sum_{i=0}^{n-1} u_0^{n-1-i} u_1^i \right) [x_0, x_1]u, \tag{12}$$

the quantity in parentheses being simply the symmetric polynomial with terms of degree $n - 1$ in $u_0 = u(x_0), u_1 = u(x_1)$. Other methods for deriving divided difference formulas will be considered in the following subsection.

Numerical experiments conducted in Pascal-XSC [10] using the arithmetic operations and standard functions formulated above show increased accuracy of differences for small h and numerical convergence of divided differences to the corresponding derivatives as $h \to 0$. This numerical convergence is essentially built into the formulas given above for arithmetic operations and standard functions. This is in sharp contrast to the calculation of divided differences by (6) for small h. In addition to the computation of differences as such, rewriting routines in real or interval f.p. arithmetic to replace subtraction by multiplication wherever possible can lead to more accuracy or narrower intervals. These "rewrite rules" can be included in a code transformation process. It should be noted that this runs counter to the process known as "strength reduction" used in optimization of compiled code with respect to speed of execution instead of accuracy [4], even though modern computers execute multiplications at essentially the same speed as subtractions.

If possible, it is desirable to set a "tolerance" on the value of $|h|$ or $|x_0 - x_1|$ and use the standard formulas if these are large enough to provide sufficient accuracy. While this gives the possibility of faster computation, it may be a problem in its own right. Although generally slower, the formulas presented above are stable over a wide range of values of the increments.

4.3   Slope Formulas

Divided differences are also called *slopes*. It follows from the chain rule (7) that one can write the *slope formula*

$$\Delta_h(fg) = S_f(g,k)\Delta_h g, \quad [x]_h(fg) = S_f(g,k)[x]_h g,$$

in terms of the slope $S_f(g,k) = [g]_k f$. Analogous formulas hold for the case x_0, x_1 are given, for example, (12) is a slope formula for $f(u) = u^n$ with

$$S_{u^n}(u_0, u_1) = \sigma_{n-1}(u_0, u_1) = \left(\sum_{i=0}^{n-1} u_0^{n-1-i} u_1^i \right),$$

the symmetric polynomial with terms of degree $n-1$ in u_0, u_1, where $\sigma_0(u_0, u_1) = 1$. If $f(u)$ has a Taylor series expansion at $u = \alpha$ which converges for $u = u_0$ and $u = u_1$, then its slope is given by

$$S_f(u_0, u_1) = \sum_{k=1}^{\infty} \frac{f^{(k)}(\alpha)}{k!} \sigma_{k-1}(u_0 - \alpha, u_1 - \alpha). \tag{13}$$

In some cases, however, slope formulas are more cumbersome than the ones given above. For example, for $f(u) = \arctan(u)$, one has

$$S_f(u,k) = \arctan_1 \left(\frac{k}{1 + u(u+k)} \right) \frac{1}{1 + u(u+k)},$$

instead of (11), where $\arctan_1(z) = \arctan(z)/z$ for $z \neq 0$, $\arctan_1(0) = 1$. The development of a slope arithmetic as such will not be pursued further here, although derivations of slopes will be presented in the next subsections.

4.4   A Taylor Series Method

It is well known from the AD method of R. E. Moore [17] that the AR of a sufficiently smooth function f can be used to generate its normalized Taylor coefficients

$$f_k = \frac{1}{k!} f^{(k)}(x) h^k, \quad k = 0, 1, 2, \dots, \tag{14}$$

of the series expansion of $f(x+h)$. Suppose that this series converges at least for small h, then one has at once

$$\Delta_h f(x) = f(x+h) - f(x) = \sum_{k=1}^{\infty} \frac{1}{k!} f^{(k)}(x) h^k = \sum_{k=1}^{\infty} f_k. \tag{15}$$

This can be written

$$\Delta_h f(x) = f_1(x,h)h, \quad f_1(x,h) = \sum_{k=1}^{\infty} \frac{1}{k!} f^{(k)}(x) h^{k-1}, \tag{16}$$

that is, $f_1(x, h) = S_f(x, h)$, the corresponding slope.

If $h = 0$, one simply obtains $\Delta_0 f(x) = 0$, $[x]_0 f = f'(x)$ by AD. For $h \neq 0$, automatic differentiation is used to compute the normalized Taylor coefficients $f_0, f_1, \ldots, f_m$ in point f.p. interval arithmetic and *interval remainder terms*

$$F_k = \frac{1}{k!} F^{(k)}(X_h) h^k, \quad k = 0, 1, \ldots, m + 1, \tag{17}$$

by f.p. interval arithmetic, where X_h denotes the closed interval with endpoints x and $x + h$ and $F^{(k)}$ is an interval extension of $f^{(k)}$. One has

$$\Delta_h f(x) = f(x + h) - f(x) \in \sum_{k=1}^{m} f_k + F_{m+1} = S_m + F_{m+1}. \tag{18}$$

An accurate summation method can be applied to obtain S_m, and the width of F_{m+1} can be reduced by the method of Corliss and Rall [6]. Since division is stable in f.p. arithmetic, one has

$$[x]_h f(x) \in (S_m + F_{m+1})/h. \tag{19}$$

The same formulas apply directly if x, h are replaced by u, k for $u = u(x)$, $k = \Delta_h u(x)$. Similar expressions also hold in x_0, x_1 and u_0, u_1 form, see (13) for example.

Equation (18) also applies if f is a function of several variables, $x = (x_1, \ldots, x_d)$, in which case $h = (h_1, \ldots, h_d)$ is the increment. A divided difference analogous to (19) in the direction of a unit vector u is formed by taking $h = \sigma u$, then

$$\partial_\sigma f(x) \in (S_m + F_{m+1})/\sigma. \tag{20}$$

4.5   Numerical Integration

Another approach to accurate, validated computation of differences is based on Taylor's theorem in integral form,

$$\Delta_h f(x) = f(x + h) - f(x) = h \int_0^1 f'(x + \theta h) d\theta. \tag{21}$$

This gives the slope formula

$$S_f(x, h) = \int_0^1 f'(x + \theta h) d\theta.$$

In this case, automatic differentiation is used to evaluate the integrand at the nodes of some rule of numerical integration, that is, for several values of θ in the interval $[0, 1]$. Once again, validation of the accuracy of the numerical integration is based on the use of an interval Taylor coefficient of f, also obtainable by automatic differentiation. This also permits the numerical integration to be adaptive, so that a difference of satisfactory accuracy can be computed with reduced effort, see Corliss and Rall [5]. From (21), it follows that

$$[x]_h f(x) = \int_0^1 f'(x + \theta h) d\theta. \tag{22}$$

Once again, the formulas above generalize immediately to forms involving u, k or x_0, x_1 and u_0, u_1.

If f is a function of several variables, (21) becomes

$$\Delta_h f(x) = \int_0^1 \nabla f(x + \theta h) \cdot h d\theta, \tag{23}$$

with the corresponding directional divided difference

$$\partial_\sigma f(x) = \int_0^1 \nabla f(x + \sigma \theta u) \cdot u d\theta. \tag{24}$$

The numerical integration method requires evaluation of only the first derivative of f at several points, the order of the interval remainder term depends on the rule of numerical integration employed.

5   Differences of Inverse Functions

Computer routines are often used to solve equations $f(x) = y$, that is, to compute values of the inverse function $x = g(y) = f^{-1}(y)$. A routine R_g for this purpose ordinarily contains a subroutine R_f for evaluation of $f(x)$. R_f is usually executed a number of times in the course of computation of a satisfactory value x of x. To find derivatives or divided differences of the inverse function g, one could simply transform the routine R_g into DR_g or $D\Delta R_g$. However, it is more efficient to transform only the subroutine R_f and then use the relationship

$$g'(y) = [f'(x)]^{-1} \tag{25}$$

for the derivative of the inverse function or

$$[y]_k g = ([x]_h f)^{-1} \tag{26}$$

for its divided difference. Thus, after $x =$ x has been obtained, it is only necessary to execute $DR_f(x)$ or $D\Delta_f(x)$ *once* to obtain the desired derivative or difference of the inverse function. Note that (25) holds in several variables for the Jacobian matrices of a transformation and its inverse.

6   Higher Order Divided Differences

Given a set of distinct points $S = \{x_0, \ldots, x_n\}$, divided differences of order up to n are defined recursively [15], for example,

$$[x_i, x_{i+1}, x_{i+2}]f = \frac{[x_i, x_{i+1}]f - [x_{i+1}, x_{i+2}]f}{x_{i+2} - x_i}, \quad i = 0, \ldots, n - 2,$$

are the second order divided differences of f. Traditional uses of higher order divided differences are the construction of interpolating polynomials or other functions. This is particularly useful if the computation of $f(x)$ is expensive and approximate values are required at a large number of points as in the case of curve rendering. Error terms for such approximations are typically expressed in terms of derivatives [15], and thus can be bounded by interval evaluation of Taylor coefficients of f. As in the case of first divided differences, problems with the accuracy of higher order divided differences arise if the spacing between at least some of the points of S becomes small. The confluent case in which some of the points of S are equal leads to Taylor coefficients. These can be computed by AD, so this case fits into the general scheme for differences, see for example [14]. The set S is sometimes referred to as the *support* of the function f.

From the standpoint of code transformation, once the routine R_f has been rewritten as a routine $D\Delta R_f$ for first divided differences, then the same rules can be applied to $D\Delta R_f$ to obtain a routine $D\Delta^2 R_f$ for second divided differences, and so on.

Another approach is based on the overloading concept, with definitions of an appropriate data type and corresponding arithmetic operators and standard functions to convert the routine R_f for f into a routine $D\Delta^n R_f$ for divided differences of f on x up to order n. The data type to be used consists of divided difference tables of order n arranged as upper triangular matrices of order $n + 1$, for example,

$$\begin{pmatrix} [x_0]f & [x_0, x_1]f & [x_0, x_1, x_2]f \\ 0 & [x_1]f & [x_1, x_2]f \\ 0 & 0 & [x_2]f \end{pmatrix}$$

for second order, where $[x_i]f = f(x_i)$, $i = 0, 1, 2$. This approach was introduced by G. Opitz [18]. The divided difference table of the support points is

$$X = \begin{pmatrix} x_0 & 1 & 0 & \cdots & 0 \\ 0 & x_1 & 1 & \cdots & 0 \\ \vdots & \ddots & \ddots & \ddots & \vdots \\ 0 & \ddots & \ddots & x_{n-1} & 1 \\ 0 & \cdots & \cdots & 0 & x_n \end{pmatrix}. \tag{27}$$

In this notation, the divided difference table for f is

$$f(X) = F = (f_{ij}),$$ (28)

where $f_{ij} = 0$ if $j < i$, and

$$f_{ij} = [x_i, \ldots, x_j]f, \quad 0 \le i \le j \le n.$$ (29)

This problem has a formal solution [18]. The eigenvalues of X are the support points $x_0, \ldots, x_n$, and the matrix of right eigenvectors of X is

$$Q = (q_{ij}), \quad q_{ij} = \begin{cases} \prod_{k=1}^{j-1}(x_j - x_k)^{-1} & \text{for } i < j, \\ 1 & \text{for } i = j, \\ 0 & \text{for } i > j. \end{cases}$$ (30)

The inverse of this matrix,

$$Q^{-1} = (\bar{q}_{ij}), \quad \bar{q}_{ij} = \begin{cases} \prod_{k=i+1}^{j}(x_i - x_k)^{-1} & \text{for } i < j, \\ 1 & \text{for } i = j, \\ 0 & \text{for } i > j, \end{cases}$$ (31)

consists of the left eigenvectors of X. Thus, if $S = \text{diag}\{x_0, \ldots, x_n\}$ is the diagonal matrix of support points, one has

$$X = QSQ^{-1},$$

and in general, for $F = \text{diag}\{f(x_0), \ldots, f(x_n)\}$,

$$f(X) = QFQ^{-1}$$ (32)

gives the divided difference table for the values of f at the support points. Thus, the operator $\Omega = Q[\ ]Q^{-1}$ constructs the divided difference table of a given set of values.

However, examination of this construction reveals that the first super-diagonal of $f(X)$ with elements $f_{i,i+1} = [x_i, x_{i+1}]f$ consists of first divided differences computed according to the usual formula (9), and similarly for higher differences [18]. Thus, (32) can yield inaccurate results. Other matrix algorithms have similar problems [16].

Analysis of propagation of error in difference tables [14] shows that errors in the first finite differences have the greatest influence. Consequently, it might be sufficient to use the method of Section 4 to compute these in order to obtain satisfactory results.

A method for accurately computing the divided difference table $f(X)$ can be obtained by transforming the routine R_f for $f(x)$ into a similar routine $\Delta^n R_f$ for $f(X)$ in the style of Section 2 and Section 4. Let u, v, w denote generic terms of the routine $R_f = \{t_1, \ldots, t_n\}$ and U, V, W the corresponding divided difference tables.

The formulas for arithmetic operations given by Opitz [18] are straightforward generalizations of Section 4. One has

$$w = c_1 u + c_2 v, \quad W = c_1 U + c_2 V, \tag{33}$$

for linear combinations, and

$$w = uv, \quad W = UV, \tag{34}$$

for products, and W is obtained by matrix multiplication. Accurate matrix multiplication in f.p. arithmetic can be done by provision of a long accumulator for sums of products [12]. For division, a formal expression is

$$w = u/v, \quad W = V^{-1}U,$$

where V is assumed to be nonsingular. However, in practice, the system of equations

$$VW = U \tag{35}$$

requires only back substitution for solution since V is upper triangular, and the method of defect correction and interval inclusion yields very accurate results [2],[10],[11]. Integral powers U^n are obtained by extended multiplication or the repeated squaring method, either of which can be done accurately. In particular, the elements of powers of X are symmetric polynomials in the support points, a generalization of (12). This can be used in a Taylor series representation of $f(X)$,

$$f(X) = \sum_{k=0}^{\infty} \frac{f^{(k)}(\alpha)}{k!}(X - \alpha I)^k, \tag{36}$$

with convergence determined by $\max |x_i - \alpha|$ [14]. This formula is essentially a generalization of (13). Of course, even if convergent, this may not be useful unless the points of S are tightly clustered around α [16].

Provision of standard functions for finite difference tables is not as straightforward as in the case of first differences. However, since the standard functions are well understood, this is easier than for functions in general. One way to do this is simply to provide routines for standard functions with sufficient precision that their difference tables can be computed in the ordinary way with full accuracy. A better variation of this "brute force" technique is to use *staggered precision*, in which the function values are stored as vectors of f.p. numbers [10],[11] to obtain results of high accuracy.

McCurdy [14] gives a "scaling and squaring" algorithm for the finite difference table of the complex exponential function based on the straightforward generalization

$$\exp(\tau U) = \sum_{k=0}^{\infty} \frac{(\tau U)^k}{k!} \tag{37}$$

of (36). With $\tau = 2^{-m}$, (37) will converge rapidly, then the result is squared m times to obtain $\exp(U)$. This "scaling and squaring" algorithm extends immediately to the sine and cosine and the corresponding trigonometric and hyperbolic functions. There are of course other accurate approximations to standard functions by rational functions or other representations which offer the possibility of increased accuracy. It is worth noting that only the first row of the finite difference table needs to be computed, that is,

$$[x_0]f, \ [x_0, x_1]f, \ \ldots, \ [x_0, x_1, \ldots, x_n]f,$$

the rest of the difference table can be "filled in" by additions [14]. This observation goes back at least to H. Briggs (1561–1631) [7], [19].

The integral representation (22) for the first divided difference generalizes to

$$[x_0, \ldots, x_m]f = \int_0^1 \int_0^{\tau_1} \cdots \int_0^{\tau_{m-1}} f^{(m)} \left(x_0 + \sum_{k=1}^m \tau_k \Delta x_{k-1} \right) d\tau_m \cdots d\tau_1,$$

$m = 2, \ldots, n$. This provides another possibility for accurate differencing, particularly for standard functions [14], and also an approach to multivariate divided differences and interpolation [3].

Formulas for higher order differences simplify drastically if the points of the *support* S of f are equally spaced, that is, $\Delta x_i = x_{i+1} - x_i = h$, $i = 0, \ldots, n - 1$.

References

1. C. Bischof, A. Carle, G. F. Corliss, and A. Griewank, ADIFOR: Automatic differentiation in a source translator environment, in Proc. of Int. Symp. on Symb. and Alg. Comp. (ISSAC 1992), ACM, New York, 1992, pp. 294–302.
2. G. Bohlender, C. Ullrich, J. Wolff von Gudenberg, and L. B. Rall, Pascal-SC, A Computer Language for Scientific Computation, Academic Press, New York, 1987.
3. C. de Boor, A multivariate divided difference, in Approximation Theory VIII, ed. by C. K. Chui and L. L. Schumaker, World Scientific, Singapore, 1995.
4. J. Cocke and J. T. Schwartz, Programming Languages and Their Compilers, Preliminary Notes, 2nd Revised Version, Courant Inst. of Math. Sci., New York University, New York, 1970.
5. G. F. Corliss and L. B. Rall, Adaptive, self-validating quadrature, SIAM J. Sci. Stat. Comput., 8 (1987), pp. 831–847.
6. G. F. Corliss and L. B. Rall, Computing the range of derivatives, in Computer Arithmetic, Scientific Computation and Mathematical Modeling, ed. by E. Kaucher, S. M. Markov, and G. Mayer, J.C. Baltzer AG, Basel, 1991, pp. 195–212.
7. H. H. Goldstine, A History of Numerical Analysis, Springer-Verlag, New York, 1977.

8. A. Griewank, Evaluating Derivatives, Principles and Techniques of Algorithmic Differentiation, SIAM, Philadelphia, 2000.

9. W. Kahan and R. J. Fateman, Symbolic computation of divided differences, Unpublished Report, Dept. of Elec. Eng. and Comp. Sci., Univ. of California-Berkeley, 1985. See http://www.cs.berkeley.edu/~fateman/papers/divdif.pdf.

10. R. Klatte, U. Kulisch, M. Neaga, D. Ratz, C. Ullrich, Pascal-XSC, Language Reference with Examples, Springer-Verlag, Berlin-New York, 1992.

11. R. Klatte, U. Kulisch, A. Wiethoff, C. Lawo, M. Rauch, C-XSC, A C++ Class Library for Extended Scientific Computation, Springer-Verlag, Berlin-New York, 1993.

12. U. W. Kulisch and W. L. Miranker, Computer Arithmetic in Theory and Practice, Academic Press, New York, 1981.

13. U. W. Kulisch and W. L. Miranker (Eds.), A New Approach to Scientific Computation, Academic Press, New York, 1983.

14. A. C. McCurdy, Accurate Computation of Divided Differences, Ph.D. diss. and Tech. Rep. UCB/ERL M80/28, Univ. of California-Berkeley, 1980.

15. W. E. Milne, Numerical Calculus, Princeton, 1949.

16. C. Moler and C. Van Loan, Nineteen dubious ways to compute the exponential of a matrix. SIAM Review **20** (1978), pp. 801–836.

17. R. E. Moore, Methods and Applications of Interval Analysis, SIAM, Philadelphia, 1979.

18. G. Opitz, Steigungsmatrizen, Zeitschrift Angew. Math. Mech. **48** (1964), pp. T52–T54.

19. R. Paige and S. Koenig, Finite differencing of computable expressions, ACM Trans. Program. Lang. Syst. **4** (1982), pp. 402–454.

20. T. W. Reps and L. B. Rall, Computational Divided Differencing and Divided-Difference Arithmetics, Tech. Rep. TR–1415, Comp. Sci. Dept., Univ. of Wisconsin-Madison, 2000.

A Comparison of Techniques for Evaluating Centered Forms

Bruno Lang

Aachen University of Technology, Institute for Scientific Computing and Computing Center, D-52056 Aachen, Germany.
lang@rz.rwth-aachen.de

Abstract. Second-order enclosures for a function's range can be computed with centered forms, which involve a so-called slope vector. In this paper we discuss several techniques for determining such vectors and compare them with respect to tightness of the resulting enclosure. We advocate that a two-stage slope vector computation with symbolic preprocessing is optimal.

1 Introduction

Enclosures for the range of functions f over parallel-epipedal domains $[x] = [\underline{x}_1, \overline{x}_1] \times \cdots \times [\underline{x}_n, \overline{x}_n]$ are required in many result-verifying algorithms, e.g., verified quadrature [4] and branch-and-bound type methods for solving non-linear systems or global optimization [6]. Severe over-estimation of the ranges can lead to a very deep recursion tree, and thus to a high running time of the algorithms, or it can even prevent the solution(s) from being located to adequate accuracy. Therefore several techniques have been devised to overcome the linear approximation error that is inherent in the standard interval evaluation of the function [1].

Most of these techniques rely on a Taylor expansion of f around some *center* $c \in \mathbb{R}^n$ (typically, but not necessarily, one chooses $c \in [x]$), and the needed enclosures for the derivatives' ranges are obtained by a combination of interval arithmetic and automatic differentiation [5,9]. For instance, the first-order expansion

$$f(x) = f(c) + \left(\frac{\partial f}{\partial x_1}(\xi), \ldots, \frac{\partial f}{\partial x_n}(\xi) \right) \cdot (x - c) ,$$

where ξ is some unknown point between c and x, immediately leads to the enclosure

$$\mathrm{rg}(f, [x]) \subseteq f(c) + \left(\frac{\partial f}{\partial x_1}[x], \ldots, \frac{\partial f}{\partial x_n}[x] \right) \cdot ([x] - c) , \tag{1}$$

which is known as the *mean value form* of f w.r.t. c, and which – under moderate assumptions – yields quadratic approximation.

Here and in the following, $\mathrm{rg}(h, [x])$ denotes the range of some function h over the domain $[x]$, $h[x]$ is *any* suitable enclosure for the range, and $h([x])$ stands for the specific enclosure obtained by standard interval evaluation.

Obviously, the gradient in (1) can be replaced with any row vector $[s]$ (called *slope vector*) such that

$$f(x) \in f(c) + [s] \cdot (x - c) \qquad \text{for all } x \in [x] \tag{2}$$

holds [7]. Doing so results in the general *centered forms*

$$\text{rg}(f, [x]) \subseteq f(c) + [s] \cdot ([x] - c) . \tag{3}$$

There are several methods for computing slope vectors and they generally lead to different vectors. By virtue of (2), each such vector must be an enclosure for the range of a *slope function* $s(x, c)$ over $[x] \times c$, where the slope function must meet the condition

$$f(x) \in f(c) + s(x, c) \cdot (x - c) \qquad \text{for all } x \in [x] .$$

In one dimension, this requirement – together with continuity – uniquely determines the slope function:

$$s(x, c) = \begin{cases} (f(x) - f(c))/(x - c), & \text{if } x \neq c \\ f'(x), & \text{if } x = c \end{cases}$$

In higher dimensions slope functions are in general non-unique.

The remainder of the paper is organized as follows. In the next section we discuss several methods for computing slope vectors, and in Sect. 3 these methods are applied to a model function f, and the resulting enclosures for the range of f are compared. The paper closes with some recommendations.

2   Methods for Computing Slope Vectors

In this section we discuss several techniques for computing slope vectors $[s]$ that fulfill (2). We will assume $c \in [x]$ in order to simplify the exposition, but this assumption is not crucial.

2.1   Off-Line Differentiation

Given an *expression for the function* f, this technique relies on the differentiation rules to recursively determine an *expression for each partial derivative* $f_i := \partial f / \partial x_i$.

$$f(x) = \text{const} \Rightarrow f_i = 0 \tag{4a}$$

$$f(x) = x_j \Rightarrow f_i = \begin{cases} 1, & \text{if } i = j \\ 0, & \text{otherwise} \end{cases} \tag{4b}$$

$$f(x) = u(x) \pm v(x) \Rightarrow f_i = u_i \pm v_i \tag{4c}$$

$$f(x) = u(x) \cdot v(x) \Rightarrow f_i = u_i \cdot v(x) + u(x) \cdot v_i \tag{4d}$$

$$f(x) = u(x)/v(x) \Rightarrow f_i = (u_i - f(x) \cdot v_i)/v(x) \tag{4e}$$

$$f(x) = e^{u(x)} \Rightarrow f_i = e^{u(x)} \cdot u_i \tag{4f}$$

$$\vdots$$

Enclosures for the derivatives' ranges over $[x]$ may then be obtained via interval evaluation $f_i([x])$ of the corresponding expressions.

The main advantage of this two-stage approach is the possibility to apply any kind of *term simplification* to the expressions f_i before the evaluation. Considering the one-variable function $f(x) = x/(1+x)$ as a simple example, straight-forward application of the differentiation rules yields the expression

$$f' = \frac{1 \cdot (1+x) - x \cdot 1}{(1+x)^2}$$

for the derivative. Simplification leads to

$$f' = \frac{1}{(1+x)^2} \, ,$$

the latter expression giving significantly sharper enclosures than the original one. In addition, the number of operations is also reduced, thus speeding up the evaluation. This aspect becomes important if the derivatives must be evaluated over many different domains, as it is the case, e.g., in the branch-and-bound algorithms.

2.2 Progressive Derivative Evaluation

Applying the mean value theorem in a component-by-component manner, we obtain the representation

$$
\begin{aligned}
f(x) \;=\; & f(c) + \frac{\partial f}{x_1}(\xi_1, c_2, \ldots, c_n) \cdot (x_1 - c_1) \\
& + \frac{\partial f}{x_2}(x_1, \xi_2, c_3, \ldots, c_n) \cdot (x_2 - c_2) \\
& + \cdots \\
& + \frac{\partial f}{x_n}(x_1, x_2, \ldots, x_{n-1}, \xi_n) \cdot (x_n - c_n) \\
\in\; & f(c) + \frac{\partial f}{x_1}([x_1], c_2, \ldots, c_n) \cdot ([x_1] - c_1) \\
& + \frac{\partial f}{x_2}([x_1], [x_2], c_3, \ldots, c_n) \cdot ([x_2] - c_2) \\
& + \cdots \\
& + \frac{\partial f}{x_n}([x_1], [x_2], \ldots, [x_n]) \cdot ([x_n] - c_n)
\end{aligned}
$$

Thus, a suitable vector $[s]$ for (3) may be computed by interval evaluation of each (simplified) expression f_i from Sect. 2.1 on the subbox $[x]^{(i)} = ([x_1], \ldots, [x_i], c_{i+1}, \ldots, c_n)$ having interval entries only in its first i places [8]. Therefore this technique, which we call *progressive derivative evaluation*, produces sharper enclosures than the standard interval evaluation discussed in Sect. 2.1.

2.3 On-Line Derivative Evaluation

In contrast to the two-stage approaches described above, the differentiation rules (4) can also be used to evaluate the gradient *during the function evaluation* by propagating not only the values of the operands, but also their gradients, through the operations that build up the expression for f.

This approach is called *automatic* or *algorithmic differentiation* (*AD*). It yields the function value and the *values* (not expressions) for the gradient at the same argument x or $[x]$.

Automatic differentiation can be applied if the function is given by some program segment, possibly involving loops, branches, and other 'nonlinear' constructs, whereas formulae for the derivatives can be determined only if the function itself is given as a formula.

If implemented with operator overloading, AD provides a very elegant way for computing derivatives. For performance reasons, however, implementations based on code-transformation are preferable. Here, the instructions for computing f are interspersed with additional code for carrying along the gradient values. For details, in particular on the highly efficient *reverse mode* of automatic differentiation, see [5]; the use of a sophisticated AD tool is described in [2].

While obtainable at low programming and evaluation cost, the enclosures computed with AD may be wider than those determined with the two-stage approaches. This is due to the fact that forward-mode AD is equivalent to the standard interval evaluation of the *non-simplified* expressions f_i.

2.4 On-Line Slope Evaluation

The differentiation rules (4) yield a vector $[s]$ containing the slopes between any pair of points in $[x]$. A narrower slope vector can be obtained by making use of the fixed expansion point c, cf. [7]. Letting $\tilde{f}_i$ denote the ith component of the slope function $s_f(x, c)$, we have

$$f(x) = \text{const} \Rightarrow \tilde{f}_i = 0 \tag{5a}$$

$$f(x) = x_j \Rightarrow \tilde{f}_i = \begin{cases} 1, & \text{if } i = j \\ 0, & \text{otherwise} \end{cases} \tag{5b}$$

$$f(x) = u(x) \pm v(x) \Rightarrow \tilde{f}_i = \tilde{u}_i \pm \tilde{v}_i \tag{5c}$$

$$f(x) = u(x) \cdot v(x) \Rightarrow \tilde{f}_i = \tilde{u}_i \cdot v(c) + u(x) \cdot \tilde{v}_i \tag{5d}$$

$$f(x) = u(x)/v(x) \Rightarrow \tilde{f}_i = (\tilde{u}_i - f(c) \cdot \tilde{v}_i)/v(x) \tag{5e}$$

$$f(x) = e^{u(x)} \Rightarrow \tilde{f}_i = e^{u(x)} \cdot \tilde{u}_i \tag{5f}$$

$$\vdots$$

Comparing with (4) we see that some occurrences of the varying x have been replaced with the fixed point c, resulting in sharper enclosures. More precisely, the interval evaluation of the derivative is equivalent to substituting $[x]$ for both the variable x *and* the fixed point c in the slope function: $f_i([x]) = s_f([x], [x])$.

If the slope computation is integrated into the function evaluation we have to carry along the function values $f([x])$, the values $\tilde{f}([x])$ for the slope vectors, *and* the values $f(c)$ because the latter are used in some of the rules, e.g., (5d) and (5e). Taking into account particular cases for values of the actual arguments, the slope vectors can be narrowed further [10].

2.5   Off-Line Slope Determination

As with differentiation, slopes may also be used in a two-stage manner, first making use of (5) to determine $2n$-variable *formulae* $\tilde{f}_i = \tilde{f}_i(x, c)$ for the components of the slope function and then evaluating these formulae at the actual arguments $[x]$ and c.

This approach preserves the general superiority of slopes based on the rules (5), as compared to the derivatives according to (4), and the option for value-dependent optimizations at evaluation time. In addition, it enables term simplifications at preprocessing time that may significantly reduce the width of the resulting vector $[s]$.

Therefore we expect this technique to yield the tightest enclosures for a function's range, and the results in the next section will confirm this claim.

3   A Numerical Example

In this section we use the methods discussed above to compute enclosures for the range of the function

$$f(x_1, x_2, x_3) = (x_1 + x_2 + x_3)^2 - 2x_1x_2 - 2x_1x_3 - 2x_2x_3 + 5$$

over the domain $[x] = [-0.125, 0.125] \times [0, 0.25] \times [0, 0.25]$. The results are given in Table 1.

First, we note that direct interval evaluation yields a sharper enclosure for the range than the derivative-based centered forms (including forward-mode automatic differentiation) *without* term simplification and progressive evaluation. This is due to the relatively large domain – on very narrow domains centered forms are always superior.

Another interesting point is that in this example progressive evaluation is effective only when the simplification is turned off. This is explained by the fact that the simplifier reduces the gradient to $(2x_1, 2x_2, 2x_3)$, which cannot benefit from progressive evaluation in the canonical order. (Changing the order in which the interval components appear in Sect. 2.2 would lead to narrower enclosures.)

Comparing the enclosures based on (4) with those based on (5) one sees that without simplification the slope arithmetic yields narrower enclosures

Table 1. Enclosures for the range of f over $[\boldsymbol{x}]$

Method	Enclosure for $\mathrm{rg}(f, [\boldsymbol{x}])$
Exact range	[5.0 , 5.140625]
Standard interval evaluation	[4.75 , 5.5515625]
Off-line differentiation, no simplification	[4.5 , 5.5625]
Off-line differentiation, no simpl., progressive evaluation	[4.6875 , 5.375]
Off-line diff. with simpl.	[4.875 , 5.1875]
Off-line diff. with simpl. and prog. eval.	[4.875 , 5.1875]
On-line differentiation (forward-mode AD)	[4.5 , 5.5625]
On-line slopes according to (5)	[4.734375, 5.328125]
Off-line slopes (5) with simpl.	[4.828125, 5.234375]
Off-line slopes (5) on simplified function	[4.921875, 5.140625]

whereas with simplification the situation is vice versa. This is due to the fact that the derivative arithmetic (4) commutes with term simplification, while the slope arithmetic (5) does not. Thus, first computing the derivatives and then simplifying them results in the same gradient $(2x_1, 2x_2, 2x_3)$ as differentiating the simplified function $f(\boldsymbol{x}) = x_1^2 + x_2^2 + x_3^2 + 5$. For the slope arithmetic the situation is different: applying it to the original function and then simplifying the expressions yields $(x_1 + c_1 + x_2 - c_2 + x_3 - c_3, -x_1 + c_1 + x_2 + c_2 + x_3 - c_3, -x_1 + c_1 - x_2 + c_2 + x_3 + c_3)$, which cannot be simplified further. By contrast, applying the slope rules to the simplified function gives $(x_1 + c_1, x_2 + c_2, x_3 + c_3)$. This problem is caused by the unsymmetry in the product formula (5d) and might be alleviated by the use of symmetric (but computationally more expensive) formulae; cf. [3].

The narrowest enclosure was obtained by applying the rules (5) to the simplified function and finally simplifying the components of the resulting slope function.

4   Summary and Recommendations

It is well-known that centered forms based on slopes yield sharper enclosures for the range of a function than those based on derivatives. Traditionally, derivatives and slopes have been evaluated on-line, i.e., simultaneously with the function itself. While this approach offers ease-of-use and higher flexibility (the functions need not be representable by closed formulae), it precludes the possibility of term simplification, which may lead to a significant reduction in the enclosure's width.

Therefore we propose a two-stage method for evaluating slope vectors: In the preprocessing stage, apply (5) or similar rules to determine formulae for the components of a slope function and then perform any width-reducing term simplifications to these formulae (e.g., minimize the number of times a variable occurs). In the evaluation phase, use standard interval evaluation with the simplified formulae to obtain the slope vector.

This approach can significantly reduce the over-estimation of the function's range and therefore speed up many algorithms for which the width of the enclosures is of paramount importance. In addition, the evaluation of the simplified formulae is often much faster than in the non-simplified case (which is equivalent to on-line slope computation), due to the decrease in the formulae's complexity.

There are two obvious cases when this technique cannot or should not be used. First, the range of applicability is restricted to functions that can be represented by a formula. Second, the preprocessing in the two-stage approach takes significantly more time than the on-line slope evaluation. This overhead does not pay if the slope function is evaluated only a few times and if we can afford using wider-than-necessary enclosures for the range.

References

1. G. Alefeld and J. Herzberger. *Introduction to Interval Computations.* Academic Press, New York, NY, 1983.
2. C. Bischof, A. Carle, P. Hovland, P. Khademi, and A. Mauer. ADIFOR 2.0 user's guide. Technical Memorandum ANL/MCS-TM-192, Argonne National Laboratory, Mathematics and Computer Science Division, 1995. Revised June 1998.
3. C. Bliek. Fast evaluation of partial derivatives and interval slopes. *Reliable Comput.*, 3:259–268, 1997.
4. G. F. Corliss and L. B. Rall. Adaptive, self-validating quadrature. *SIAM J. Sci. Stat. Comput.*, 8(5):831–847, 1987.
5. A. Griewank. *Evaluating Derivatives: Principles and Techniques of Algorithmic Differentiation.* Frontiers in Applied Mathematics. SIAM, Philadelphia, PA, 2000.
6. R. B. Kearfott. *Rigorous Global Search: Continuous Problems.* Kluwer Academic Publishers, Dordrecht, The Netherlands, 1996.
7. A. Neumaier. *Interval Methods for Systems of Equations.* Cambridge University Press, Cambridge, UK, 1990.
8. J. B. Oliveira. New slope methods for sharper interval functions and a note on Fischer's acceleration method. *Reliable Comput.*, 2(3):299–320, 1996.
9. L. B. Rall. *Automatic Differentiation: Techniques and Applications.* Lecture Notes in Computer Science, No. 120. Springer-Verlag, Berlin, Germany, 1981.
10. D. Ratz. *Automatic Slope Computation and its Application in Nonsmooth Global Optimization.* Shaker Verlag, Aachen, Germany, 1998. Habilitationsschrift.

On the Limit of the Total Step Method in Interval Analysis

Günter Mayer[1] and Ingo Warnke[2]

Universität Rostock, Fachbereich Mathematik,
D–18051 Rostock, Germany.
[1] guenter.mayer@mathematik.uni-rostock.de
[2] ingo@sun4.math.uni-rostock.de

Abstract. We derive a linear system for the midpoint and the radius of the limit $[x]^*$ of the interval total step method $[x]^{k+1} = [A][x]^k + [b]$ provided that $\rho(|[A]|) < 1$. The coefficients of this system are formed by lower and upper bounds of the input intervals, their choice depends on the position of the components of $[x]^*$ with respect to zero. For particular input data this choice can be made without knowing $[x]^*$. For nonnegative $[A]$ the coefficients are determined by solving at most $n + 1$ real linear systems.

1 Introduction

We consider the interval iteration

$$[x]^{k+1} = [A][x]^k + [b], \quad k = 0, 1, \ldots \tag{1}$$

with an $n \times n$ interval matrix $[A] = ([a]_{ij})$ and an interval vector $[b] = ([b]_i)$ with n components. This iteration is usually called (interval) *total step method* (cf. [1], e.g.). It is well–known that (1) is globally convergent if and only if $\rho(|[A]|) < 1$ where $\rho(\cdot)$ denotes the spectral radius and $|[A]|$ is the absolute value of $[A]$. The limit $[x]^*$ of (1) is then the unique fixed point of the interval function

$$[f]([x]) := [A][x] + [b] \tag{2}$$

and contains the solution set

$$S := \{\, x \in \mathbb{R}^n \mid x = Ax + b, \ A \in [A], \ b \in [b] \,\}. \tag{3}$$

These results go back to O. Mayer – cf. [6], [1] or [7] – and are repeated in the subsequent Theorem 1. In the case $\rho(|[A]|) \geq 1$ the fixed points of $[f]$ are extensively studied in [5]. For particular $[A]$ and $[b]$ with $\rho(|[A]|) < 1$ the limit $[x]^*$ can be expressed by means of the bounds of $[a]_{ij}$, $[b]_i$ as can be seen from [2] and [4]. For general $[A]$, $[b]$ the shape of $[x]^*$ is unknown up to now. We will fill this gap by the present paper in the following sense: We will derive a real $2n \times 2n$ linear system for the midpoint and the radius of $[x]^* = ([x]_i^*)$. The coefficients of this linear system can again be expressed by means of the bounds of $[a]_{ij}$ and $[b]_i$. Their choice depends on the position of $[x]_i^* = [\underline{x}_i^*, \overline{x}_i^*]$ with respect to zero and is fixed for the whole i–th column if $\overline{x}_i^* \leq 0$ or $\underline{x}_i^* \geq 0$ (cf. Theorem 2). Although the coefficients cannot be fixed a priori in the general case, the linear system gives insight in the structure of $[x]^*$. Its coefficients can be determined in advance for all particular cases known to the authors from the literature (cf. Theorem 3). For nonnegative interval matrices $[A]$, i.e., $\underline{a}_{ij} \geq 0$ for all i, j, we show that at most $n+2$ real linear $n \times n$ systems have to be solved in order to represent $[x]^*$. For the proof we use ideas of Barth and Nuding in [2]. Note that there $2n$ linear systems seem to be required for $[x]^*$. Modifying, however, slightly the algorithm given in [2] yields also to the number $n + 2$.

2   Notations

We denote the set of real compact intervals $[a] = [\underline{a}, \overline{a}]$ by $I(\mathbb{R})$, the set of real interval vectors by $I(\mathbb{R}^n)$, the set of real $n \times n$ interval matrices by $I(\mathbb{R}^{n \times n})$. We use $[A] = [\underline{A}, \overline{A}] = ([a]_{ij}) = ([\underline{a}_{ij}, \overline{a}_{ij}]) \in I(\mathbb{R}^{n \times n})$ simultaneously without further reference, and we apply a similar notation for interval vectors. Sometimes we use $\min[a]$ instead of $\underline{a}$ and $\max[a]$ instead of $\overline{a}$ and an analogous notation for interval matrices. Degenerate intervals $[a] = [a, a]$ are identified with its unique element writing $[a] \equiv a$ for simplicity. We call $[a] \in I(\mathbb{R})$ symmetric if $[a] = -[a]$, i.e., if $[a] = [-a, a]$ with some real number $a \geq 0$. For a general interval $[a]$ we introduce the midpoint $\check{a} := (\underline{a} + \overline{a})/2$, the absolute value $|[a]| := \max\{|\underline{a}|, |\overline{a}|\}$, and the radius $r_a := (\overline{a} - \underline{a})/2$. For interval vectors and interval matrices these quantities are defined entrywise, for instance $|[A]| := (|[a]_{ij}|) \in \mathbb{R}^{n \times n}$. For elementary formulas dealing with these auxiliary quantities see, e.g., the introductory chapters of [1] or [7]. In order to facilitate the formulation of our results in Section 3 we define the boundary $\partial[A] \subseteq [A]$ of an interval matrix $[A] \in I(\mathbb{R}^{n \times n})$ by

$$\partial[A] := \{\, A = (a_{ij}) \mid a_{ij} \in \{\underline{a}_{ij},\ \overline{a}_{ij}\} \text{ for } i, j = 1, \ldots, n \,\}.$$

We equip $\mathbb{R}^{n \times n}$ by the entrywise defined natural partial ordering '$\leq$' and call $A \in \mathbb{R}^{n \times n}$ nonnegative if $O \leq A$ or, equivalently, $A \geq O$. If T is a set with finitely many elements we denote their number by $\#(T)$.

3   Results

Before we present our own results we recall a result of O. Mayer [6] which regulates the convergence of the total step method (1). Its proof is finally based on Banach's fixed point theorem together with an appropriate scaled maximum norm, and on the inclusion monotonicity of the interval arithmetic. It is easily accessible, e.g., in [1] and [7].

Theorem 1.
For every starting vector $[x]^0 \in I(\mathbb{R}^n)$ the iteration (1) converges to the same vector $[x]^* \in I(\mathbb{R}^n)$ if and only if $\rho(|[A]|) < 1$. In this case $[x]^*$ contains the solution set S from (3) and is the unique fixed point of $[f]([x]) = [A][x] + [b]$.

Our next result has purely auxiliary character. It is tacitly used in several of our proofs and is itself part of the Perron and Frobenius theory on nonnegative matrices. (Cf. [7] or [8], e.g., for a proof.)

Lemma 1.
Let A, $B \in \mathbb{R}^{n \times n}$ with $O \leq A$. If $|B| \leq A$ then $\rho(B) \leq \rho(|B|) \leq \rho(A)$.

As is well–known the usual interval arithmetic satisfies the subset property

$$[a]([b] + [c]) \subseteq [a][b] + [a][c] \quad \text{for } [a], [b], [c] \in I(\mathbb{R})$$

which was called subdistributivity (cf. [1], e.g.). Equality holds only in selected cases, for instance if $[a]$ is degenerate. This observation forms the basis of the following lemma which is crucial for our main result in Theorem 2. The lemma admits a particular representation of the product between two intervals, and, finally, between an interval matrix and an interval vector.

Lemma 2.
Let $[a], [b] \in I(\mathbb{R})$, $[x] \in I(\mathbb{R}^n)$, $[A] \in I(\mathbb{R}^{n \times n})$.

a) There exist $\underset{\sim}{a}, \tilde{a} \in \{\underline{a}, \overline{a}\}$ such that

$$\min([a][b]) = \min(\underset{\sim}{a}[b]) = \underset{\sim}{a}\check{b} - |\underset{\sim}{a}|r_b,$$
$$\max([a][b]) = \max(\tilde{a}[b]) = \tilde{a}\check{b} + |\tilde{a}|r_b.$$

b) With $\underset{\sim}{a}, \tilde{a}$ from a) we get the representation

$$[a][b] = [\underset{\sim}{a}\check{b}, \tilde{a}\check{b}] + [-|\underset{\sim}{a}|, |\tilde{a}|]r_b = \begin{cases} [\underset{\sim}{a}, \tilde{a}]\check{b} + [-|\underset{\sim}{a}|, |\tilde{a}|]r_b, & \text{if } \check{b} \geq 0, \\ [\tilde{a}, \underset{\sim}{a}]\check{b} + [-|\underset{\sim}{a}|, |\tilde{a}|]r_b, & \text{if } \check{b} < 0. \end{cases}$$

c) There exist $\underset{\sim}{A}, \tilde{A} \in \partial[A]$ such that

$$[A][x] = [\underset{\sim}{A}\check{x}, \tilde{A}\check{x}] + [-|\underset{\sim}{A}|, |\tilde{A}|]r_x. \tag{4}$$

The entries $\underset{\sim}{a}_{ij}$, $\tilde{a}_{ij}$ are (not necessarily uniquely) determined by

$$\min([a]_{ij}[x]_j) = \min(\underset{\sim}{a}_{ij}[x]_j) \quad \text{and} \quad \max([a]_{ij}[x]_j) = \max(\tilde{a}_{ij}[x]_j),$$

respectively, following the lines in a).

Proof.

a) Due to the definition of the real interval arithmetic there is an element $\underset{\sim}{a} \in \{\underline{a}, \bar{a}\}$ such that

$$\min([a][b]) = \min(\underset{\sim}{a}[b]) = \min(\underset{\sim}{a}(\check{b} + [-1, 1]r_b))$$
$$= \min(\underset{\sim}{a}\check{b} + [-1, 1]|\underset{\sim}{a}|r_b)) = \underset{\sim}{a}\check{b} - |\underset{\sim}{a}|r_b.$$

The remaining part of a) follows analogously.

b) Due to a) we have still to show that $\underset{\sim}{a}\check{b} \leq \tilde{a}\check{b}$ holds. To this end we assume the converse, i.e., $\underset{\sim}{a}\check{b} > \tilde{a}\check{b}$.

Case 1: $|\underset{\sim}{a}| \leq |\tilde{a}|$.

Here we get the contradiction

$$\min(\tilde{a}[b]) \geq \min([a][b]) = \underset{\sim}{a}\check{b} - |\underset{\sim}{a}|r_b > \tilde{a}\check{b} - |\tilde{a}|r_b = \min(\tilde{a}[b]).$$

Case 2: $|\underset{\sim}{a}| > |\tilde{a}|$.

Similarly as above we get the contradiction

$$\max(\underset{\sim}{a}[b]) \leq \max([a][b]) = \tilde{a}\check{b} + |\tilde{a}|r_b < \underset{\sim}{a}\check{b} + |\underset{\sim}{a}|r_b = \max(\underset{\sim}{a}[b]).$$

c) is a direct consequence of a) and b). If $[x]_j = -[x]_j$ and $[a]_{ij} = -[a]_{ij}$ one can choose $\underset{\sim}{a}_{ij} = \tilde{a}_{ij} = \underline{a}_{ij}$ or $\underset{\sim}{a}_{ij} = \tilde{a}_{ij} = \bar{a}_{ij}$ showing that the choice for $\underset{\sim}{a}_{ij}$, $\tilde{a}_{ij}$ is not necessarily unique.

□

In order to get insight in the choice of $\underset{\sim}{a}$, $\tilde{a}$ we start with a look at Table 1 in which the multiplication $[a] \cdot [b]$ of two intervals is expressed by means of their bounds.

Table 1. Multiplication $[a] \cdot [b]$

	$\bar{b} \leq 0$	$\underline{b} < 0 < \bar{b}$	$0 \leq \underline{b}$
$\bar{a} \leq 0$	$[\bar{a}\underline{b}, \underline{a}\bar{b}]$	$[\underline{a}\bar{b}, \underline{a}\underline{b}]$	$[\underline{a}\bar{b}, \bar{a}\underline{b}]$
$\underline{a} < 0 < \bar{a}$	$[\bar{a}\underline{b}, \underline{a}\underline{b}]$	$[\min\{\bar{a}\underline{b}, \underline{a}\bar{b}\}, \max\{\underline{a}\underline{b}, \bar{a}\bar{b}\}]$	$[\underline{a}\bar{b}, \bar{a}\bar{b}]$
$0 \leq \underline{a}$	$[\bar{a}\underline{b}, \underline{a}\bar{b}]$	$[\bar{a}\underline{b}, \bar{a}\bar{b}]$	$[\underline{a}\underline{b}, \bar{a}\bar{b}]$

The case

$$\underline{a} < 0 < \bar{a}, \quad \underline{b} < 0 < \bar{b} \tag{5}$$

of this table can be further resolved. Since here $r_a > 0$, $r_b > 0$ the inequalities in (5) are equivalent to

$$-1 < \frac{\check{a}}{r_a} < 1, \quad -1 < \frac{\check{b}}{r_b} < 1,$$

Table 2. Multiplication $[a] \cdot [b]$ in the case $\underline{a} < 0 < \bar{a}$, $\underline{b} < 0 < \bar{b}$

	$-1 < \dfrac{\check{a}}{r_a} \le -\dfrac{\check{b}}{r_b} < 1$	$-1 < -\dfrac{\check{b}}{r_b} < \dfrac{\check{a}}{r_a} < 1$
$-1 < \dfrac{\check{a}}{r_a} \le \dfrac{\check{b}}{r_b} < 1$	$[\underline{a}\bar{b}, \underline{a}\bar{b}]$	$[\underline{a}\bar{b}, \bar{a}\bar{b}]$
$-1 < \dfrac{\check{b}}{r_b} < \dfrac{\check{a}}{r_a} < 1$	$[\bar{a}\underline{b}, \underline{a}\underline{b}]$	$[\bar{a}\underline{b}, \bar{a}\bar{b}]$

Table 3. $[a] \cdot [b] = [\underline{a}\check{b}, \tilde{a}\check{b}] + [-|\underline{a}|, |\tilde{a}|]r_b$

	$\bar{b} \le 0$		$\underline{b} < 0 < \bar{b}$		$0 \le \underline{b}$	
	$\underline{a}$	$\tilde{a}$	$\underline{a}$	$\tilde{a}$	$\underline{a}$	$\tilde{a}$
$\bar{a} \le 0$	$\bar{a}$	$\underline{a}$	$\underline{a}$	$\underline{a}$	$\underline{a}$	$\bar{a}$
$\underline{a} < 0 < \bar{a}$	$\bar{a}$	$\underline{a}$	$\otimes$		$\underline{a}$	$\bar{a}$
$0 \le \underline{a}$	$\bar{a}$	$\underline{a}$	$\bar{a}$	$\bar{a}$	$\underline{a}$	$\bar{a}$

and we get Table 2 as the result.

Depending on the cases in these tables one can immediately deduce the choices for $\underline{a}$, $\tilde{a}$. They are summarized in Table 3.

According to Table 2 the circled cross $\otimes$ in the middle of Table 3 has to be replaced by the cases in Table 4.

Table 4. $[a] \cdot [b] = [\underline{a}\check{b}, \tilde{a}\check{b}] + [-|\underline{a}|, |\tilde{a}|]r_b$ in the case $\underline{a} < 0 < \bar{a}$, $\underline{b} < 0 < \bar{b}$

	$-1 < \dfrac{\check{a}}{r_a} \le -\dfrac{\check{b}}{r_b} < 1$		$-1 < -\dfrac{\check{b}}{r_b} < \dfrac{\check{a}}{r_a} < 1$	
	$\underline{a}$	$\tilde{a}$	$\underline{a}$	$\tilde{a}$
$-1 < \dfrac{\check{a}}{r_a} \le \dfrac{\check{b}}{r_b} < 1$	$\underline{a}$	$\underline{a}$	$\underline{a}$	$\bar{a}$
$-1 < \dfrac{\check{b}}{r_b} < \dfrac{\check{a}}{r_a} < 1$	$\bar{a}$	$\underline{a}$	$\bar{a}$	$\bar{a}$

Based on Lemma 2 we are now able to prove the following Theorem 2 as the main result of our paper.

Theorem 2.
Let $[A] \in I(\mathbb{R}^{n \times n})$, $[b] \in I(\mathbb{R}^n)$, $\rho(|[A]|) < 1$.
If $[x] \in I(\mathbb{R}^n)$ *satisfies the linear system*

$$\check{x} = \frac{1}{2}(\tilde{A} + \underline{A})\check{x} + \frac{1}{2}(|\tilde{A}| - |\underline{A}|)r_x + \check{b} \tag{6}$$

$$r_x = \frac{1}{2}(\tilde{A} - \underline{A})\check{x} + \frac{1}{2}(|\tilde{A}| + |\underline{A}|)r_x + r_b \tag{7}$$

or, equivalently,

$$\underline{x} = \frac{1}{2}(\underline{A} + |\underline{A}|)\underline{x} + \frac{1}{2}(\underline{A} - |\underline{A}|)\overline{x} + \underline{b} \tag{8}$$

$$\overline{x} = \frac{1}{2}(\tilde{A} - |\tilde{A}|)\underline{x} + \frac{1}{2}(\tilde{A} + |\tilde{A}|)\overline{x} + \overline{b} \tag{9}$$

with $\underline{A}, \tilde{A} \in \partial[A]$ *as in Lemma 2c) then* $[x]$ *is the unique limit* $[x]^*$ *of the total step method (1).*

Conversely, the limit $[x]^*$ *of (1) satisfies (6), (7) or, equivalently (8), (9) with* $[x] = [x]^*$ *and corresponding choices for* $\underline{A}, \tilde{A} \in \partial[A]$.

Proof.
Subtracting (7) from (6) yields to

$$\underline{x} = \check{x} - r_x = \underline{A}\check{x} - |\underline{A}|r_x + \check{b} - r_b$$
$$= \underline{A}(\underline{x} + \overline{x})/2 - |\underline{A}|(\overline{x} - \underline{x})/2 + \underline{b}$$
$$= \frac{1}{2}(\underline{A} + |\underline{A}|)\underline{x} + \frac{1}{2}(\underline{A} - |\underline{A}|)\overline{x} + \underline{b}$$

which is (8). Adding (6) to (7) results analogously in (9). Thus (6), (7) imply (8), (9). In order to prove the converse, multiply the sum of (8) and (9) by 0.5 and replace $\underline{x}$, $\overline{x}$ by $\check{x} - r_x$, $\check{x} + r_x$, respectively. This leads to (6). By the same substitution in half of the difference between (9) and (8) one gets (7).

Let the midpoint and the radius of $[x]$ now satisfy (6) and (7). Then the difference and the sum of these two systems result in

$$\underline{x} = \check{x} - r_x = \underline{A}\check{x} - |\underline{A}|r_x + \check{b} - r_b = \min([A][x]) + \underline{b}$$

and, similarly,

$$\overline{x} = \check{x} + r_x = \tilde{A}\check{x} + |\tilde{A}|r_x + \check{b} + r_b = \max([A][x]) + \overline{b}$$

where we exploited the definition of $\underline{A}$ and $\tilde{A}$, respectively, according to Lemma 2. This leads to $[x] = [A][x] + [b]$, and by the uniqueness of the fixed point (see Theorem 1) we get the assertion.

Let, conversely, $[x] = [x]^*$ be the limit of (1) and define $\underline{A}, \tilde{A}$ according to (4). From there we get $\overline{x} = \tilde{A}\check{x} + |\tilde{A}|r_x + \overline{b}$ and $\underline{x} = \underline{A}\check{x} - |\underline{A}|r_x + \underline{b}$. Multiplying these equations by 0.5 and adding/subtracting them yields to (6) and (7), respectively.

□

It is worth noticing that the linear systems (6), (7) and (8), (9), respectively, are uniquely solvable: From Theorem 1 we know that the limit $[x]^*$ of (1) exists. According to Theorem 2 it solves the above–mentioned linear systems. If there is another solution of these systems it will be a limit of (1) contradicting the uniqueness guaranteed by Theorem 1.

Although Theorem 2 does not provide the matrices $\underset{\sim}{A}$, $\tilde{A} \in \partial[A]$ in advance it shows that the midpoint, the radius and the bounds of $[x]^*$ can be represented by means of elements from the boundary of $[A]$ in a rather structured way. If one solves any particular system $x = Ax + b$ with $A \in [A]$, $b \in [b]$, for instance the midpoint equation $x = \check{A}x + \check{b}$, then the sign of the component x_j determines in many cases already at least one of the two entries $\underset{\sim}{a}_{ij}$, $\tilde{a}_{ij}$ for $i = 1, \ldots, n$. If, for instance, $x_j > 0$ then necessarily $\overline{x}_j > 0$. Therefore, only the two cases (i) $0 \leq \underline{x}_j$ or (ii) $\underline{x}_j < 0 < \overline{x}_j$ can occur. Hence $\underset{\sim}{a}_{ij} = \underline{a}_{ij}$ if $\overline{a}_{ij} \leq 0$ and $\tilde{a}_{ij} = \overline{a}_{ij}$ if $0 \leq \underline{a}_{ij}$, independently of the precise position of $\underline{x}_j$ – cf. Table 3 with $[a] = [a]_{ij}$ and $[b] = [x]_j$. In the case (i) this table implies that one has to choose $\underset{\sim}{a}_{ij} = \underline{a}_{ij}$ in the *whole* column of $\underset{\sim}{A}$, and, similarly, $\tilde{a}_{ij} = \overline{a}_{ij}$ in the *whole* column of $\tilde{A}$. The case (ii) causes more trouble if $\underline{a}_{ij} < 0 < \overline{a}_{ij}$ for some value of i because now Table 4 regulates the choice, and this requires more knowledge on $[x]_j$. Unfortunately, all cases can in fact occur as the one–dimensional example in Table 5 shows. Here, $[x]^* = [a][x]^* + [b]$ is listed for different choices of $[A] \equiv [a]$ and $[b]$.

We want to show now how Theorem 2 can be applied to particular input data $[A]$ and $[b]$. As a preparation for this we prove the following Lemma 3.

Lemma 3.

Let $[A] \in I(\mathbb{R}^{n \times n})$, $[b] \in I(\mathbb{R}^n)$, $\rho(|[A]|) < 1$ and let $[x]^*$ be the limit of (1).

a) $[x]^* = -[x]^*$ if and only if $[b] = -[b]$.
b) $0 \in [b]$ implies $0 \in [x]^*$ but not vice versa.

Proof.

a) Let $[x]^* = -[x]^*$ hold. Multiplying $[x]^* = [A][x]^* + [b]$ by -1 yields to $-[x]^* = [A](-[x]^*) - [b]$, and by the assumption for $[x]^*$ we get $[A][x]^* - [b] = [x]^* = [A][x]^* + [b]$ whence $-[b] = [b]$. Let, conversely, $[b] = -[b]$ hold. This implies $-[x]^* = [A](-[x]^*) - [b] = [A](-[x]^*) + [b]$ hence $-[x]^*$ is another fixpoint of $[f]$ from (2). Since, by Theorem 1, this fixed point is unique we end up with $[x]^* = -[x]^*$.

b) Choose $b = 0 \in [b]$, $A \in [A]$. Then $x = 0$ solves $x = Ax + 0$, hence $x = 0 \in S \subseteq [x]^*$. The one–dimensional example $[A] \equiv [a] = [-\frac{3}{4}, \frac{1}{2}]$, $[b] \equiv b = 2$ implies $[x]^* = [-1, 4]$ and shows that $0 \in [x]^*$ does not always imply $0 \in [b]$.

$\square$

Table 5. $[x]^* = [a][x]^* + [b]$ for different choices of $[a]$, $[b]$

$[a]$	$[x]^*$	$[b]$	$[a]$	$[x]^*$	$[b]$
$-\frac{1}{2}$	-2	-3	$\frac{1}{2}$	-6	-3
$-\frac{1}{2}$	$[-2, 2]$	$[-1, 1]$	$\frac{1}{2}$	$[-6, 6]$	$[-3, 3]$
$-\frac{1}{2}$	2	3	$\frac{1}{2}$	6	3

$[a]$	$[x]^*$	$[b]$	$\frac{\check{a}}{r_{\check{a}}}$	$\frac{\check{x}^*}{r_{\check{x}^*}}$
$[-\frac{1}{4}, \frac{1}{2}]$	$[3, 12]$	6		
$[-\frac{1}{2}, \frac{1}{4}]$	$[-8, 6]$	$[-5, 2]$	$-\frac{1}{3}$	$-\frac{1}{7}$
$[-\frac{1}{4}, \frac{1}{2}]$	$[-3, 12]$	$[0, 6]$	$\frac{1}{3}$	$\frac{3}{5}$
$[-\frac{1}{4}, \frac{1}{2}]$	$[-12, 3]$	$[-6, 0]$	$\frac{1}{3}$	$-\frac{3}{5}$
$[-\frac{1}{4}, \frac{1}{2}]$	$[-8, 6]$	$[-4, 3]$	$\frac{1}{3}$	$-\frac{1}{7}$
$[-\frac{1}{4}, \frac{1}{2}]$	$[-12, -3]$	-6		

Theorem 3.
Let $[A] \in I(\mathbb{R}^{n \times n})$, $[b] \in I(\mathbb{R}^n)$, $\rho(|[A]|) < 1$, and let $[x]^*$ be the limit of (1).

a) In each of the cases
   i) $[A] \equiv A \in \mathbb{R}^{n \times n}$
   ii) $[A] = -[A]$
   iii) $[b] = -[b]$
   we obtain the representation

$$[x]^* = \hat{x} + (I - |[A]|)^{-1}(r_A|\hat{x}| + r_b)[-1, 1]. \tag{10}$$

with $\hat{x} = (I - \check{A})^{-1}\check{b}$.

b) If $\underline{A} \geq O$ then

$$\underline{x}^* = (I - \underline{A})^{-1}\underline{b}, \qquad \overline{x}^* = (I - \tilde{A})^{-1}\overline{b} \tag{11}$$

with A, $\tilde{A} \in \partial[A]$ as in Theorem 2. In particular, $[x]^*$ is the interval hull of S from (3). The choice $\underline{a}_{ij} = \underline{a}_{ij}$ or $\underline{a}_{ij} = \overline{a}_{ij}$ remains the same in the *whole j-th column. An analogous statement holds for $\tilde{a}_{ij}$.*

*If $\hat{x}$ is the solution of any linear system $x = Ax + b$ with $A \in [A]$, $b \in [b]$
then*

$$\underline{a}_{ij} = \overline{a}_{ij} \text{ if } \hat{x}_j \leq 0, \qquad \tilde{a}_{ij} = \overline{a}_{ij} \text{ if } \hat{x}_j \geq 0, \tag{12}$$

i.e., $\hat{x}$ determines at least n^2 entries of $\underline{A}$, $\tilde{A}$.

c) *If $\underline{A} \geq O$ then*

$$[x]^* = \begin{cases} [(I - \overline{A})^{-1}\underline{b}, (I - \underline{A})^{-1}\overline{b}], & \text{if } \overline{b} \leq 0, \\ [(I - \overline{A})^{-1}\underline{b}, (I - \overline{A})^{-1}\overline{b}], & \text{if } \underline{b} \leq 0 \leq \overline{b}, \\ [(I - \underline{A})^{-1}\underline{b}, (I - \overline{A})^{-1}\overline{b}], & \text{if } 0 \leq \underline{b}. \end{cases}$$

Proof.

a) i)   Since $[A] \equiv A$ we get $\underline{A} = \tilde{A} = \check{A} = A$. The equations (6), (7)
read then $\check{x} = A\check{x} + \check{b}$, $r_x = |A|r_x + r_b$. They imply $\check{x} = (I - A)^{-1}\check{b} =$
$(I - \check{A})^{-1}\check{b} = \hat{x}$, $r_x = (I - |A|)^{-1}r_b$, and (10) follows from $r_A = O$.
ii)   From $[A] = -[A]$ we get $\frac{1}{2}(|\check{A}| - |\underline{A}|) = O$, $\frac{1}{2}(|\tilde{A}| + |\underline{A}|) = |[A]|$. If
$\check{x}_j^* \geq 0$ then $\underline{a}_{ij} = \underline{a}_{ij}$, $\tilde{a}_{ij} = \overline{a}_{ij}$ and $\frac{1}{2}(\tilde{a}_{ij} - \underline{a}_{ij})\check{x}_j^* = r_{a_{ij}}\check{x}_j^* = r_{a_{ij}}|\check{x}_j^*|$. If
$\check{x}_j^* < 0$ then $\underline{a}_{ij} = \overline{a}_{ij}$, $\tilde{a}_{ij} = \underline{a}_{ij}$ and $\frac{1}{2}(\tilde{a}_{ij} - \underline{a}_{ij})\check{x}_j^* = -r_{a_{ij}}\check{x}_j^* = r_{a_{ij}}|\check{x}_j^*|$.
This implies $\frac{1}{2}(\underline{A} + \tilde{A}) = \check{A} = O$ and $\frac{1}{2}(\tilde{A} - \underline{A})\check{x}^* = r_A|\check{x}^*|$. From (6),
(7) we therefore get $\check{x}^* = \check{b} = \hat{x}$ and $r_{x^*} = r_A|\check{x}^*| + |[A]|r_{x^*} + r_b$ whence
$r_{x^*} = (I - |[A]|)^{-1}(r_A|\check{x}^*| + r_b) = (I - |[A]|)^{-1}(r_A|\hat{x}| + r_b)$. This proves
(10).
iii)   The assumption together with Lemma 3 imply

$$[x]^* = -[x]^*. \tag{13}$$

Therefore, $[A][x]^* = |[A]|[x]^* = |\dot{A}|[x]^* = \dot{A}[x]^*$ for each $\dot{A} \in [A]$ with
$|\dot{A}| = |[A]|$, hence one can choose $\underline{A} = \tilde{A} = \check{A}$. The equations (6) and
(7) read $\check{x} = \check{A}\check{x}$ and $r_x = |\dot{A}|r_x + r_b = |[A]|r_x + r_b$ and have the unique
solutions $\check{x}^* = \check{x} = 0$ (in conformity with (13)) and $r_x = (I - |[A]|)^{-1}r_b$.
This proves (10) when taking into account $\check{b} = 0$ and $\hat{x} = 0$.
b) For $\underline{A} \geq O$ the equations (8), (9) are more appropriate than (6), (7).
Since $|\underline{A}| = \underline{A}$, $|\tilde{A}| = \tilde{A}$ they yield to

$$\underline{x} = \underline{A}\underline{x} + \underline{b}, \qquad \overline{x} = \tilde{A}\overline{x} + \overline{b}$$

whence (11) follows immediately. The choice (12) results from $\underline{x}_j^* \leq \hat{x}_j \leq$
0 and $0 \leq \hat{x}_j \leq \overline{x}_j^*$, respectively, using the last row of Table 3 and the
remarks after the proof of Theorem 2. The columnwise choice for $\underline{A}$ and
$\tilde{A}$ follows again from the last row of Table 3.
c) Since $O \leq \underline{A} \leq \overline{A}$ the Neumann series shows $(I - \overline{A})^{-1} \geq O$ and, similarly,
$(I - \tilde{A})^{-1} \geq O$. If $\overline{b} \leq 0$ then $\overline{x}^* = (I - \tilde{A})^{-1}\overline{b} \leq 0$ whence $\underline{A} = \overline{A}$, $\tilde{A} = \underline{A}$
according to Table 3. The assertion follows now from (11). The remaining
part of c) can be seen analogously.   □

We mention that the parts a) i) – a) iii) of Theorem 3 were stated in partly equivalent forms in [4] and proved there without using Theorem 2 which at that time was unknown to the authors. The unified representation (10) of the three cases i) – iii) is given for the first time in the present paper. The parts b) and c) were stated (slightly modified) in [2] and proved there in a different way; see also [1], p. 144 ff and [3]. The matrices A, $\tilde{A}$ can be computed by an iterative process which was first given in [2]. We shortly review this method showing by the way that $n + 1$ linear systems are sufficient to end up with A, $\tilde{A}$. For the proof we need the following lemma.

Lemma 4.
Let $A \in \mathbb{R}^{n \times n}$, $b \in \mathbb{R}^n$, $A \geq O$, $\rho(A) < 1$, $x^* = Ax^* + b$. If $x \leq Ax + b$ then $x \leq x^*$; if $x \geq Ax + b$ then $x \geq x^*$.

Proof.
Start the iteration $x^{k+1} = Ax^k + b$ with $x^0 = x$. Then, by the first assumption, $x^0 \leq Ax^0 + b = x^1$ whence, by induction, $x^0 \leq x^k$ for $k = 1, 2, \ldots$. With $k \to \infty$ one gets $x \leq x^*$. The second assertion is proved analogously.

$\square$

Theorem 4.
Let $[A] \in I(\mathbb{R}^{n \times n})$, $\underline{A} \geq O$, $\rho(|[A]|) < 1$, $[b] \in I(\mathbb{R}^n)$. Then A, $\tilde{A}$ from (11) can be constructed by solving at most $n + 1$ real linear systems of size $n \times n$. In order to compute the limit $[x]^*$ of (1) one has to solve at most $n + 2$ such linear systems.

Before we prove this theorem we remark that for A, $\tilde{A}$ the last row in Table 3 will be used in a slightly modified form which we list in Table 6.

Table 6. $[a] \cdot [b] = [\underline{ab}, \tilde{ab}] + [-|\underline{a}|, |\tilde{a}|]r_b$ in the case $0 \leq \underline{a}$

	$\overline{b} < 0$		$\underline{b} \leq 0 \leq \overline{b}$		$0 < \underline{b}$	
	$\underline{a}$	$\tilde{a}$	$\underline{a}$	$\tilde{a}$	$\underline{a}$	$\tilde{a}$
$0 \leq \underline{a}$	$\overline{a}$	$\underline{a}$	$\overline{a}$	$\overline{a}$	$\underline{a}$	$\overline{a}$

This table together with (12) should be kept in mind during the whole proof. The proof itself starts by choosing some matrix $A^{(1)} \in \partial[A]$ and a 'corresponding' vector $x^{(1)} \in S$. The matrix $A^{(1)} = (a_{ij}^{(1)})$ will be changed in its j-th column to another matrix $A^{(2)} \in \partial[A]$ in an appropriate way whenever the products $a_{ij}^{(1)} x_j^{(1)}$, $i = 1, \ldots, n$, can be increased by this change in the case of finding $\tilde{A}$, and decreased in the case of finding $\underset{\sim}{A}$, respectively. It will turn out later on from Table 6 that these replacements occur if and only if $x_j^{(1)} < 0$ in the first case and $x_j^{(1)} > 0$ in the second one. By $n^{(1)}$ we will count the number of these 'false' columns of $A^{(1)}$. The process will be repeated iteratively ending up with $\tilde{A}$ and $\underset{\sim}{A}$, respectively. It is described by the Steps 1 – 4 in the subsequent proof and represents a slight modification of the algorithm in [2].

Proof of Theorem 4.

We first describe the iterative process mentioned above which leads to $\tilde{A}$, $\underset{\sim}{A}$ by solving at most $n + 1$ linear systems of equations.

Initialization for $\tilde{A}$

Let $\tilde{A}^{(1)} := \overline{A}$.

Step 1
Solve $x = \tilde{A}^{(1)}x + \overline{b}$, denote the solution by $\tilde{x}^{(1)}$ and define

$$\tilde{n}^{(1)} := \#(\{\, j \mid \tilde{x}_j^{(1)} < 0 \,\}),$$

$$\tilde{A}^{(2)} = (\tilde{a}_{ij}^{(2)}) \in \mathbb{R}^{n \times n} \text{ with } \tilde{a}_{ij}^{(2)} := \begin{cases} \overline{a}_{ij} & \text{if } \tilde{x}_j^{(1)} \geq 0, \\ \underline{a}_{ij} & \text{if } \tilde{x}_j^{(1)} < 0. \end{cases}$$

Step 2
If $\tilde{A}^{(1)} = \tilde{A}^{(2)}$ or $\tilde{n}^{(1)} = 1$ define $\tilde{A} := \tilde{A}^{(2)}$ and stop.
Otherwise repeat the Steps 1 and 2 with the upper indices being increased by one.

Initialization for $\underset{\sim}{A}$

Define

$$A^{(1)} = (\underset{\sim}{a}_{ij}^{(1)}) \in \mathbb{R}^{n \times n} \text{ by } \underset{\sim}{a}_{ij}^{(1)} := \begin{cases} \overline{a}_{ij} & \text{if } \tilde{x}_j^{(1)} \leq 0, \\ \underline{a}_{ij} & \text{if } \tilde{x}_j^{(1)} > 0. \end{cases}$$

If $\tilde{n}^{(1)} = n$ define $\underset{\sim}{A} := \overline{A}$ and stop.

Step 3
Solve $x = \underset{\sim}{A}^{(1)}x + \underset{\sim}{b}$ and denote the solution by $\underset{\sim}{x}^{(1)}$. Define

$$\underset{\sim}{n}^{(1)} := \#(\{\, j \mid \underset{\sim}{x}_j^{(1)} > 0 \,\}),$$

$$\underset{\sim}{A}^{(2)} = (\underset{\sim}{a}_{ij}^{(2)}) \in \mathbb{R}^{n \times n} \text{ with } \underset{\sim}{a}_{ij}^{(2)} := \begin{cases} \overline{a}_{ij} & \text{if } \underset{\sim}{x}_j^{(1)} \leq 0, \\ \underline{a}_{ij} & \text{if } \underset{\sim}{x}_j^{(1)} > 0. \end{cases}$$

Step 4

If $\underset{\sim}{A}^{(1)} = \underset{\sim}{A}^{(2)}$ or $\underset{\sim}{n}^{(1)} = 0$ define $\underset{\sim}{A} := \underset{\sim}{A}^{(2)}$ and stop.

Otherwise repeat the Steps 3 and 4 with the upper indices being increased by one.

We now comment on these steps.

CASE 1: $\tilde{A}^{(1)} = \tilde{A}^{(2)}$

By the definition of $\tilde{A}^{(2)} \geq O$ we have

$$\tilde{A}\tilde{x}^{(1)} = \tilde{A}^{(2)}\tilde{x}^{(1)} = \tilde{A}^{(1)}\tilde{x}^{(1)} = \max([A]\tilde{x}^{(1)}),$$

hence $\tilde{x}^{(1)} = \tilde{A}\tilde{x}^{(1)} + \bar{b} = \max([A]\tilde{x}^{(1)} + [b])$.

CASE 2: $\tilde{A}^{(1)} \neq \tilde{A}^{(2)}$

Let $\tilde{x}^{(2)}$ be the solution of $x = \tilde{A}^{(2)}x + \bar{b}$. By the definition of $\tilde{A}^{(2)}$ we obtain

$$\tilde{x}^{(1)} = \tilde{A}^{(1)}\tilde{x}^{(1)} + \bar{b} \leq \tilde{A}^{(2)}\tilde{x}^{(1)} + \bar{b}, \tag{14}$$

hence Lemma 4 guarantees

$$\tilde{x}^{(1)} \leq \tilde{x}^{(2)}, \tag{15}$$

in particular, $\tilde{x}_j^{(2)} \geq 0$ whenever $\tilde{x}_j^{(1)} \geq 0$. Therefore, $\tilde{n}^{(2)} \leq \tilde{n}^{(1)}$. Moreover,

$$\tilde{A}^{(1)}\tilde{x}^{(1)} \leq \tilde{A}^{(2)}\tilde{x}^{(1)} \leq \tilde{A}^{(2)}\tilde{x}^{(2)}$$

as can be seen from (14) and (15). If $\tilde{n}^{(1)} = \tilde{n}^{(2)}$ then $\tilde{A}^{(2)} = \tilde{A}^{(3)}$, and the algorithm terminates in the next step. Otherwise $\tilde{n}^{(2)} < \tilde{n}^{(1)}$.

In the particular case $\tilde{n}^{(1)} = 1$ either $\tilde{n}^{(2)} = 0$ or $\tilde{n}^{(2)} = 1$ is possible. If $\tilde{n}^{(2)} = 0$ then $\tilde{x}^{(2)} \geq 0$, hence $\tilde{x}^{(2)} = \tilde{A}^{(2)}\tilde{x}^{(2)} + \bar{b} \leq \overline{A}\tilde{x}^{(2)} + \bar{b}$. Since $\tilde{x}^{(1)}$ solves $x = \overline{A}x + \bar{b}$, Lemma 4 guarantees $\tilde{x}^{(2)} \leq \tilde{x}^{(1)}$, and with (15) we get $\tilde{x}^{(2)} = \tilde{x}^{(1)} \geq 0$ contradicting $\tilde{n}^{(1)} = 1$. Therefore, $\tilde{n}^{(2)} = 1 = \tilde{n}^{(1)}$, and (15) yields to $\tilde{A}^{(3)} = \tilde{A}^{(2)} = \tilde{A}$. Knowing this one needs not compute $\tilde{x}^{(2)}$ if one is only interested in $\tilde{A}$ and not in the fixed point $[x]^*$ itself; $\tilde{x}^{(2)}$ is needed, however, for $\overline{x}^*$, which means that one additional linear system has to be solved for $\overline{x}^*$.

These arguments can be repeated for each sweep of the Steps 1 and 2. The matrix $\tilde{A}$ will finally be defined by $\tilde{A} = \tilde{A}^{(k)}$ for some $k \leq \tilde{n}^{(1)} + 1$, where

$$\text{at most}\quad \max(1, \tilde{n}^{(1)})\quad \text{linear systems have to be solved.} \tag{16}$$

The maximum is necessary because in the case $\tilde{n}^{(1)} = 0$ Step 1 has to be executed once. For the solution $\tilde{x}^{(k)}$ of $x = \tilde{A}x + \bar{b}$ one needs solving at most $\tilde{n}^{(1)} + 1$ linear systems. It satisfies

$$\tilde{x}^{(1)} \leq \tilde{x}^{(k)}\quad \text{and}\quad \tilde{A}^{(1)}\tilde{x}^{(1)} \leq \tilde{A}\tilde{x}^{(k)} = \max([A]\tilde{x}^{(k)}). \tag{17}$$

Before we comment on the Steps 3 and 4 we note that

$$\tilde{x}^{(1)} = \tilde{A}^{(1)}\tilde{x}^{(1)} + \bar{b} \geq \underline{A}^{(1)}\tilde{x}^{(1)} + \underline{b}$$

whence

$$\underline{x}^{(1)} \leq \tilde{x}^{(1)} \qquad (18)$$

again by Lemma 4. Therefore,

$$\underline{n}^{(1)} \leq n - \tilde{n}^{(1)} \qquad (19)$$

and

$$\underline{A}^{(1)}\underline{x}^{(1)} \leq \underline{A}^{(1)}\tilde{x}^{(1)} \leq \tilde{A}^{(1)}\tilde{x}^{(1)} \qquad (20)$$

where we used the definition of $\underline{A}^{(1)}$ for the last inequality.

If $\tilde{n}^{(1)} = n$ as in the initialization step for $\underline{A}$, then (18), (19) result in $\underline{x}^{(1)} \leq 0$, $\underline{n}^{(1)} = 0$, in particular $\underline{A} = \overline{A}$ without any further computation. Together with (16), (11) and the arguments at the end of this proof we get the assertions of the theorem in the case $\tilde{n}^{(1)} = n$ with at most n linear systems for $\underline{A}$, $\tilde{A}$.

CASE 1: $\underline{A}^{(1)} = \underline{A}^{(2)}$
Here we get

$$\underline{A}\underline{x}^{(1)} = \underline{A}^{(2)}\underline{x}^{(1)} = \underline{A}^{(1)}\underline{x}^{(1)} = \min([A]\underline{x}^{(1)}),$$

hence $\underline{x}^{(1)} = \underline{A}\underline{x}^{(1)} + \underline{b} = \min([A]\underline{x}^{(1)} + [b])$.

CASE 2: $\underline{A}^{(1)} \neq \underline{A}^{(2)}$
Let $\underline{x}^{(2)}$ be the solution of $x = \underline{A}^{(2)}x + \underline{b}$. Then by the same arguments as above one gets

$$\underline{x}^{(2)} \leq \underline{x}^{(1)}, \qquad \underline{A}^{(2)}\underline{x}^{(2)} \leq \underline{A}^{(1)}\underline{x}^{(1)} \quad \text{and} \quad \underline{n}^{(2)} \leq \underline{n}^{(1)}.$$

If $\underline{n}^{(2)} = \underline{n}^{(1)}$ then $\underline{A}^{(2)} = \underline{A}^{(3)} = \underline{A}$; otherwise $\underline{n}^{(2)} < \underline{n}^{(1)}$.

These arguments can be repeated for each sweep of the Steps 3 and 4. The iterative process terminates with $\underline{A} = \underline{A}^{(\ell)}$ for some $\ell \leq \underline{n}^{(1)} + 2$, where at most $\underline{n}^{(1)} + 1$ linear systems have to be solved.

We now assume

$$\tilde{n}^{(1)} < n, \quad \underline{n}^{(1)} = n - \tilde{n}^{(1)} \quad (\neq 0). \qquad (21)$$

Then (18) implies $\underline{x}_j^{(1)} > 0$ if and only if $\tilde{x}_j^{(1)} > 0$, and the construction of $\underline{A}^{(1)}$ yields to

$$\underline{A}^{(1)} = \underline{A}^{(2)} = \underline{A}.$$

Therefore, we have to solve at most $\max(1, \tilde{n}^{(1)}) + 1$ linear systems in order to obtain $\underline{A}$, $\tilde{A}$. The inequality

$$\max(1, \tilde{n}^{(1)}) + 1 \leq n + 1 \tag{22}$$

is obvious, and by virtue of the first assumption in (21) we get equality in (22) if and only if $n = 1$. In this case (21) implies $\tilde{n}^{(1)} = 0$, hence we need at most $n + 2 = 3$ linear equations for $\bar{x}^* = \tilde{x}^{(1)}$, $\underline{x}^* = \underline{x}^{(2)}$, which proves the theorem in this particular case. Otherwise strict inequality holds in (22), and by (11) together with the arguments at the end of this proof we obtain the assertions of the theorem under the assumptions (21).

We next assume

$$\tilde{n}^{(1)} < n, \quad \underline{n}^{(1)} < n - \tilde{n}^{(1)}.$$

Here we get $\underline{n}^{(1)} + 1 \leq n - \tilde{n}^{(1)}$ whence

$$\max(1, \tilde{n}^{(1)}) + (\underline{n}^{(1)} + 1) \leq \begin{cases} n + 1, & \text{if} \quad \tilde{n}^{(1)} = 0, \\ n, & \text{if} \quad 0 < \tilde{n}^{(1)} < n. \end{cases}$$

Therefore, if $\tilde{n}^{(1)} = 0$ then $(\tilde{n}^{(1)} + 1) + (\underline{n}^{(1)} + 2) = \max(1, \tilde{n}^{(1)}) + (\underline{n}^{(1)} + 2) \leq n + 2$, and if $0 < \tilde{n}^{(1)} < n$ then $(\tilde{n}^{(1)} + 1) + (\underline{n}^{(1)} + 2) \leq n + 2$, too. This means that $n + 1$ is an upper bound for the maximal number of linear systems of equations to be solved for $\underline{A}$ and $\tilde{A}$, and at most $n + 2$ such systems are needed to compute $\tilde{x}^{(k)}$, $\underline{x}^{(\ell)}$ with $\underline{x}^{(\ell)} = \underline{A}^{(\ell)}\underline{x}^{(\ell)} + \underline{b}$. Analogously to (17) one obtains

$$\underline{x}^{(\ell)} \leq \underline{x}^{(1)} \quad \text{and} \quad \min([A]\underline{x}^{(\ell)}) = \underline{A}\underline{x}^{(\ell)} \leq \underline{A}^{(1)}\underline{x}^{(1)}. \tag{23}$$

Combining (17), (18), (20) and (23) finally leads to $\underline{x}^{(\ell)} \leq \tilde{x}^{(k)}$ and $\underline{A}\underline{x}^{(\ell)} \leq \tilde{A}\tilde{x}^{(k)}$. Therefore, $[x] := [\underline{x}^{(\ell)}, \tilde{x}^{(k)}]$ is an interval vector which satisfies $[A][x] = [\underline{A}\underline{x}^{(\ell)}, \tilde{A}\tilde{x}^{(k)}]$, and Theorem 2 shows via (8), (9) that $[x]$ is the limit of (1). $\qquad \square$

We note that Theorem 4 is mainly a theoretical result. It does not mean that one should really solve the (at most) $n + 2$ linear systems mentioned there in order to obtain the limit $[x]^*$ of (1).

We conclude our paper with an example in which we show that the bounds in Theorem 4 for the number of linear systems are sharp if one computes $\underline{A}$, $\tilde{A}$, $[x]^*$ as in the proof of this theorem.

Example 1.

a) Let $n = 1$, $[A] = [0, \frac{1}{2}]$, $[b] = [-2, 4]$. The Steps 1 and 2 lead to

$$\tilde{x}^{(1)} = 8, \ \tilde{n}^{(1)} = 0, \ \tilde{A}^{(2)} = \overline{A} = \tilde{A}^{(1)} = \tilde{A}.$$

The Steps 3 and 4 result in

$$\underline{A}^{(1)} = \underline{A}, \ \underline{x}^{(1)} = -2, \ \underline{n}^{(1)} = 0, \ \tilde{A}^{(2)} = \overline{A} = \underline{A}.$$

Hence $n + 1 = 2$ equations were solved for $\underline{A}$, $\tilde{A}$, $\overline{x}^* = \tilde{x}^{(1)}$, and another one is needed for $\underline{x}^* = \underline{x}^{(2)} = -4$.

If, however, Theorem 3 c) is applied (which is always possible in the one–dimensional case) no equation is needed for $\underline{A}$, $\tilde{A}$ and only two equations have to be solved for $[x]^*$.

b) Let $n = 2$, $[A] = \begin{pmatrix} [\frac{1}{8}, \frac{1}{4}] & [0, \frac{1}{4}] \\ [0, \frac{1}{2}] & [\frac{1}{8}, \frac{1}{4}] \end{pmatrix}$, $[b] = \begin{pmatrix} [-10, -1] \\ [2, 10] \end{pmatrix}$.

Then $[x]^* = \begin{pmatrix} [-16, 4] \\ [-8, 16] \end{pmatrix}$, and the Steps 1 and 2 lead to

$$\tilde{x}^{(1)} = \begin{pmatrix} 4 \\ 16 \end{pmatrix}, \ \tilde{n}^{(1)} = 0, \ \tilde{A}^{(2)} = \overline{A} = \tilde{A}^{(1)} = \tilde{A}.$$

The Steps 3 and 4 result in

$$\underline{A}^{(1)} = \underline{A}, \ \underline{x}^{(1)} = \begin{pmatrix} -\frac{80}{7} \\ \frac{16}{7} \end{pmatrix}, \ \underline{n}^{(1)} = 1, \ \underline{A}^{(2)} = \begin{pmatrix} \frac{1}{4} & 0 \\ \frac{1}{2} & \frac{1}{8} \end{pmatrix}, \ \underline{x}^{(2)} = \begin{pmatrix} -\frac{40}{3} \\ -\frac{16}{3} \end{pmatrix},$$

$$\underline{n}^{(2)} = 0, \ \underline{A}^{(3)} = \overline{A} = \underline{A}.$$

Hence $n + 1 = 3$ equations were solved for $\underline{A}$, $\tilde{A}$, $\overline{x}^* = \tilde{x}^{(1)}$, and another one is needed for $\underline{x}^* = \underline{x}^{(3)} = (-16, -8)^T$.

$\square$

References

1. Alefeld, G., Herzberger, J. (1983) Introduction to Interval Computations. Academic Press, New York
2. Barth, W., Nuding, E. (1974) Optimale Lösung von Intervallgleichungssystemen. Computing **12**, 117–125
3. Kulisch, U. (1969) Grundzüge der Intervallrechnung. In: Laugwitz, L. (Ed.) Überblicke Mathematik 2. Bibliographisches Institut, Mannheim, 51–98

4. Mayer, G, Warnke, I. (2001) On the Shape of the Fixed Points of $[f]([x]) = [A][x] + [b]$. In: Alefeld, G., Rohn, J., Rump, S. M., Yamamoto, T. (Eds.): Symbolic Algebraic Methods and Verification Methods – Theory and Applications. Springer, Wien, to appear

5. Mayer, G, Warnke, I., On the Fixed Points of the Interval Function $[f]([x]) = [A][x] + [b]$. Submitted for publication.

6. Mayer, O. (1968) Über die in der Intervallrechnung auftretenden Räume und einige Anwendungen. Ph.D. Thesis, Universität Karlsruhe, Karlsruhe

7. Neumaier, A. (1990) Interval Methods for Systems of Equations. Cambridge University Press, Cambridge

8. Varga, R. S. (1962) Matrix Iterative Analysis. Prentice–Hall, Englewood Cliffs, N. J.

How Fast can Moore's Interval Integration Method Really be?

Jürgen Herzberger

University of Oldenburg, Department of Mathematics,
D-26111 Oldenburg, Germany.

Abstract. Moore's interval integration method which uses only function evaluations of the integrant is interpreted as an approximation of the fundamental integration by Riemann-sums. In this way we can estimate the speed of convergence of this method and it is shown that its convergence factor of its linear convergence rate is bounded to below. This lower bound is shown to be sharp in the sense that there is a wide class of functions for which it cannot be improved. In particular this is true for all rational functions only considered by Moore.

1 Introduction

In this note we are considering real functions

$$f : [a, b] \ni x \longrightarrow f(x) \in \mathbb{R}$$

which do have an interval extension. This means, that we can plug in an interval $[x]$ for the variable x in $f(x)$ and when all operations in $f(x)$ ar done in interval arithmetic (see Moore [5] or Alefeld and Herzberger [1]) we get an interval as result. Moore [5] mainly focused on rational expressions for which such an interval extension is guaranteed by the four basic interval operations $+, -, \cdot, /$.

One of the main features of using interval arithmetic is that, when properly applied, it can automatically give bounds for the error of an algorithmic procedure for approximating the solution of mathematical problems. These bounds take into account the disretization errors of the method used as well as the accumulated rounding errors during its execution in floating-point arithmetic.

One basic mathematical task is to calculate the finite integral

$$\int_a^b f(x)dx$$

and following our considerations we now want to get an including interval $[i] \in I(\mathbb{R})$ with

$$\int_a^b f(x)dx \in [i].$$

Here the width of $[i] = [\underline{i}, \overline{i}]$ should be narrow, this means

$$w([i]) = \overline{i} - \underline{i} \le \varepsilon.$$

Usually one constructs such an interval $[i]$ by calculating a sequence of supersets $[i_n] \supseteq [i]$, $n \ge 1$ represented by intervals $[i_n]$ with the following basic properties:

(i) $\int_a^b f(x)dx \in [i_n]$, $n \ge 1$,

(ii) $\lim_{n \to \infty} [i_n] = \int_a^b f(x)dx$,

(iii) and here we are considering only those $[i_n]$ which require just (interval-) function evaluations and no derivatives.

In floating-point arithmetic (ii) will only guarantee that the procedure under consideration has an interval $[i]$ as result due to the rounding errors during the calculations.

2   Moore's Algorithm

In his book of 1966, Moore proposed in a very natural way the following formula for $[i_n]$:

$$[i_n] = h \cdot \sum_{k=0}^{n-1} f\left([x_k, x_{k+1}]\right), \qquad n \ge 1, \tag{1}$$

where

$$a = x_0 < x_1 \cdots < x_n = b \quad \text{and} \quad x_{k+1} - x_k = h, \quad 0 \le k \le n - 1.$$

is a subdivision of the interval $[a, b]$.

For this method Moore proved the following theorem:

Theorem (Moore). *Let $f(x)$ be a rational expression, then the intervals $[i_n]$ do have the following properties:*

(i) $\int_a^b f(x)dx \in [i_n]$, $n \ge 1$,

(ii) $w\left([i_n]\right) \le C \cdot h$, $C \in \mathbb{R}$,

and from (i) and (ii) we get necessarily

(iii) $\lim_{n \to \infty} [i_n] = \int_a^b f(x)dx$.

The Theorem of Moore implicitly states that the order of convergence of the sequence $\{[i_n]\}$ towards the integral $\int_a^b f(x)dx$ is at least linear. But then it remains the important question: "How small can the constant C in (ii) be chosen?" If we were able to give an estimation with C_h and $C_h \longrightarrow 0$

as h tends to 0 for a nontrivial class of functions than the method would be superlinearly convergent at least for this class. But we will show that for all nontrivial functions there is a lower bound $C > 0$ for this constant. Thus the method is only linearly convergent and we can give for a wide class of functions the smallest possible constant C in estimation (ii).

In order to prove our result we first have to analyze Moore's method and to point out to which well-known methods it is related to.

Given a subdivision of the interval $[a, b]$ by

$$a = x_0 < x_1 \ldots < x_n = b \quad \text{with} \quad x_{k+1} - x_k = h, \quad 0 \leq k \leq n - 1$$

we construct in the usual way (see Courant and John [3]) the lower and upper Riemann-sums for $f(x)$ on $[a, b]$ by

$$\check{i}_n = h \cdot \sum_{k=0}^{n-1} \underline{f}_k \quad \text{and} \quad \hat{i}_n = h \cdot \sum_{k=0}^{n-1} \overline{f}_k ,$$

where

$$\underline{f}_k = \min \left\{ f(x) : x \in [x_k, x_{k+1}] \right\} \quad \text{and} \quad \overline{f}_k = \max \left\{ f(x) : x \in [x_k, x_{k+1}] \right\} .$$

It is well-known from calculus that

$$\check{i}_n \leq \int_a^b f(x) dx \leq \hat{i}_n, \quad n \geq 1$$

and

$$\lim_{n \to \infty} \check{i}_n = \lim_{n \to \infty} \hat{i}_n = \int_a^b f(x) dx$$

holds true.

Now, we slightly rewrite the integration by Riemann-sums:

$$[i_n] = [\check{i}_n, \hat{i}_n] = h \cdot \sum_{k=0}^{n-1} [\underline{f}_k, \overline{f}_k], \quad n \geq 1$$

with

$$\int_a^b f(x) dx \in [i_n] \quad \text{and} \quad \lim_{n \to \infty} [i_n] = \int_a^b f(x) dx .$$

Moore's idea is to approximate the range-intervals $[\underline{f}_k, \overline{f}_k]$, the calculation of which requires in general to solve an optimization problem, by the supersets of the function interval evaluations $f([x_k, x_{k+1}])$. The inclusion property of interval arithmetic (Moore [5]) guarantees that

$$[\underline{f}_k, \overline{f}_k] \subseteq f([x_k, x_{k+1}]), \quad 0 \leq k \leq n - 1 .$$

Thus he can define the outer approximation of the intervals defined by the Riemann-sums $\left[\breve{i}_n, \hat{i}_n\right]$ by the interval expression

$$[i_n] = h \cdot \sum_{k=0}^{n-1} f\left([x_k, x_{k+1}]\right), \quad n \geq 1. \tag{2}$$

It clearly follows immediately by construction that

$$\int_a^b f(x)dx \in \left[\breve{i}_n, \hat{i}_n\right] \subseteq [i_n]$$

holds for every $n \geq 1$ when $[i_n]$ is defined by (2).

We want to illustrate this view of Moore's integration method by the following Figure 1 concerning the subinterval $[x_k, x_{k+1}]$:

3   Estimation of the Integration Error

From our derivation in the last Section 2, it is evident that for Moore's integration formula (2) we have the representation

$$[i_n] = \int_a^b f(x)dx + [e_n]$$

where

$$[e_n] = [e_{n_1}] + h \cdot [e_{n_2}] =$$

$$= [\text{integration error of Riemann-sums}]+$$

$$+ h \cdot [\text{approximation error of interval evaluations.}]$$

First, let us consider the approximation error of using interval evaluations instead of ranges of values:

$$[e_{n_2}] = \sum_{k=0}^{n-1} \left[e_{n_2}^{(k)}\right],$$

where $\left[e_{n_2}^{(k)}\right]$ is the approximation error of $f\left([x_k, x_{k+1}]\right)$.

It is well-known (see Moore [5] or Alefeld and Herzberger [1]) that

$$q\left(\left[\underline{f}_k, \overline{f}_k\right], f\left([x_k, x_{k+1}]\right)\right) \leq \gamma \cdot w\left([x_k, x_{k+1}]\right), \quad \gamma \text{ independent of } k,$$

holds true for the Hausdorff-metric $q(\cdot, \cdot)$ in $I(\mathbb{R})$.

Summarizing this we finally get the estimation

$$w\left([e_{n_2}]\right) = \sum_{k=0}^{n-1} w\left([e_{n_2}]\right) \leq \gamma \cdot (b-a)/n \cdot \sum_{k=0}^{n-1} 1 = \gamma \cdot (b-a).$$

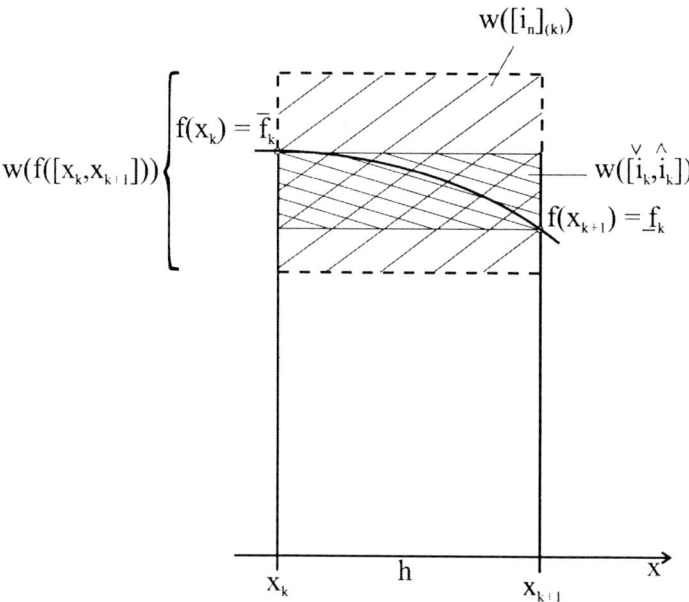

Fig. 1. Integration error for Riemann-sums and Moore's method

Now we need an estimation for the integration error of the Riemann-sums. For this we can refer to Drager [4]. He proves the following Theorem, for which we give a version matching our requirements.

Drager considers a special class of real functions fulfilling the

Condition (FMC). There are points

$$a = x_0 < x_1 \ldots x_K < x_{K+1} = b$$

so that f is weakly monotone on $\left[x_{j-1}, x_j\right]$ for $1 \leq j \leq K+1$ and no x_j can be removed without destroying this property.

The quantity

$$V = \sum_{j=1}^{K+1} \left| f\left(x_{j-1}\right) - f\left(x_j\right) \right|$$

usually is called the total variation of f on $[a, b]$.

Now we are able to formulate Drager's result in our setting.

Theorem (Drager 1987). *Let Δ_h be the difference of the upper and lower Riemann-sums*

$$\Delta_h = \hat{i}_n - \check{i}_n \geq 0$$

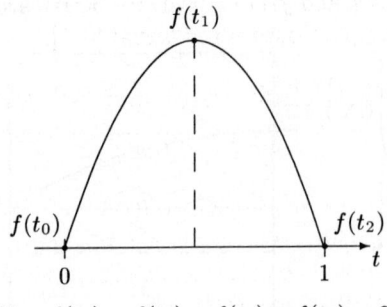

$$V = f(t_1) - f(t_0) + f(t_1) - f(t_2) = 2$$

Fig. 2. Total variation for a parabolic function

for an equally-spaced subdivision of the interval $[a, b]$ into n subintervals. If f satisfies property (FMC) and is Lipschitz-continuous in $[a, b]$, then

$$\Delta_h \leq V \cdot (b - a)/n \,.$$

If f has a continuous second deriviative, this can be improved to

$$|\Delta_n - V \cdot (b - a)/n| \leq C/n^3 \quad \text{with} \quad C > 0 \,,$$

which indicates that the constant V cannot be improved.

Remark. All rational expressions $r(x)$ – considered in Moore's book of 1966 – satisfy the assumptions of the theorem above and thus the theorem is true for them.

If we now apply the statement of the theorem above to our error estimation by assuming that our function f satisfies the assumptions of the last theorem, then we get our final estimation

$$w([e_n]) \leq V \cdot (b - a)/n + \gamma \cdot (b - a)^2/n = V \cdot (b - a)/n \cdot (1 + \gamma \cdot (b - a))$$

The constant γ in this estimation can be calculated in some way by bounds for the partial derivatives of expressions related to $f(x)$ (see Alefeld and Herzberger [1]).

4   Conclusions

In order to discuss the error estimation 3 we have a look at the constant γ in 3.

(i) First, this constant γ can be arbitrarily large.

Example 1. $f(x) = x$ and $f_1(x) = 501 \cdot x - 500 \cdot x$ are rational expressions for the identity mapping onto the interval $[0, 1]$.

$$q\big([\underline{f}_1, \overline{f}_1], f_1([0, 1])\big) \leq 501 \cdot w\big([0, 1]\big) + 500 \cdot w\big([0, 1]\big) - w\big([0, 1]\big)$$
$$= 1000 \cdot w\big([0, 1]\big).$$

(ii) The constant γ in (3) can be equal to 0.

This is true, for example, for the class of expressions, in which the variable x is occuring only once. Then after Apostolatos and Kulisch [2] its interval evalution is equal to its range of values.

Example 2. $f(x) = 2/\sqrt{\ln(1 + x^2) + 2}$.

As a consequence of case (ii) we immediately get from (3) an answer to our question posed in the introduction.

"The constant C in Moore's estimation (ii) of the integration error is always greater or equal than V if f belongs to the class (FMC) and is Lipschitz-continuous in $[a, b]$. Furthermore, if f has a continuous second derivative then V cannot be replaced by a smaller value. In particular for all rational expressions this statement is true."

References

1. Alefeld, G., Herzberger, J. (1974) Einführung in die Intervallrechnung. BI-Wissenschaftsverlag, Wiesbaden 1974; English edition: Introduction to Interval Computations, Academic Press, New York 1983
2. Apostolatos, N., Kulisch, U. (1967) Grundlagen einer Maschinenintervallarithmetik. Computing **2** (1967), 89–104
3. Courant, R., John, F. (1989) Introduction to Calculus and Analysis I. Springer-Verlag, Berlin
4. Drager, L.D. (1987) A simple Theorem on Riemann integration, based on classroom experience. In: Mathematical Modelling - Classroom Notes in Applied Mathematics, M.S. Klamkin (ed.), SIAM, Philadelphia, pp. 188–192
5. Moore, R.E. (1966) Interval Analysis. Prentice-Hall Inc., Englewood-Cliffs N.J.
6. Neumaier, A. (1990) Interval Methods for Systems of Equations. Cambridge University Press, Cambridge

Numerical Verification and Validation of Kinematics and Dynamical Models for Flexible Robots in Complex Environments

Wolfram Luther, Eva Dyllong, Daniela Fausten, Werner Otten, and Holger Traczinski

Gerhard–Mercator–University of Duisburg,
D-47048 Duisburg, Germany
{luther, dyllong, fausten, otten, traczinski}@informatik.uni-duisburg.de

Abstract. We give a survey on well-known and new interval methods and algorithms with result verification in the field of robotics. In particular we present optimal linear controller design, reliable geometric computations for distances between a point and a non-convex polyhedron or a NURBS curve, path planning, and failure detection with fault tree logic for flexible robots in complex environments. We also present an extension of a multi body modelling and simulating tool, which provides error propagation control and reliable numerical algorithms.

1 Introduction

The estimation or measurement of the gravitational, static friction, and joint torque parameters, joint angles and rates can be performed only with inaccuracies within the framework of simulating and controlling flexible robots. Uncertainty in sensor measurement and discretization approaches contribute to a significant error accumulation during the computation of collision free arm postures, joint angles or the payload of the end-effector. There are similar problems in the context of robot localisation and the local path planning or parameter estimation and identification of a non-linear controller for multi body systems with many degrees of freedom. After modelling these errors, all inaccuracies have to be estimated and controlled [15,16]. Some of these problems which have their origins in methods from classical modelling and simulating tools are investigated in [26,4,11]. Therefore, one goal of our approach is to innovate methods to extend the object oriented programming package MOBILE [14] with numerical routines using interval data type, controlled rounding and the precise dot product.

In this context a new extension for MOBILE designed for modelling multi body systems has been developed [25]. For reliable results it supports interval arithmetic and error calculus. In conjunction with correct rounding a verification of MOBILE results is possible. Interval arithmetic is also usable to cope with uncertain data, then inputs are intervals which contain the correct value. Especially in this respect we are interested in computation errors we make, regardless of which value of the input is taken. For that purpose a self acting error calculus due to Krämer [19] is implemented in the MOBILE extension. Another advantage of this package is the easy handling of the new features because interval arithmetic and error calculus are used automatically.

An important part in the framework of control and navigation problems play matrix and vector differential equations with a square non-linearity. One of the emphasis of our activities lies in the employment of validation tools and in the development of verification algorithms to analyse the stationary solutions and the end-value-problems of these differential equations. In the context of optimal control we have developed and implemented algorithms for the verified calculation of solutions of continuous and discrete-time algebraic Riccati equations [23,24]. These equations are to be solved when we want to find the steady state solutions of matrix Riccati differential or difference equations with constant coefficients, respectively, which arise in the theory of automatic control and linear filtering. In these algorithms we can handle uncertain data, too, and we get guaranteed enclosures of the exact solution.

Another centre of our interest forms the development of efficient and accurate algorithms for distance calculation between a flexible robot and a target or obstacles in the complex environment and for resulting contact problems. Robust solutions to this problem are also used in the collision-free path planning if a given end-effector is moving amid a collection of known obstacles from an initial to a desired final position. For simulation the obstacles are taken to be a collection of polyhedral or free-formed objects like NURBS surfaces. Accurate algorithms have been developed, based on suitable projections and using controlled rounding and the precise dot product whereby a verified enclosure of the solution is ensured [9]. If the end-effector or the sensor is taken to be a single moving point an efficient distance algorithm, which doesn't rely on convex properties and thus is applicable to non-convex polyhedral surfaces, has been developed [7]. Under the same assumption the problem has been solved for the more difficult case of NURBS-defined solids based on subdivision techniques and using an algorithm for the solution of non-linear polynomial systems by Sherbrooke and Patrikalakis. The extension of this algorithm introduces interval arithmetic, the interval version of the convex hull, and a modified Simplex algorithm and the new solver allows a verification of obtained consequences [8] using some new criteria to guarantee the existence of zeros within the calculated inclusions [10].

Besides, we try to find out a functional description of solids from visual censoring. Such a description leads to a considerable reduction of data, which makes an effective storing and processing possible. An approach, like a hierarchical representation with octrees, Gauß–pyramid or ray tracing methods is still a challenging task of our access to find validated enclosures of obstacles.

Robot manipulators are a critical mixture of mechanical, electrical, and electronic components. Consequently, there are numerous failure modes inherent in a typical manipulator system. Our work is focused on redundant robots with multiple sensors. A five arm manipulator mounted on a rotatory base and driven by hydraulic actuators by means of a transmission mechanism is discussed in more detail. We develop an accurate algorithm using interval computations to propagate given input probability densities by fault tree logic (typically AND or OR gates) up to the root to estimate the overall robot failure probability density [22]. In the future, we will include a graphical interface to introduce the input distributions and control parameters and the fault tree logic. Furthermore, integration into the object-oriented programming package MOBILE and an interface to a database with reliability data is planned.

2   Error Propagation Control and Reliable Numerical Algorithms in MOBILE

In various fields of robotics, the multi body system approach gives the basis for the needed modelling and simulation software. A prototype is specified using an alphanumeric editor or a graphical user interface. The governing equations of the constrained motion are generated. Thus, fast and reliable numerical algorithms for time integration and solving non linear equations describing all constrains are required to analyse the behaviour of the model. A further objective is to develop methods for an accurate treatment of (overdetermined) differential-algebraic equations. During the last years, great efforts were made to provide efficient libraries often in Fortran or C++ double precision standard to solve the equations of motions by enhanced explicit, half-explicit or implicit Runge–Kutta methods and Differential-Algebraic System Solvers. The multi body system package MBSPACK [28] proposes a collection of several well-known algorithms together with a suitable interface to support versatile and competitive simulation tools. Another package for numerics is Scilab [13], which also provides a linkage to computer algebra systems Maple and MuPAD.

We have presented a framework for the object-oriented programming package MOBILE designed for modelling of multi body systems to handle uncertain or imprecise inputs and to get verified results. A self acting error calculus from Krämer [19] is added to calculate the maximum absolute computational error. In a second step a verifying numerical integration tool based on Lohner's PASCAL XSC program AWA [21] as well as a solver of non linear polynomial equation systems will be integrated in MOBILE.

The main idea of the new extension package is that a user familiar with MOBILE does not have to learn a new computer language. Of course, there are new objects for mathematics and kinetostatic transmission elements providing interval arithmetic and error calculus. These objects work in the same way as the corresponding well-known objects. Only the names of the new objects are different from the names of the well-known objects. All names of MOBILE objects starting with Mo are cloned by interval objects beginning with MoInterval, generally speaking, MoIntervalxxx complements Moxxx.

The basic class is `MoInterval`. It contains a variable of the type `INTERVAL` of the interval arithmetic package PROFIL/BIAS by Knüppel [17,18] and a variable of the type `double` for the absolute error.

Here we give a simple program listing concerning a two-linked manipulator [14]. It consists of two revolute joints R1 and R2 with non-orthogonal, non-intersecting axes. The first joint rotates around the fixed z-axis while the second joint is aligned with the x-axis of the moving frame. Joints and links are defined by two local frames K_i, K_j, an angle and an axis or a length, respectively. The links are assumed to point in direction of the fixed z-axis and the moving y-axis, respectively. The objective is to evaluate the joint torques when a load is applied to the tip of the manipulator in direction of the z-axis of the last frame K_4.

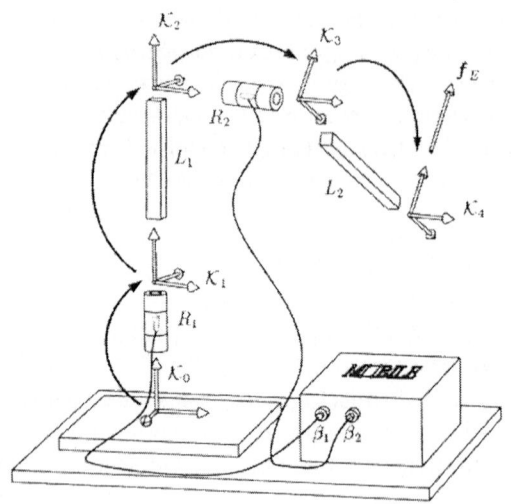

Fig. 1. Two-linked manipulator

```
#include <Mobile/MoIntervalElementaryJoint.h>
#include <Mobile/MoIntervalRigidLink.h>
#include <Mobile/MoIntervalMapChain.h>
main() {
  MoIntervalFrame K0, K1, K2, K3, K4;
  MoIntervalAngularVariable beta1, beta2;
  MoIntervalVector l1, l2;
  MoIntervalElementaryJoint R1(K0,K1, beta1, zAxis);
  MoIntervalElementaryJoint R2(K2,K3, beta2, xAxis);
```

```
MoIntervalRigidLink L1(K1, K2, l1);
MoIntervalRigidLink L2(K3, K4, l2);
MoIntervalMapChain Manipulator;
Manipulator << R1 << L1 << R2 << L2;
l1 = MoIntervalVector(0,0,1);
l2 = MoIntervalVector(0,1,0);
int nsteps = 10;
MoInterval forceMagnitude = 1;
beta1.q = 90 * INTERVAL_DEG_TO_RAD;
beta2.q =-45 * INTERVAL_DEG_TO_RAD;
for ( int i = 0; i < nsteps; i++) {
  Manipulator.doMotion(DO_INTERVAL_POSITION);
  Manipulator.cleanUpForces();
  K4.f = forceMagnitude * MoIntervalVector(0,0,1) * K4.R;
  Manipulator.doForce();
  cout << "Torque joint 1 = " << beta1.Q
       << "\nTorque joint 2 = " << beta2.Q << "\n";
  beta1.q += 90 * INTERVAL_DEG_TO_RAD / float(nsteps);
  beta2.q += 180 * INTERVAL_DEG_TO_RAD / float(nsteps);
}
}
```

Note that constants INTERVAL_DEG_TO_RAD and DO_INTERVAL_POSITION of the interval version have to be chosen. To assign an interval value to a variable, e.g. forceMagnitude, put

```
MoInterval forceMagnitude = MoInterval(0.8, 1.2);
```

The output of this interval MOBILE program is a list of guaranteed intervals for torque values of both joints, and a maximum absolute error is given for each interval. Any maximum error is smaller than $3.9 \cdot 10^{-15}$, and a minimum of 13 exact digits is ensured for any torque value. Thus, for the first time it is possible to use an automatic error control in modelling and simulating multi body systems, even if precise data are not known.

Further work will provide an automatic transcription of common MO-BILE programs into programs working with intervals and providing accurate results and error estimations.

As another application of error calculus coming from control theory we have studied the identification and simulation of a hydraulic differential cylinder with servo valve. Here denote s the position of the piston rod. The two pressures are p_A and p_B and u is the control voltage of the servo valve. The stroke of the cylinder is 200 mm, the diameter of the piston is 25 mm, and the diameter of the rod is 18 mm. Actuators of this kind are used in elastic hydraulic robots. We are going to identify as model for the position s of the piston depending on the control voltage u and the pressure p_B.

As model structure we choose a Sugeno–Takagi fuzzy model. For a detailed discussion of this approach see [1] or [20]. The parameters of the model are

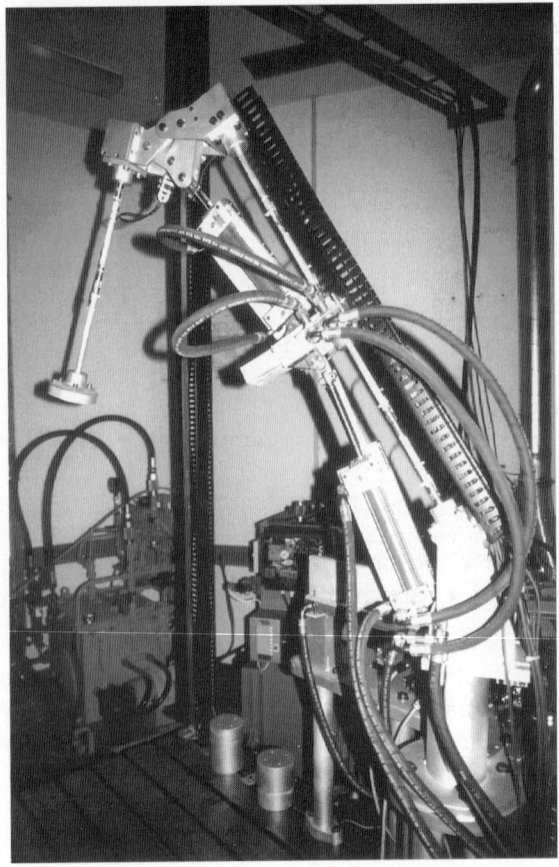

Fig. 2. Elastic hydraulic robot (HyRob)

identified with the FIMO MATLAB package [2]. The method starts with a
time discrete system with m inputs $u_1(k), \ldots, u_m(k)$ and one output $s(k)$.
The non-linear difference equation joining u_i and s is approximated by several
auto-regressive models of the form

$$s(k) = c + \sum_{i=1}^{n} a_i s(k-i) + \sum_{i=1}^{m} \sum_{j=0}^{\nu_i} b_{ij} u_i(k-j).$$

The task to identify the real coefficients a_i, b_{ij}, c is well known and can
be solved with the least squares method. However, such a simple model is
only a good approximation of the non-linear system in a bounded domain
of the given data. Therefore, it would be advantageous to combine several
local models. The partition of the set of data can be found by the Fuzzy-c-
Means algorithm or by the Gustafson–Kessel algorithm [3]. Once a partition

found, we have to combine the subsystems to obtain an overall system using multivariate Shepard's weight-functions and blending techniques.

We present the equations of the Sugeno–Takagi fuzzy model for the position s with inputs p_B, and u and use four local models with the following structure for the position

$$s(k) = c + a_1 s(k - 1) + b_{11} p_B(k - 1) + b_{24} u(k - 4).$$

Here the data set has dimension three. The further constants in the linear difference equation are omitted. The parameters of the models are determined according to Table 1. The prototypes for the position model are given in Table 2.

Table 1. Local Models

Parameter	Model 1	Model 2	Model 3	Model 4
c	$7.564 \cdot 10^{-5}$	$-9.587 \cdot 10^{-5}$	$1.552 \cdot 10^{-5}$	$-9.991 \cdot 10^{-5}$
a_1	-1.000	-1.001	$-9.997 \cdot 10^{-1}$	-1.000
b_{11}	$-1.542 \cdot 10^{-11}$	$3.091 \cdot 10^{-12}$	$-4.465 \cdot 10^{-13}$	$6.650 \cdot 10^{-11}$
b_{24}	$5.749 \cdot 10^{-5}$	$9.316 \cdot 10^{-5}$	$8.324 \cdot 10^{-5}$	$8.285 \cdot 10^{-5}$

Table 2. Prototypes

Parameter	1	2	3	4
s	$1.0474199 \cdot 10^{-1}$	$7.3378302 \cdot 10^{-5}$	$7.3225103 \cdot 10^{-2}$	$1.1336200 \cdot 10^{-1}$
p_B	$5.4239085 \cdot 10^{6}$	$3.1518045 \cdot 10^{6}$	$2.1645755 \cdot 10^{6}$	$1.3875264 \cdot 10^{6}$
u	-4.5641599	$2.2579128 \cdot 10^{-1}$	$4.0656880 \cdot 10^{-1}$	4.6694999

For the input s, p_B, and u of the model for the position s we use maximum relevant intervals $s = [4 \cdot 10^{-6}, 0.1945]$, $p_B = [9.375 \cdot 10^{5}, 6.1523 \cdot 10^{6}]$, $u = [-9.7, 9.7]$ to get an absolute error bound. With these intervals an error estimation is not possible. We divided the intervals into smaller ones and calculate all the boxes coming from this process. We do this recursively and get a verification in 99,999821 % of all combinations. This number represents the ratio of the volume of boxes where a verification is possible, to the volume of the box of maximum relevant intervals mentioned above. For the p_B-interval $[2164575.4, 2164575.6]$ and the u-interval $[0.3, 0.51]$ we can not verify the position s. This situation happens in simulation with the robot, and a developed

Sugeno–Takagi fuzzy model for pressure p_A yields bad results in this case, too. So simulation and verification results are in good correspondence, even more, error calculus indicates a serious model error besides the numerical error.

In more detail, we verify the position s resulting from measurements of s, p_A, p_B, and u on the robot every 2 msec using 8000 data samples in each parameter. We start our calculation with the input $s(4)$, $p_B(4)$, and $u(1)$ from data to determine an interval for $s(5)$ and the error. From now on we take the input-position in conjunction with the error from previous calculation and the other inputs from data and compute the position s and the maximum absolute error after 8000 steps. The difference from the measured position was $6.13 \cdot 10^{-4}$. Even if we do the computations with uncertain inputs for p_B of order $0.1\,\%$ and for u of order $1\,\%$ simulating a uniformly distributed error of measurement, the maximum absolute error for s is smaller than $2.86 \cdot 10^{-9}$.

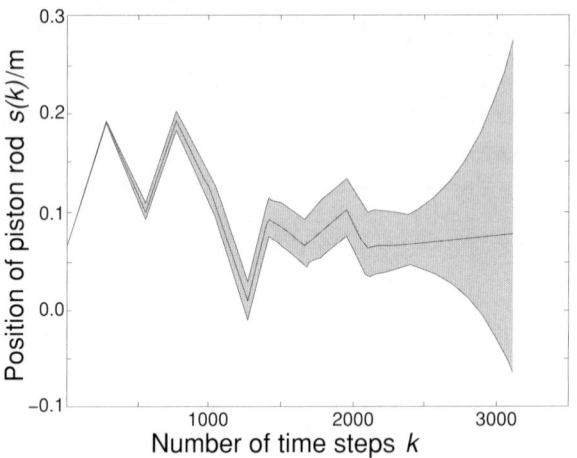

Fig. 3. Enclosure of position s with uncertainty in all inputs

Thus we have shown that a verification is possible in the cases where the simulation yields good results, in cases where a verification fails the model has to be improved. Besides a validation of the identified parameters of the Sugeno–Takagi fuzzy model was obtained.

3   Verified Calculation of the Solution of Discrete-Time Algebraic Riccati Equation

The discrete-time algebraic matrix Riccati equation (DTARE) is to be solved when we want to find steady-state solutions of the matrix Riccati difference equations with constant coefficients which arises in the theory of automatic control and linear filtering.

The discrete-time linear dynamical system is described by $x(t + 1) = Ax(t) + Bu(t)$ for the state vector x and the control input vector u and appropriate real matrices A and B together with a quadratic cost functional given by a sum over terms $x^T(t_i)Qx(t_i)$ and $u^T(t_i)Ru(t_i)$ with symmetric positive (semi)-definite matrices Q and R. The unique solution of this problem is given by the state-feedback control $u(t) = Hx(t)$, where $H = -(R + B^T X B)^{-1}B^T X A$, and X satisfies the algebraic matrix equation

$$P(X) := A^T X A - X - A^T X B (R + B^T X B)^{-1} B^T X A + Q = 0.$$

Then there is the following verification algorithm [24]:

1. With the aid of a classical solution scheme compute a first approximation W of the interval DTARE-problem together with the matrices $\text{mid}(A), \text{mid}(B), \text{mid}(Q)$, and $\text{mid}(R)$, so that the residual $Y := P(W)$ is sufficiently small in a norm, e.g. the maximum row sum matrix norm.
2. Blow up the matrix W to $W + \Delta$ by an error matrix Δ, choose for example $\Delta := [-\delta, \delta]W$ or $\Delta := [-\delta, \delta](1, 1, \ldots, 1)^T(1, 1, \ldots, 1)$ with $\delta > 0$.
3. Transform the matrix equation $0 = CXD - X + K$ (*) with suitable matrices C, D, K obtained by developing $P(X + \Delta) - P(X)$ in terms of order one and two in Δ to a linear equation system $(I - \Delta^T \otimes C)\text{vec } X = \text{vec } K$ using the Kronecker product $\otimes$ and the vec-function of a matrix Y as a compound column of all columns of Y. We interpret a solution X of (*) as an image under the mapping $T : \Delta \to X$. If U is a fixed point of T, we find that $P(W + U) = 0$.
4. Use an appropriate interval equation solver to solve this linear equation system and get a solution interval matrix $[X]$.
5. If the error matrix Δ encloses $[X]$, an application of Brouwer's fixed point theorem ensures that the interval matrix $W + [X]$ encloses $W + U$ as a solution of DTARE.

The enclosure can be optimised by introducing an adaptive error matrix Δ. Furthermore, inner dependencies in the linear system whereof the verified solution has been calculated can be used to reduce the dimension of the system and to speed up the algorithm. Applying our result to control theory, we have to check if each symmetric matrix in the real enclosure $W + [X]$ is positive semi-definite which can be done by a LU-decomposition of this interval matrix. If the diagonal elements of the upper triangular matrix U only contain positive values, then every symmetric matrix in $W + [X]$ is positive definite. As a numerical example the nine dimensional discrete-time tubular ammonia reactor model described in [29, p. 258] in point and interval matrix form is discussed. Analogous results for the continuous-time algebraic

Riccati problem were presented at the Scan–98 and can be found in [23]. By the way we emphasise that this approach provides a complete stability theory for fuzzy–Sugeno controllers.

4   Accurate Distance Calculation Algorithms

In several papers we give efficient and accurate algorithms to calculate the distance between convex polyhedra, a point and a non-convex polyhedron, and a point and a NURBS curve or surface [9,7,8]. The distance calculation is an essential component of robot motion planning and control to steer the robot away from its surrounding obstacles or to work on a target surface. The obstacles may be polyhedral objects, quadratic surfaces, which include spherical and cylindrical surfaces or more general surface types like non-uniform rational B-splines (NURBS). There is an abundance of literature to calculate the distance between convex and non-convex objects. However, in most cases, the non-convex objects are decomposed into convex components which are pre-processed to a hierarchical representation. Then an iteration process finds the distance between both objects. However, our scope is different from these results. Obstacles are often modelled or reconstructed from sonar and visual data leading to uncertain information. We are interested in simple algorithms to calculate the distance between two objects like points or (non-) convex polyhedra or NURBS-surfaces with interval vertices. Therefore, we only use accurate operations and perform the calculation of the Euclidean distance by a sequence of subdivisions and projections into certain planes or edges with help of the precise dot product.

Our algorithm to compute the distance between a point and a non-convex polyhedron (Fig. 4) doesn't need to decompose the polyhedron into convex parts, uses no iteration and yields the result with high accuracy. Explicit absolute resp. relative errors of a real distance point X and the distance D to the (non-)convex polyhedron and the calculated approximations x and d are derived.

The accurate algorithm shown in Fig. 5 for the distance calculation between a point Q and a NURBS curve $C(u)$ of degree p with control points P_i, weights $w_i, i = 0, \ldots, n$, and a knot sequence U, and its extension to the case of a point and a surface is crucially based on some appropriate decomposition into n_p rational Bézier segments with control points P_{jk} and weights w_{jk}, subdivision to obtain flat sub-segments and projection techniques. Then some evaluations of suitable scalar products decide on a further subdivision.

The number of necessary subdivision steps was calculated in advance. The inherent distance calculations are processed by projection or are transformed in terms of the solution of the polynomial equation $(Q - C(u))C'(u) = 0$ in the variable u. We calculate the roots of this equation using either the well known (interval) Newton method, or one of the recently implemented solution methods, the projected-polyhedron (PP) and the linear program-

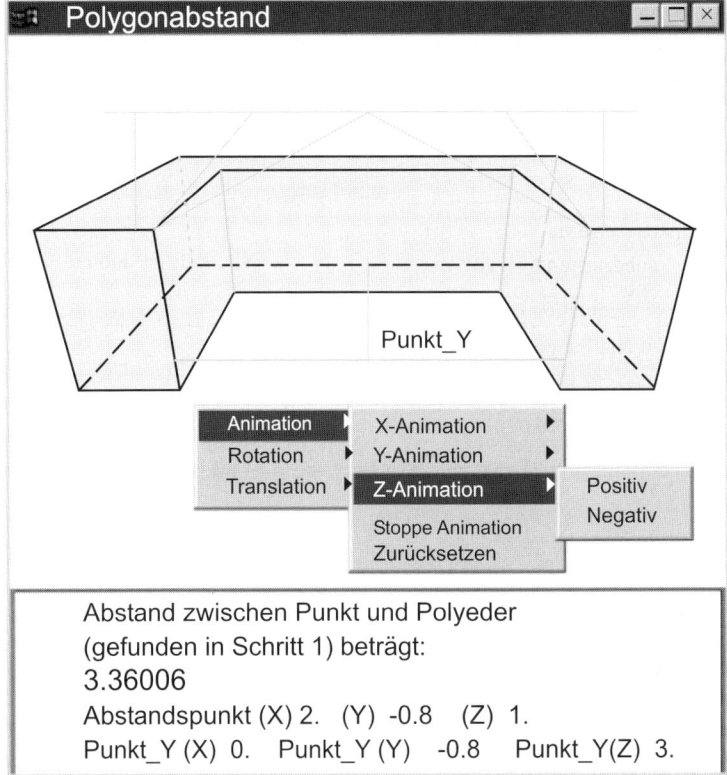

Fig. 4. Distance calculation point on non-convex polyhedron

ming (LP) technique, developed by Sherbrooke and Patrikalakis [27]. They rely on the representation of polynomials in the multivariate Bernstein basis, the convex hull property and on subdivision or linear programming. We have developed an interval version of the PP/LP algorithm using interval arithmetic and considering a correct handling of roots of order two, suitable modifications of Graham's scan algorithm, the revised simplex method by Gass [12], and an adapted interval-based subdivision by de Casteljau. The solver has been implemented in C++ using the library Profil/BIAS. This improves the robustness of the distance algorithm, assures an interval enclosure of the solution, and makes it suitable for verification of off-line tasks in the path planning.

5   Accurate Robot Reliability Estimation

Recently, we have developed an accurate algorithm using interval computations to propagate given input probability densities by fault tree logic (typi-

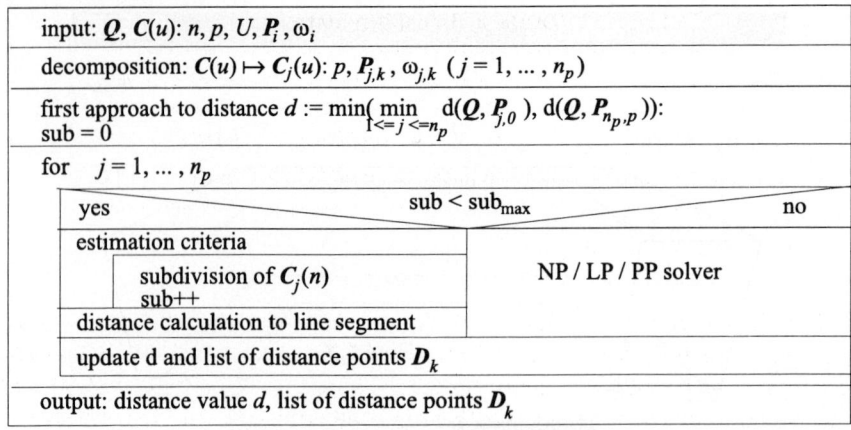

Fig. 5. Calculation of the distance between a point and a NURBS curve

cally AND or OR gates) up to the root to estimate the overall robot failure probability density.

A three joint robot was deployed in hazardous environments for remediation of highly radioactive waste, which make reliability analysis and fault tolerance of most critical importance. Furthermore, a five arm manipulator mounted on a rotatory base and driven by hydraulic actuators by means of a transmission mechanism is discussed in more detail. Both robots are kinematically redundant and so they have sufficient dexterity to avoid obstacles.

Failure logic for a three joint planar robot with redundant sensors, actuators and dual position sensors (position and velocity) can be represented by a fault tree. To improve the fault tolerance of the system, redundant sensors are added at each joint. It is technically not so easy to add an additional motor to drive a joint.

The fault tree has three binary subtrees. On the lowest level of the subtrees there are sensor leafs and AND-gates, a combined sensor node and an actuator node are inputs for OR-gates, and finally joint-failure nodes are introduced into OR-gates (see Fig. 6). Thus, the failure of one joint-combination 1,2, 1,3 and 2,3 yields the general failure of the robot. Starting with failure probabilities for electric motors and sensors we can compute the failure probability p of the top event and the reliability R of the robot is $R = 1 - p$.

However, there is a lack of reliability data for robots. Table 3 shows reliability interval data derived from the 1995 Nonelectronic Parts Reliability Data (NPRD–95) reported in the literature [6]. In an older approach the input data was treated as 'fuzzy'. They were represented as trapezoids over the intervals $[u, x]$ with inner points v and w. These points correspond to an uncertainty in statistics, i.e. 68 per cent of cases will be between these bounds and 90 per cent between the outer bounds.

Table 3. Reliability data

Component	$u = p_{ll}$	$v = p_l$	$w = p_r$	$x = p_{rr}$
Electric Motor	0.000739	0.00203	0.0416	0.11
Optical Encoder Sensor	0.00124	0.00341	0.0698	0.184

For the electric motor the base failure probability is $p = 0.00924$ and for the optical encoder sensor the failure probability is $p = 0.0155$. It was observed that 68 percent of component failure rates λ will be between 0.22 and 4.5 times the reported value p. Furthermore, 90 percent of failure rates will be between 0.08 and 11.9 times the reported value. The total mission operation time for the robot is assumed to be on the order of 1000 hours. An approximate probability of failure was determined as λt, where $t = 1000$ hours.

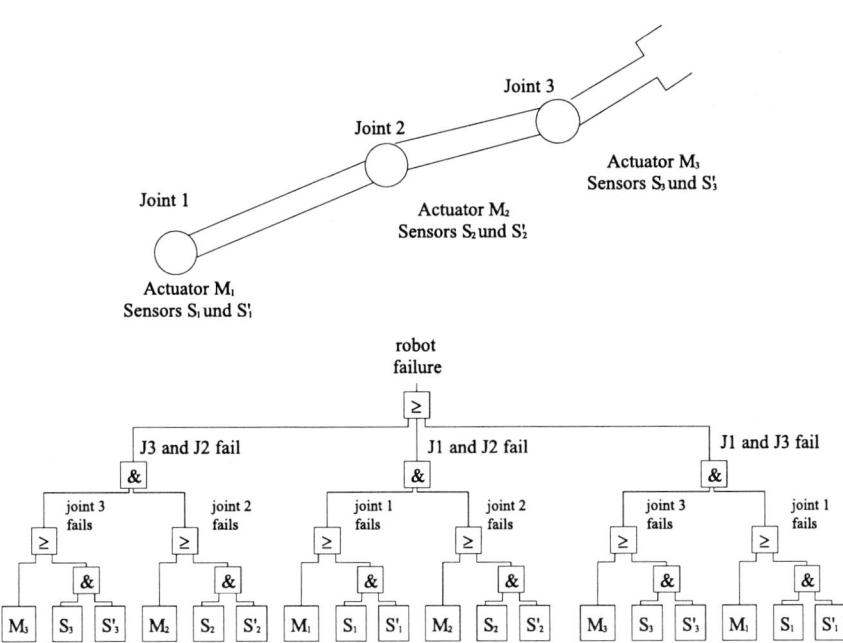

Fig. 6. Fault tree for three joint planar robot with redundant sensors [5]

To compute the reliability estimates for the robot, the trapezoidal estimates were propagated through the fault tree, where the output distributions are computed pairwise from the corners of the input trapezoidal distributions:

Multiplication: $(u, v, w, x)_{\text{new}} = (u_1 u_2, v_1 v_2, w_1 w_2, x_1 x_2)$

Addition: $(u, v, w, x)_{\text{new}} = (u_1 + u_2, v_1 + v_2, w_1 + w_2, x_1 + x_2)$.

This approach allows the incorporation of some additional uncertainty information in the output distribution, beyond the simple propagation of the means. However, there is no theoretical foundation for this approach and the trapezoidal output distribution seems to overestimate the reliability.

The robot reliability problem can be viewed as a particular case of the general problem of combining of a sequence of mutually dependent arithmetic functions obtained here from the probability density estimates of the initial input variables. In fact, these densities can naturally be expressed in the form of histograms, thus an interval approach is quite adequate.

Two major difficulties are encountered when using an interval approach to solve this problem. One arises from the fact that the arithmetic function involved are mutually dependent.

The other problem is that if no approximations are taken into account, the number of intervals that can be generated from the computation can grow up to a point that the computation using intervals has similar complexity than other traditional simulation methods based on discretized input values.

A new interval based treatment was introduced by Carreras et al. [5]. However, the algorithm is difficult to re-implement, the accuracy is not very clear, and the globally fine discretization is very time expensive.

Our algorithm calculates by a fault tree analysis the probability distribution of robot failure. We discuss multiple joint robots with redundant electric motors and optical encoder sensors. The failure probabilities of both components are given by two polygonal splines reported in the literature. General step functions and other types of actuators or sensors are supported. Probabilities are propagated by fault tree logic (typically AND or OR gates) up to the root applying standard interval calculus to obtain the output failure probability density for the robot manipulator. Great emphasis is laid to guarantee accurate results and to optimise time complexity. Therefore, the basic interval $[0, 1]$ is divided into a finer scale near zero and a coarser one near one.

We apply our scenario to a five arm manipulator with an electric motor pair at the bottom and optical sensor pairs at each joint. Thus, the robot supports failure of one electric motor at the bottom, one sensor in each optical encoder sensor pair and three of five motors at the joints. We attach great importance to give reliable results by a systematic small overestimation of failure probability throughout the tree.

Now, we want to explain our algorithm. The algorithm has different parts and uses several procedures. First, control parameters and input distributions are initialised and conditions for robot failure are defined. Then the failure

probabilities are propagated through the tree up to root to give the top event
probability distribution as a step function over both scales.

- Introduce the existing reliability data for electric motors and optical encoder sensor by the knot points in form of a polygonal spline.
- Define two scales as finer and coarser subdivision of the interval $[0, 1]$ into
  two overlapping intervals $[0, 2sc]$ and $[sc, 1]$ and divide these intervals into
  kN, resp. N subranges of equal size.
- Map the input distribution uniformly to the small intervals. Thus, each
  interval supports a fraction of failure probability and summing up over
  all disjoint intervals we obtain 1.
- Introduce the fault tree logic in form of minimal cut sets, i.e. combinations
  of component failures are determined which lead to robot failure.

After an initialisation several gate operations are needed: Gate 'AND'
multiplication, gate 'OR' addition, gate operation with more than two inputs
and a further procedure to evaluate repeated multiplications and additions
(cascaded AND and OR gates) in an efficient way.

The input distributions are propagated through the whole fault tree and
we obtain the output failure distribution as a step function over the two scales
or a bound p for the accumulated distribution. (x per cent of the probability
mass of robot failure is located in the interval $[0, p]$.)

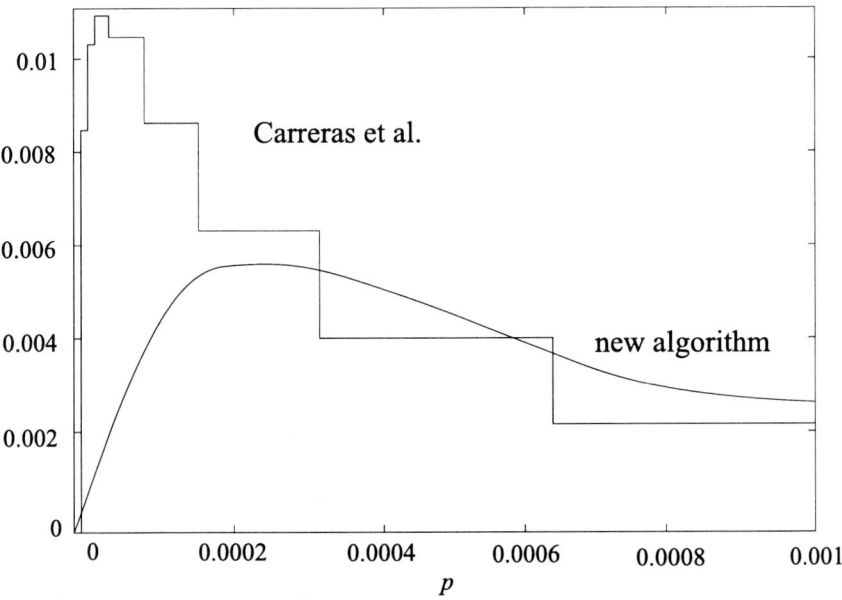

Fig. 7. Output distributions of three joint robot

The output failure probability distribution obtained by combining two input intervals with uniformly distributed probabilities has a peak form. To guarantee correct bounds the peak has to be translated to the right adjacent interval.

Our result gives approximately the same concentration of the density between 25 % and 50 % as the result by Carreras et al. However, the time complexity is quite better and the method supports more complex configurations. A reduction of discretization steps shows a remarkable stability of the output result.

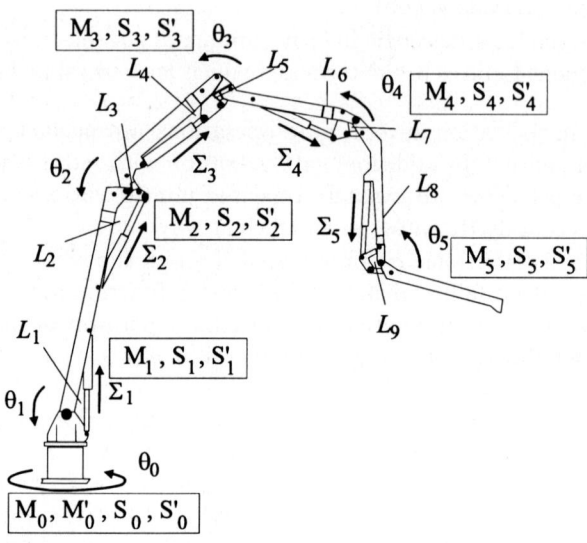

Fig. 8. Five arm manipulator

The following tables and Fig. 8 show the output distribution of robot failure and the five arm manipulator:

Table 4. Trapezoidal output distribution

5 %	16 %	84 %	95 %	p
$2.084 \cdot 10^{-6}$	$1.575 \cdot 10^{-5}$	$6.65 \cdot 10^{-3}$	$5.024 \cdot 10^{-2}$	$3.257 \cdot 10^{-4}$

The trapezoidal output distribution is to optimistic near zero but surprisingly good otherwise. A small user interface permits to introduce the

Table 5. Output distribution, scale 1: 0.2510^{-5}, scale 2: 0.510^{-3}, $sc = 10^{-3}$

5 %	16 %	20 %	30 %	40 %
$8.0 \cdot 10^{-5}$	$1.85 \cdot 10^{-4}$	$2.325 \cdot 10^{-4}$	$3.85 \cdot 10^{-4}$	$6.05 \cdot 10^{-4}$

50 %	70 %	84 %	95 %	99 %
$9.175 \cdot 10^{-4}$	$3 \cdot 10^{-3}$	$6.5 \cdot 10^{-3}$	$2.35 \cdot 10^{-2}$	$8.2 \cdot 10^{-2}$

structure of the manipulator (the number of arms, one or two electric motors at a joint) and to parameterise both scales and the accuracy of calculation.

In the future, we will include a graphical interface to insert the input distributions, control parameters and the fault tree logic. Furthermore, integration into the object-oriented programming package MOBILE is planned. An interface to a database with reliability data is desirable.

6   Further Work

In Fig. 9 the main topics of our work are summarised. Within the new research period 2001–2003 of "Sonderforschungsbereich 291: Elastische Handhabungssysteme für schwere Lasten in komplexen Operationsbereichen" further work will be focused on the integration of reliable numerical algorithms supporting directed rounding, interval data types and the precise dot product in the multi body toolkit MOBILE. Classical user programs are automatically translated in programs with error control and verified results. The distance calculation algorithms are extended to the dynamic case with moving obstacles and the data transfer is adapted to the MechaSTEP standard. Furthermore, offset and probabilistic occupancy octree models will be examined.

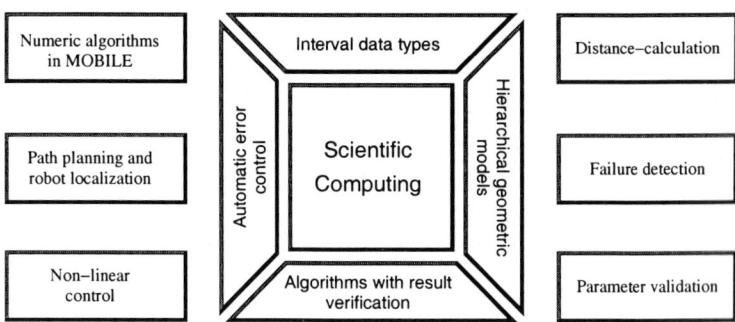

Fig. 9. Centre of interests

7 Acknowledgement

The authors thank Prof. Dr. W. Krämer, who initiated parts of this work, for helpful discussions and suggestions.
This work is supported by "Deutsche Forschungsgemeinschaft" within the scope of "Sonderforschungsbereich 291".

References

1. Baluska, R. (1998) Fuzzy Modeling for Control. Kluwer Academic Publisher, Boston.
2. Bernd, Th., Kroll, A. (1998) FIMO 8.1: Ein Programmpaket zur rechnergestützten Fuzzy-Modellierung nichtlinearer Prozesse. 8. Workshop Fuzzy-Control, Dortmund, GMA–Fachausschuss 5.22, 154–167.
3. Bezdek, J. C. (1981) Pattern Recognition with Fuzzy Objective Function Algorithms. Plenum Press, New York.
4. Büdding, G., Fausten, D., Krämer, W., Luther, W., Möllers, Th., Otten, W., Traczinski, H. (1999) Verifikation von Lösungen ausgewählter Probleme aus der Modellierung von Manipulatoren. Technical Report, Gerhard–Mercator–Universität Duisburg SM–DU–433.
5. Carreras, C., Walker, I. D., Nieto, O., Cavallaro, J. R. (1999) Robot reliability estimation using interval methods. In: MISC'99: Workshop on Applications of Interval Analysis to Systems and Control, Girona, Spain, 371–385.
6. Denson, W., Chandler, G., Crowell, W., Clark, A., Jaworski, P. (1994) Nonelectronic Parts Reliability Data. Technical Report NPRD–95, Reliability Analysis Center, Rome, NY.
7. Dyllong, E., Luther, W. (2000) An accurate computation of the distance between a point and a polyhedron. In: Berveiller, M., Louis, A. K., Fressengeas, C. (Eds.) ZAMM, Vol. 80 of GAMM 99 Annual Meeting, Metz, France, April 12–16, WILEY–VCH, Berlin, S771–S772.
8. Dyllong, E., Luther, W. (2000) Distance-Calculation Between a Point and a NURBS Surface. In: Laurent, P.-J., Sablonnière, P., Schumaker, L. L. (Eds.) Curve and Surface Design, Saint Malo, 1999, Vanderbilt University Press, Nashville, TN, ISBN 0-8265-1356-5.
9. Dyllong, E., Luther, W., Otten, W. (1999) An accurate distance-calculation algorithm for convex polyhedra. Reliable Computing 3 (5), 241–254.
10. Fausten, D., Luther, W. (2000) Verifizierte Lösungen von nichtlinearen polynomialen Gleichungssystemen. Technical report, Gerhard–Mercator–Universität Duisburg SM–DU–477.
11. Fausten, D., Möllers, Th., Traczinski, H. (1999) Verified simulation of a hydraulic drive. In: RoMoCo'99: IEEE Workshop On Robotic Motion And Control, 23–27.
12. Gass, S. I. (1985) Linear Programming: Methods and Applications. Mc Graw Hill, New York.
13. Gomez, C. (Ed.) (1999) Engineering and Scientific Computing with Scilab. Birkhauser.
14. Kecskeméthy, A. (1996) MOBILE Version 1.2 User's Guide and Reference Manual.

15. Kieffer, M., Walter, E. (1998) Interval analysis for guaranteed nonlinear parameter estimation. In: Atkinson, A. C., Pronzato, L., Wynn, H. P. (Eds.) MODA 5 – Advances in Model-Oriented Data Analysis and Experiment Design, Physica, Heidelberg, 115–125.

16. Kieffer, M., Jaulin, L., Walter, E., Meizel, D. (1999) Guaranteed mobile tracking using interval analysis. In: MISC'99: Workshop on Applications of Interval Analysis to Systems and Control, Girona, Spain, 347–360.

17. Knüppel, O. (1993) Bias – basic interval arithmetic subroutines. Technical Report 93.3, TU Hamburg–Harburg.

18. Knüppel, O. (1993) PROFIL – Programmer's Runtime Optimized Fast Interval Library. Technical Report 93.4, TU Hamburg–Harburg.

19. Krämer, W. (1998) A Priori Worst Case Error Bounds for Floating-Point Computations. IEEE Trans. on Computers 47 (7), 750–756.

20. Kroll, A. (1997) Fuzzy-Systeme zur Modellierung und Regelung komplexer technischer Systeme, Ph. D. thesis, Universität Duisburg.

21. Lohner, R. (1989) Einschließungen bei Anfangs- und Randwertaufgaben gewöhnlicher Differentialgleichungen. In: Kulisch, U. W. (Hrsg.) Wissenschaftliches Rechnen mit Ergebnisverifikation, Vieweg, Braunschweig, 183–223.

22. Luther, W. (1999) Accurate robot reliability estimation by fault tree techniques. Technical report, Gerhard–Mercator–Universität Duisburg SM–DU–459.

23. Luther, W., Otten, W. (1999) Verified calculation of the solution of algebraic Riccati equation. In: Cendes, T. (Ed.) Developments in Reliable Computing, 105–119.

24. Luther, W., Otten, W., Traczinski, H. (1999) Verified calculation of the solution of discrete-time algebraic Riccati equation. In: MISC'99: Workshop on Applications of Interval Analysis to Systems and Control, Girona, Spain, 411–421.

25. Luther, W., Traczinski, H. (1999) Error propagation control in MOBILE: extended basic mathematical objects and kinetostatic transmission elements. In: Kecskeméthy, A., Schneider, M., Woernle, C. (Eds.) Advances in Multibody Systems and Mechatronics, TU Graz, 267–276.

26. Morales, D., Son, T. C. (1998) Interval Methods in Robot Navigation. Reliable Computing 4, 55–61.

27. Sherbrooke, E. C., Patrikalakis, N. M. (1993) Computation of the solution of nonlinear polynomial systems. Computer Aided Geometric Design 10, 379–405.

28. Simeon, B. (1995) MBSPACK – numerical integration software for constrained mechanical motion. Surv. Math. Ind., 169–202.

29. Sima, V. (1996) Algorithms for linear-quadratic optimization. Marcel Dekker, New York.

On the Ubiquity of the Wrapping Effect in the Computation of Error Bounds

Rudolf J. Lohner

Institut für Angewandte Mathematik, Universität Karlsruhe (TH),
D-76128 Karlsruhe, Germany.
rudolf.lohner@math.uni-karlsruhe.de

Abstract. Historically, the wrapping effect was discovered and named in the context of solving ordinary initial value problems in interval arithmetic. Its explanation was obviously geometric: rotations of interval vectors enclosing the set of solutions catch excessive points into the enclosure which may eventually 'explode' exponentially. Also discrete dynamical systems share this undesirable behaviour. In the literature the wrapping effect has been discussed primarily in this context.

However, the wrapping effect is not confined to the computation of bounds for dynamical systems – it is much more a phenomenon which occurs concealed within many other problems such as difference equations, linear systems with full or with banded and even triangular matrix, similarly in non-linear systems and even in automatic differentiation. It appears that almost any algorithm which computes rigorous error bounds in some 'iterative' or 'recurrent' fashion may become a victim of the wrapping effect. This paper gives an overview on many such wrapping prone problems as well as on old and new methods designed to eliminate or at least diminish the wrapping effect.

1 Introduction or What is the Wrapping Effect?

The wrapping effect was discovered and named by R.E. Moore [14,5] in the context of solving ordinary initial value problems in interval arithmetic. The classical model problem for the explanation of the wrapping effect is the harmonic oscillator with initial values taken in some box:

$$u" + u = 0, \ u(0) \in [u_0], \ u'(0) \in [u_1].$$

The exact solution set in the phase plane is indicated in Fig. 1 by the dashed square rotating clockwise around the origin. A straight forward interval enclosure method takes some time step size $h > 0$ and successively computes enclosures at $t_j = jh$ by starting with the enclosure at t_{j-1} and wrapping the propagated solution set to an interval vector at t_j as is also sketched in Fig. 1.

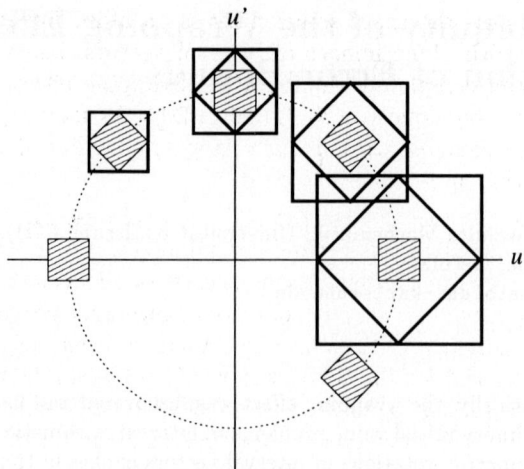

Figure 1: Wrapping effect for the harmonic oscillator.

Thus the explanation for the rapid growth of the enclosure set is geometrically obvious: rotations of interval vectors enclosing the solution set catch excessive points into the enclosure which may eventually 'explode' exponentially. Another observation is that no numerical errors such as roundoff or discretization errors are involved in the wrapping operation – it is solely due to the enclosure in interval vectors.

Therefore, we do not even need the context of a differential equation to examine the wrapping effect: A simple matrix vector iteration shows all important details.

If for the rotation A with angle ϕ and initial condition $[x_0]$

$$A = \begin{pmatrix} \cos\phi & \sin\phi \\ -\sin\phi & \cos\phi \end{pmatrix}, \quad [x_0] = \begin{pmatrix} [1-\varepsilon, 1+\varepsilon] \\ [1-\varepsilon, 1+\varepsilon] \end{pmatrix}$$

we perform the interval iteration

$$[x_{n+1}] = A[x_n].$$

then the sequence of the diameter vectors $d_n = d([x_n])$ satisfies

$$d_{n+1} = |A|d_n = \begin{pmatrix} |\cos\phi| & |\sin\phi| \\ |\sin\phi| & |\cos\phi| \end{pmatrix} d_n = \underbrace{(|\sin\phi| + |\cos\phi|)^n}_{>1} \underbrace{\begin{pmatrix} 2\varepsilon \\ 2\varepsilon \end{pmatrix}}_{d_0}.$$

The d_n diverge exponentially. For small ϕ this occurs with a blow up factor $\approx e^{2\pi} \approx 535$ per revolution.

There are other examples where the solution has no rotations in the phase plane but still the wrapping effect occurs massively (e.g. $u_1' = u_1 - 2u_2, u_2' = 3u_1 - 4u_2$, see [10,11]).

In the linear case the exact solution set is an affine image of the initial set. Computing with affine images of interval vectors, i.e. computing with parallelepipeds can completely eliminate the wrapping effect, however, only if all computations are done exactly ([4,10,11,20]). As soon as computational errors have to be taken into account the wrapping effect is still a serious issue. We will discuss this in Section 3.

For nonlinear discrete or continuous dynamical systems the situation is even less favourable. Here the nonlinear dependence on the initial data prevents the a priori choice of some 'optimal' class of enclosure sets. The exact solution set may develop a very complicated shape and become non-convex or even multiply connected. Thus parallelepipeds or even any class of convex sets may be unsuitable for the computation of enclosures. With the recently developed 'Taylor models' ([3,13]) the dependency problem can be handled to a large degree very satisfactorily, however, the computational cost may increase rapidly with the number of variables and the wrapping effect will not be completely eliminated (see Section 3).

We take this as a reason to define only loosely what we mean with the wrapping effect: It is the undesirable overestimation of a solution set of an iteration or recurrence which occurs if this solution set is replaced by a superset of some 'simpler' structure and this superset is then used to compute enclosures for the next step which may eventually lead to an exponential growth of the overestimation. Although not very precisely defined, this notion of the wrapping effect covers the observations made with the simple model problem as well as being useful for more complicated classes of enclosure sets which have been introduced by several authors [3,7–11,13,19,22].

The focus of this paper will not be the dependency problem of nonlinear problems but the fact that in each step of a 'marching algorithm' we have to take into account additional local errors (roundoff, discretization etc.) which force overestimations already in the linear case where we cannot benefit from nonlinear tools.

In Section 2 we will identify many different kinds of problems which are prone to the wrapping effect. The reason for this is basically that these problems can be reformulated as a linear or nonlinear recurrence which reveals where and how they are vulnerable. The enumeration of such problems is by far not complete. Still it gives an impression of how widespread such problems are which might suffer from the wrapping effect if enclosures for their solutions are to be computed. Section 3 gives an overview on different appoaches to eliminate or at least to reduce the wrapping effect. Most of the methods which were proposed by authors in the last four decades are of an intuitive geometric nature. However, also purely algebraic approaches have been considered. Conlusions in Section 4 discuss the present situation especially with a look on the cost of the methods proposed in Section 3.

Throughout the paper we assume that the reader has a basic knowlegde in interval analysis as presented e.g. in [1] and [18]. For brevity of the presentation we implicitly assume that necessary conditions such as smoothness of functions, existence of interval evaluations etc. are fulfilled wherever used but not stated explicitly.

2   Where does the Wrapping Effect appear?

2.1   Matrix-Vector Iterations

As in the model problem from Section 1 we consider matrix vector iterations

$$[x_{n+1}] = A_n[x_n] + b_n, \quad [x_0] \in \mathrm{I\!I\!R} .$$

Here we must expect overestimations, since for the spectral radii of a matrix A and the matrix of its absolute values $|A|$ we have

$$\rho(A) \leq \rho(|A|)$$

and for the diameter vectors $d_n = d([x_n])$ there holds

$$d_{n+1} = |A_n|d_n.$$

If A_n and b_n are allowed to contain intervals too, then

$$d(A_n)|[x_n]| + |A_n|d_n + d(b_n) \geq d_{n+1} \geq |A_n|d_n + d(b_n) .$$

If for example

$$\rho(A) < 1 \text{ and } \rho(|A|) > 1$$

then the solution set

$$\{x_{n+1} = A_n x_n + b_n | x_0 \in [x_0]\}, n \geq 0,$$

may shrink to a point whereas the interval iterates $[x_n]$ diverge.

The same may (and does) happen in matrix-products, matrix-powers as well as in more complicated matrix and matrix-vector expressions.

2.2   Discrete Dynamical Systems

For a nonlinear iteration with $f : \mathrm{I\!R}^n \to \mathrm{I\!R}^n$ sufficiently smooth

$$x_{n+1} = f(x_n), \quad x_0 \text{ given}$$

or with intervals

$$[x_{n+1}] = f([x_n]), \quad x_0 \in [x_0] \text{ given}$$

we can apply a mean-value form which gives tighter enclosures as long as $d([x_n])$ remains small:

$$[x_{n+1}] = f(\tilde{x}_n) + f'([x_n])([x_n] - \tilde{x}_n), \quad \tilde{x} \in [x_n].$$

Formally this is a 'linear' iteration scheme as in the previous subsection, however, with an interval matrix $A_n = f'([x_n])$ depending on the iteration index n. Therefore we must expect the same kind of difficulties which can occur in the linear case: Iterations may diverge even though the exact solution set stays bounded.

In the mean-value representation the nonlinearity of the original iteration is hidden inside the functional matrix $A_n = f'([x_n])$ whose diameter additionally grows if $d([x_n])$ does.

2.3   Continuous Dynamical Systems (ODEs)

For an ordinary initial value problem with $g : \mathbb{R}^{n+1} \to \mathbb{R}^n$

$$x'(t) = g(t, x(t)), \quad x(t_0) = x_0,$$

a numerical one step method which also takes into account all roundoff and discretization errors leads to a discrete system again which basically has the form

$$[x_{n+1}] = [x_n] + h\Phi([x_n], t_n) + [z_{n+1}] .$$

Here h is the step size, $t_n = t_0 + nh$, $x_n = x(t_n) \in [x_n]$, Φ comes from the one step method and $[z_{n+1}]$ is an interval vector containing all local errors. Since this is the type of problem for which the wrapping effect has been studied most intensively, we refer to the literature for further details ([3,4,10,11,14–16]).

As with linear and nonlinear discrete dynamical systems we see that we potentially run into problems with the wrapping effect. Any counter measures that work in the discrete case should also work in the continuous case and vice versa.

2.4   Difference Equations

Whereas the previously mentioned problems all were problems for systems in $\mathbb{R}^n$ the following linear difference equation

$$\begin{cases} a_0 z_n + a_1 z_{n+1} + \cdots + a_m z_{n+m} + a_{m+1} z_{n+m+1} = b_n, \quad n \geq 0 \\ z_0, z_1, \ldots, z_m \text{ given} \end{cases}$$

is a recurrence equation for scalar values z_n only. There seems to be no reason to expect a behaviour similar to the previous cases.

However, rewriting the scalar equation as a matrix-vector iteration by use of $x_n := (z_n, z_{n+1}, \ldots, z_{n+m})^T \in \mathrm{IR}^{m+1}$:

$$
x_{n+1} = \begin{pmatrix} z_{n+1} \\ z_{n+2} \\ \vdots \\ z_{n+m+1} \end{pmatrix} = \begin{pmatrix} 0 & 1 & 0 & & 0 \\ 0 & 0 & \ddots & & \vdots \\ \vdots & & \ddots & \ddots & 0 \\ 0 & \cdots & & 0 & 1 \\ \dfrac{-a_0}{a_{m+1}} & & \cdots & \dfrac{-a_{m-1}}{a_{m+1}} & \dfrac{-a_m}{a_{m+1}} \end{pmatrix} x_n + \begin{pmatrix} 0 \\ 0 \\ \vdots \\ 0 \\ \dfrac{b_n}{a_{m+1}} \end{pmatrix}
$$

it can be seen immediately that both formulations are equivalent not only theoretically but also computationally: Solving the scalar difference equation for z_{n+m+1} requires precisely the same computation as doing one step of the matrix-vector form.

Unexpected overestimations in the scalar computation are often attributed to data dependency. In the vector formulation, however, it becomes evident that such overestimations are due to the wrapping effect in its classical geometric appearance. Data dependence of the initial values is linear only and can be dealt with by the use of parallelepipeds as we will see in Section 3. The techniques described there can be applied only to the matrix-vector formulation however, not to the scalar formulation. This suggests that difference equations should really be treated as matrix-vector iterations if tight error bounds are to be computed.

As an example, consider the recurrence equations for the Chebycheff polynomials

$$
T_0(x) = 1, \quad T_1(x) = x, \quad T_{n+1}(x) - 2xT_n(x) + T_{n-1}(x) = 0 .
$$

If we wish to compute an enclosure of $T_n(x_0)$ for some value x_0 and a large value of n then we can use this difference equation and compute the desired value in interval arithmetic. However, this computation is nothing else but forward computation in interval arithmetic for the matrix-vector iteration

$$
\begin{pmatrix} T_n(x_0) \\ T_{n+1}(x_0) \end{pmatrix} = \begin{pmatrix} 0 & 1 \\ -1 & 2x_0 \end{pmatrix} \begin{pmatrix} T_{n-1}(x_0) \\ T_n(x_0) \end{pmatrix} .
$$

Even though we have point data at the beginning interval floating-point arithmetic will introduce roundoff errors. Then a very fast blow up of the enclosures will occur due to the wrapping effect. If the argument x_0 is not a floating-point number, then such intervals will enter the computation right from the beginning. If we take e.g. $x_0 = 0.99$ then we obtain the following output from a short PASCAL-XSC program (IEEE double precision) for the values of n and the enclosures for $T_n(0.99)$:

n	T[n](0.99)	
2	[9.601999999999999E-001,	9.602000000000007E-001]
3	[9.11195999999999E-001,	9.11196000000002E-001]
4	[8.43968079999998E-001,	8.43968080000004E-001]
5	[7.5986079839999E-001,	7.5986079840001E-001]
6	[6.6055630083198E-001,	6.6055630083202E-001]
7	[5.480406772473E-001,	5.480406772474E-001]
8	[4.245642401176E-001,	4.245642401179E-001]
9	[2.925965181856E-001,	2.925965181861E-001]
10	[1.54776865889E-001,	1.54776865891E-001]
15	[-5.246430527E-001,	-5.246430525E-001]
20	[-9.52088247E-001,	-9.52088240E-001]
25	[-9.222663E-001,	-9.222657E-001]
30	[-4.496E-001,	-4.494E-001]
35	[2.3E-001,	2.5E-001]
40	[6.9E-001,	9.3E-001]
45	[-8.0E+000,	1.0E+001]
50	[-7.1E+002,	7.2E+002]

Of course the values of the Chebychev polynomials can be computed in a much easier way, since $T_n(x) = \cos(n \arccos x)$. This example should only demonstrate the difficulties that can arise with difference equations even as simple as this one. This example is also important since many other well known functions can be computed by use of certain difference equations such as three term recursions for orthogonal polynomials and special functions like Lagrange polynomials, Bessel functions and many others.

Similarly, nonlinear difference equations can be treated in an obvious way by rewriting them in an equivalent nonlinear vector formulation. This can then be evaluated by the mean-value form as was done for nonlinear discrete dynamical systems. The two representations are theoretically equivalent but no longer computationally. Nevertheless, for small diameters of the iterates we should expect better enclosures from the mean-value formulation however, this is a matrix-vector iteration again (with interval data) and such a problem has already been identified as being susceptible to the wrapping effect.

2.5   Linear Systems with (Banded) Triangular Matrix

Forward and backward substitution in interval arithmetic for triangular matrices are known to be mostly unstable processes resulting in large overestimations.

That this is also due to the wrapping effect becomes evident e.g. in the case of banded triangular matrices since there is an obvious equivalence with linear difference equations which in turn are equivalent to matrix-vector iterations as we have demonstrated in the previous subsection.

This is precisely the reason why verification methods for linear systems $Ax = b$ which use an LU-factorization usually break down for medium to high system dimensions unless special methods against the wrapping effect are employed.

The following example is taken from [19]: The system $Ax = b$ with three band lower triangular matrix

$$A = \begin{pmatrix} 1 & 0 & 0 & 0 & \cdots & 0 \\ 1 & 1 & 0 & 0 & & 0 \\ 1 & 1 & 1 & 0 & & 0 \\ 0 & 1 & 1 & 1 & \ddots & \vdots \\ \vdots & \ddots & \ddots & \ddots & \ddots & 0 \\ 0 & \cdots & 0 & 1 & 1 & 1 \end{pmatrix}$$

and right hand side $(b_1, 0, 0, \ldots, 0)^T$, $b_1 = [-\varepsilon, \varepsilon]$ can be reformulated as a linear second order difference equation:

$$\begin{cases} x_1 = b_1, \ x_2 = -b_1 \\ x_{n+1} + x_n + x_{n-1} = 0, \ n \geq 2. \end{cases}$$

Computing the solution of this difference equation by solving for x_{n+1} in interval arithmetic is computationally equivalent with interval forward substitution with the original matrix.

For any fixed b_1 the exact solution is:

$$x_1 = b_1, \quad x_2 = -b_1, \quad x_n = \begin{cases} -b_1 & \text{for } n = 3k - 1 \\ 0 & \text{for } n = 3k \\ b_1 & \text{for } n = 3k + 1 \end{cases}$$

Interval forward substitution, however, yields for $b_1 = [-\varepsilon, \varepsilon]$:

$$x_1 = [-\varepsilon, \varepsilon], \ x_2 = [-\varepsilon, \varepsilon], \ x_n = [-a_n\varepsilon, a_n\varepsilon], \ n \geq 3$$

where the a_n are the Fibonacci numbers $a_1 = a_2 = 1$, $a_{n+1} = a_n + a_{n-1}$. Therefore, the diameters $d(x_n)$ diverge exponentially:

$$d(x_n) = 2a_n\varepsilon > \text{const} \cdot 1.62^n.$$

On the other hand the optimal solution set stays bounded:

$$x_n = \begin{cases} 0 & \text{for } n = 3k \\ [-\varepsilon, \varepsilon] & \text{for } n = 3k + 1. \end{cases}$$

Again we have seen from the theoretical and computational equivalence of two problems, that susceptibility for the wrapping effect in one problem class induces the same risk for the other problem class. Since the bandwidth itself does not play any role and also the resulting difference equations need not necessarily have constant coefficients this observation holds for any triangular matrix.

2.6   Automatic Differentiation

The computation of Taylor coefficients of a scalar function by automatic differentiation (see [10,11,15,18,21]) often involves forward/backward substitution for a triangular matrix. As we know from the previous subsection we must be suspicious and expect to get difficulties here also. Examples are division, square root and virtually all elementary function.

To compute Taylor coefficients $(w)_k$ of a quotient $w = u/v$ of two functions u, v with Taylor coefficients $(u)_k, (v)_k$ we have to solve the triangular Toeplitz system

$$\begin{pmatrix} (v)_0 & 0 & 0 & \cdots & 0 \\ (v)_1 & (v)_0 & 0 & & 0 \\ (v)_2 & (v)_1 & (v)_0 & & 0 \\ \vdots & \ddots & \ddots & \ddots & 0 \\ (v)_n & \cdots & (v)_2 & (v)_1 & (v)_0 \end{pmatrix} \begin{pmatrix} (w)_0 \\ (w)_1 \\ (w)_2 \\ \vdots \\ (w)_n \end{pmatrix} = \begin{pmatrix} (u)_0 \\ (u)_1 \\ (u)_2 \\ \vdots \\ (u)_n \end{pmatrix}$$

for $(w)_0, \ldots (w)_n$. Here, in general, all quantities are intervals due to roundoff errors.

The way how this is usually done is by interval forward substitution which results in the well known and widely used recurrence formulae (see [21]). Here often large overestimations are produced as can be seen from the simple example $f(x) = \sin(e^x)/e^{-x}$. Computing an enclosure of the 30th Taylor coefficient $(f)_{30}$ at $x = -8$ by the recurrence formula in interval arithmetic (IEEE double floating-point) yields $(f(-8))_{30} \in [-6, 6] \cdot 10^{-17}$.

If instead we first precondition the triangular system with an approximate inverse of the mid-point matrix and solve the resulting system by interval forward substitution, then we get practically no wrapping effect and as an enclosure of the 30th Taylor coefficient at $x = -8$ we get $(f(-8))_{30} \in [-8.708693, -8.708691] \cdot 10^{-30}$.

In general this preconditioning technique is quite expensive. However, here we are dealing with a Toeplitz matrix the consequence of which is that all matrix operations can be done at a cost being only quadratic in the system dimension.

3   How can we Reduce the Wrapping Effect?

In Section 2 we demonstrated that the straight forward interval solution of many important classes of problems may yield bounds which are too bad by an exponentially growing overestimation. In order to get satisfactory bounds new algorithms have to be designed which are much less vulnerable to the wrapping effect. In this section we give a short overview over different such techniques which have been created by different authors.

We formulate all methods for the special case of matrix-vector iterations. As we have seen in the previous section this is the central problem class in the sense that once we have a means to reduce the wrapping effect for this special problem class, then we can reformulate other problem classes in order to take advantage from the method for matrix-vector iterations.

We note that a typical property of many of these methods is, that they try to separate (up to a certain degree) the computation of the operator (i.e. the matrix product) and its action on the initial set x_0. A successful separation of both has a good chance to avoid many unnecessary wrapping operations – at least if the operator itself does not involve large perturbations (i.e. interval entries with large diameter).

3.1   Rearranging Expression Evaluation

We start with a purely algebraic approach to reduce the wrapping effect, see Gambill and Skeel [5]. It is based on the idea of minimizing the depth of the computational graph induced by the matrix-vector iteration thereby having the (somewhat vague) intention to reduce the number of operations with dependent variables:

Instead of computing

$$x_{n+1} = A_n(A_{n-1}(A_{n-2}(\cdots A_1(A_0 x_0)\cdots)))$$

which would be the normal order of operations we compute the product of the matrices first and do this in such a way that the expression is not very deeply nested. If $n = 2^m$ then this can be achieved for example by grouping pairs of factors recursively:

$$x_{n+1} = (\cdots((A_n A_{n-1})(\cdots))\cdots((A_3 A_2)(A_1 A_0))\cdots)x_0$$

Modifications for other values of n can be obtained easily. Also many variations are possible such as taking groups of more than two factors or even varying the number of factors in different groups.

Numerical experiments often show astonishingly narrow enclosures. One reason for this behaviour is, that the method first computes the operator, i.e. the matrix product, before it is applied to the argument x_0. For initial values with large diameter and matrices with very small diameters (ideally point matrices) this is a favorable situation.

Nevertheless, the arbitrariness in grouping the terms gives rise to the possibility to construct counterexamples for each given grouping strategy such that the method yields exponentially growing overestimations. For grouping in pairs a simple counterexample is given with all matrices equal to the 2×2 rotation matrix with 30° rotation: The group pairing will always compute rotations whose angles are a multiple of 30° but never a multiple of 90°.

For nonlinear problems this method does not behave satisfactorily since now the matrices are interval matrices which depend on the iterates. This causes the matrix products to blow up rather soon in most cases.

3.2   Coordinate Transformations

The oldest approach which also has been continuously modified and developed further by different authors uses enclosures in suitably transformed coordinate systems instead of just taking interval vectors in the original Cartesian system.

A more geometrical interpretation of this approach is that now the enclosing sets are suitably chosen parallelepipeds (i.e. affine images of interval vectors). As already discussed in Section 1 this class of sets is optimal in the linear case if the initial set belongs already to this class and if computations are done exactly. Because of this reason this approach is rather successful also if computation is not exact and generates additional error intervals in each iteration.

A parallelepiped can be represented as $P = \{c + Bx | x \in [x]\}$ with a point vector c, a point matrix B and an interval vector $[x]$.

Then the original iteration

$$x_{n+1} = A_n x_n + b_n$$

is replaced by the equivalent one:

$$\begin{cases} x_{n+1} = B_{n+1} y_{n+1}, & B_0 := I, \quad y_0 := x_0 \\ y_{n+1} = (B_{n+1}^{-1} A_n B_n) y_n + B_{n+1}^{-1} b_n \end{cases} .$$

In interval arithmetic:

$$\begin{cases} [x_{n+1}] = B_{n+1} [y_{n+1}], & B_0 := I, \quad [y_0] := [x_0] \\ [y_{n+1}] = (B_{n+1}^{-1} A_n B_n) [y_n] + B_{n+1}^{-1} b_n \end{cases} .$$

The crucial point in this approach is the choice of basis matrices B_{n+1}. Several possibilities will be discussed now.

Coordinate Transformations – non orthogonal   Choose the basis matrices as good approximations of the iteration map [14,15,4,10,11] .

This means $B_{n+1} \approx A_n A_{n-1} \cdots A_1 A_0$, which can be achieved by choosing $B_{n+1} \approx A_n B_n$ in each step.

An advantage of this choice is that the edges of the enclosing set are roughly parallel to those of the solution set itself (at least in the linear case). This however is at the same time also a disadvantage: If there are dominating eigenvalues (e.g. all A_i are constant with a largest simple eigenvalue) then B_{n+1} may become ill conditioned and usually becomes even singular numerically. In this case the method breaks down.

Unfortunately this is the most common situation: the method behaves similar to the power method for the computation of the dominant eigenvector in B_{n+1} with the consequence that the columns of B_{n+1} become linearly dependent very soon. Then B_{n+1} rapidly becomes ill-conditioned or even singular such that B_{n+1}^{-1} will cause the enclosures to blow up.

This limited applicability may be considerably broadened if we restrict the pairwise angles between the columns of the basis matrix B_n to stay above a minimal prescribed angle. The resulting matrices have bounded condition numbers and offer a great flexibility in choosing an appropriate basis. Up to now this variant has not yet been tested thoroughly, so there is not yet practical experience with this modification.

Coordinate Transformations – orthogonal Choose orthogonal basis matrices to keep them well conditioned and invertible [10,11].

The choice of an orthogonal matrix B_n guarantees that it has much more favorable properties than with the previous choice. A suitable orthogonal matrix can be obtained from a QR-factorization of $A_n B_n$: Compute $A_n B_n = Q_n R_n$ and choose $B_{n+1} \approx Q_n$. (If computations are done exactly then $B_{n+1}^{-1} A_n B_n = R_n$ is upper triangular.)

It is advisable to apply a pivoting strategy prior to the QR-factorization by sorting the columns of $A_n B_n$ in descending order according to the lengths of the columns of $A_n B_n \text{diag}(d([y_n]))$. The columns of this matrix span a good approximation of the exact solution set. The advantage of this pivoting becomes clear from Figure 2: If a_1, a_2 are the columns which span an approximation of the solution set (dashed) and which are to be orthogonalized then obviously proceeding in decreasing order of their length (right picture) yields much narrower enclosures than taking a different order (left picture).

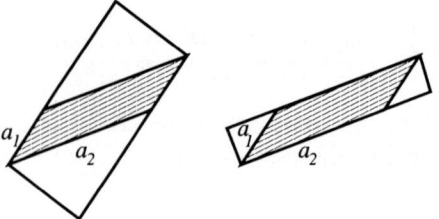

Figure 2: Sorting edges prior to orthogonalization.

A big advantage of this method is that it never breaks down because of a singular matrix. In practical computations is has proven to be very robust and mostly delivers astonishingly narrow enclosures.

However, also this method is not infallible. Kühn [7] has constructed an example where this choice of orthogonal basis matrices leads to an exponential overestimation of the solution set:

$$A_0 = \begin{pmatrix} 1 & 1 \\ 0 & 1 \end{pmatrix}, \quad A_1 = A_0^{-1}, \quad A_2 = \begin{pmatrix} 2 & 0 \\ 0 & 1 \end{pmatrix}, \quad A_3 = A_0^T,$$

$$A_4 = A_3^{-1}, \quad A_5 = A_2^{-1}, \quad A_{n+6} = A_n, \quad n \geq 5 .$$

After each six steps this iteration is the identity. However, starting the iteration with $[x_0] = ([-1,1],[-1,1])^T$ and choosing orthogonal basis matrices as described above doubles the initial interval after each six iterations eventually producing an exponential blow up whereas the exact solution set stays bounded.

Coordinate Transformations – Constant Matrix Case ($A_n = A$) If the iteration matrices A_n are all identical, $A_n = A$, then it is possible to obtain some quantitative information about its behaviour. A very detailed discussion in this direction can be found in Nedialkov and Jackson [17] in this volume.

Because of the general rule $d(A[x]) \geq |A|d([x])$ the growth of the diameter of $[x_n]$ for the non orthogonal choice of the matrices B_n is dictated by the spectral radius $\rho(|A|)$ which may be considerably larger than $\rho(A)$.

The situation is much more complicated in the case of orthogonal matrices B_n. Here, writing $C_{n+1} := B_{n+1}^{-1} A B_n$, the growth of the diameters of $[x_n], [y_n]$ is dictated by $\rho(|C_{n+1}|)$. Since C_{n+1} is upper triangular we have

$$\rho(|C_{n+1}|) = \rho(C_{n+1}) \ .$$

Moreover, since B_n and B_{n+1} are orthogonal it follows that

$$C_{n+1}^T C_{n+1} = B_n^T A^T A B_n \Rightarrow \sigma(C_{n+1}) = \sigma(A)$$

where $\sigma(\cdot)$ denotes the spectral norm (i.e. the largest singular value).

For orthogonal A we trivially obtain always $\rho(|C_{n+1}|) = 1$ which demonstrates that this choice of B_n is ideal in this case.

For symmetric A we obtain the result

$$\rho(|C_{n+1}|) = \rho(C_{n+1}) \leq \sigma(C_{n+1}) = \sigma(A) = \rho(A)$$

which shows that here the orthogonal choice of B_n also yields optimal results concerning the growth of the enclosure. For more and deeper results see [17].

3.3   Ellipsoids

Neumaier [19] proposes the use of ellipsoids as enclosure sets. An ellipsoid is represented as

$$E(z, L, r) := \{z + Lx | x \in \mathbb{R}^n, \ ||x||_2 \leq r\}$$

where the $n \times n$-matrix L is chosen to be lower triangular.

In [19] an algorithm is developed which computes enclosures for x_n as ellipsoids. Some numerical examples for constant 2×2 (interval) iteration matrices A demonstrate the applicability of the algorithm. The results are mostly superior to naive interval arithmetic. There do not, however, exist comparisons with methods that use affine coordinate transformations as discussed earlier. The algorithm for computing with ellipsoids is more complicated than computing with parallelepipeds.

An advantage of the use of ellipsoids could be the fact that ellipsoids have a smooth boundary which make them perhaps easier to handle analytically. Also some applications can be treated in a more natural way as with parallelepipeds (e.g. stability regions and confidence regions, [19]).

3.4 Zonotopes

Proposing zonotopes as enclosing sets Kühn has introduced another promising class of sets into the arena against the wrapping effect [7–9]. Zonotopes are the Minkowski sum of line segments and thus are convex polyhedrons. The sum of three line segments and the resulting zonotope is depicted in Figure 3:

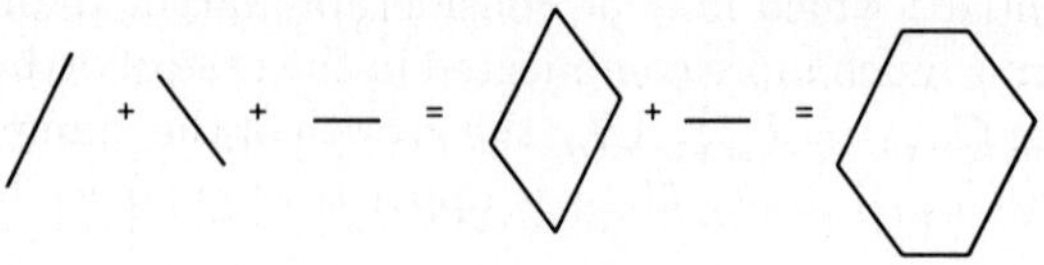

Figure 3: Zonotope as sum of three line segments.

A parallelepiped in $\mathrm{I\!R}^n$ is a zonotope: The Minkowski sum of n line segments. This is used by Kühn to represent zonotopes as a sum of m parallelepipeds

$$z := \sum_{k=1}^{m} B_k[z_k] := \{\sum_{k=1}^{m} B_k z_k | z_k \in [z_k], k = 1 \ldots m\}$$

which enables comfortable computation:

$$Az + b = \sum_{k=1}^{m} AB_k[z_k] + b = \sum_{k=1}^{m} B'_k[z_k] + b \ .$$

The number m of terms to represent a zonotope is kept constant by wrapping the smallest term in the sum into an interval vector and adding it to the next larger term. Kühn discusses many possible strategies how this can be done in detail. The key point is that wrapping operations are done only with small terms which effectively delays or almost eliminates a blow up of the enclosing zonotope.

The method has a large degree of flexibility due to the free choice of the number m of terms in a zonotope and many possible strategies of wrapping and combining different terms in the representation of a zonotope. For satisfactory results m must not be too small, usually $m = 5, \ldots, 10$.

An advantage over the coordinate transformation approach is that no inverse matrices have to be computed. On the other hand, however, the zonotope and coordinate transformation approaches can be combined by using coordinate transformations whenever several terms in the zonotope representation have to be added together. Then the necessary wrapping operations can be performed with much less overestimation than would be obtained by pure interval arithmetic.

3.5   Taylor Models

Taylor models were introduced by Berz [3] and Makino/Berz [13]. A Taylor model is an enclosure of a function $f(x)$ by use of a polynomial $T(x)$ with floating-point coefficients and a corresponding remainder term $[I_f]$ which is an interval vector.

$$f(x) \in T(x) + [I_f]$$

Usually $T(x)$ is a (very close) approximation of the Taylor polynomial of $f(x)$ at some specified expansion point and $[I_f]$ contains the corresponding remainder term and all other kinds of errors (round-off, approximation).

Berz and Makino report very successful applications of Taylor models to different types of problems such as computing range of values, univariate and multivariate verified integration, ordinary initial value problems and others.

In the linear case (i.e. $T(x)$ is linear) Taylor models represent a special zonotope, the sum of a parallelepiped plus an interval vector:

Computing a matrix-vector iteration $x_{n+1} = Ax_n$ by use of such a linear Taylor model yields

$$
\begin{aligned}
x_0 &= \tilde{x}_0 + (x - \tilde{x}_0), \quad x \in [x_0], \\
x_1 &= Ax_0 = A(\tilde{x}_0 + (x - \tilde{x}_0)) = A\tilde{x}_0 + A(x - \tilde{x}_0) \\
&= \tilde{x}_1 + A_1(x - \tilde{x}_0) + [I_1], \quad x \in [x_0], \\
x_2 &= Ax_1 = A(\tilde{x}_1 + A_1(x - \tilde{x}_0) + [I_1]) \\
&= \tilde{x}_2 + A_2(x - \tilde{x}_0) + [I_2], \quad x \in [x_0], \\
&\ \vdots \\
x_{n+1} &= Ax_n = A(\tilde{x}_n + A_n(x - \tilde{x}_0) + [I_n]) \\
&= \tilde{x}_{n+1} + A_{n+1}(x - \tilde{x}_0) + [I_{n+1}], \quad x \in [x_0],
\end{aligned}
$$

where $\tilde{x}_k$ and A_k are floating-point quantities, $A_0 = I$, and the interval vectors $[I_{k+1}] = A[I_k] + (A\tilde{x}_k - \tilde{x}_{k+1}) + (AA_k - A_{k+1})(x - \tilde{x}_0)$, $[I_0] = 0$ contain all errors. We see that the wrapping effect is not completely eliminated: It is just moved to the smallest term $A[I_k]$ within the Taylor model representation which here at the same time is a zonotope representation.

Taylor models have their strong point in nonlinear problems. To a high degree they solve the dependency problem which we have identified in Section 1 because they approximate this dependence up to the degree of the approximating polynomial $T(x)$. However, the cost for computing such Taylor models can be very high as compared to other methods. Also, as can be seen in the linear case that the wrapping effect is still not completely eliminated. Nevertheless, because of the successful treatment of the dependency problem Taylor models are a valuable new tool.

4   Conclusion

We have identified many situations where the wrapping effect may severely blow up computed error bounds if no suitable measures are taken against it. Unfortunately these situations can be problems in which it is not at all obvious that the wrapping effect may be the cause of large overestimations (as e.g. in automatic differentiation).

On the other hand we also gave an overview over different methods which can be used with more or less success to reduce or almost eliminate the influence of the wrapping effect. However, at the present stage no method can be recommended to be always superior to all others or even just to be successful in all applications.

Whereas coordinate transformations and parallelepipeds have proven to be rather robust for some time now, new candidates such as zonotopes and Taylor models are an interesting and promising step ahead.

In summary we state that among the presently available methods there is no general rule available to choose the 'right' one. All have their advantages and disadvantages. There is one thing. however, that is common to all the methods: The cost for computing tight enclosures for an n-dimensional vector iteration is at least of order $O(n^3)$ whereas the cost for simple floating-point computation and naive (but usually unacceptable) interval arithmetic is only $O(n^2)$.

Therefore, the main problem which still is unsolved is the question if there are any methods that compute tight bounds without or with only negligible wrapping effect whose cost is only of order $O(n^2)$.

References

1. Alefeld, G., Herzberger, J.: *An Introduction to Interval Computations*. Academic Press, New York, 1983.
2. Anguelov, R., Markov, S.: *Wrapping effect and wrapping function*. Reliab. Comput. 4, No.4, 311-330 (1998).
3. Berz, M.: *Verified integration of ODEs and flows using differential algebraic methods on high-order Taylor models*. Reliab. Comput. 4, No.4, 361-369 (1998).
4. Eijgenraam, P.: *The Solution of Initial Value Problems Using Interval Arithmetic*. Math. Centre Tracts 144, Mathematisch Centrum, Amsterdam (1981).
5. Gambill, T.N. Skeel, R.D.: *Logarithmic Reduction of the Wrapping Effect with Applications to Ordinary Differential Equations*. SIAM J. Numer. Anal. 25, 153–162 (1988).
6. Jackson, L.W.: *Interval Arithmetic Error-Bounding Algorithms*. SIAM J. Numer. Anal. 12, 223–238 (1975).
7. Kühn, W.: *Rigorously Computed Orbits of Dynamical Systems Without the Wrapping Effect*. Computing 61, No.1, 47-67 (1998).
8. Kühn, W.: *Zonotope Dynamics in Numerical Quality Control*. In: Hege, H-Ch. (ed.) et al.: Mathematical visualization. Algorithms, applications, and numerics. International workshop Visualization and mathematics, Berlin, Germany, September 16-19, 1997. Berlin: Springer. 125-134 (1998).

9. Kühn, W.: *Towards an Optimal Control of the Wrapping Effect.* In: Csendes, Tibor (ed.), Developments in reliable computing. SCAN-98 conference, 8th international symposium on Scientific computing, computer arithmetic and validated numerics. Budapest, Hungary, September 22-25, 1998. Dordrecht: Kluwer Academic Publishers. 43-51 (1999).

10. Lohner, R.J.: *Enclosing the Solutions of Ordinary Initial- and Boundary-Value Problems.* In: Kaucher, E., Kulisch, U., Ullrich Ch. (eds.): Computerarithmetic, pp. 225–286, Teubner Stuttgart (1987).

11. Lohner, R.J.: *Einschliešung der Lösung gewöhnlicher Anfangs- und Randwertaufgaben und Anwendungen.* Dissertation, University of Karlsruhe (1988).

12. Lohner, R.J.: *Verified Computing and Programs in Pascal-XSC.* Habilitationsschrift, University of Karlsruhe (1994).

13. Makino, K., Berz, M.: *Efficient control of the dependency problem based on Taylor model methods.* Reliab. Comput. 5, No.1, 3-12 (1999).

14. Moore, R.E.: *Automatic local coordinate transformations to reduce the growth of error bounds in interval computation of solutions of ordinary differential equations.* Error in Digital Comput. 2, Proc. Symp. Madison 1965, 103-140 (1965).

15. Moore, R.E.: *Interval Analysis.* Englewood Cliffs, N.J. Prentice-Hall (1966).

16. Nedialkov, N.S., Jackson, K.R., Corliss, G.F.: *Validated solutions of initial value problems for ordinary differential equations.* Appl. Math. Comput. 105, No.1, 21-68 (1999).

17. Nedialkov, N.S., Jackson, K.R.: *A New Perspective on the Wrapping Effect in Interval Methods for Initial Value Problems for Ordinary Differential Equations.* This Volume. (2001)

18. Neumaier, A.: *Interval methods for systems of equations.* Encyclopedia of Mathematics and its Applications, 37. Cambridge etc.: Cambridge University Press. xvi, (1990).

19. Neumaier, A.: *The Wrapping Effect, Ellipsoidal Arithmetic, Stability and Confidence Regions,* Computing, Suppl. 9, 175–190 (1993).

20. Nickel, K.: *How to fight the wrapping effect.* Lect. Notes Comput. Sci. 212, 121-132 (1986).

21. Rall, L.B.: *Automatic Differentiation: Techniques and Applications.* Lecture Notes in Computer Science, No. 120, Springer-Verlag, Berlin, 1981.

22. Stewart, N.P.: *A heuristic to reduce the wrapping effect in the numerical solution of* $x' = f(t, x)$. BIT 11, 328-337 (1971).

A New Perspective on the Wrapping Effect in Interval Methods for Initial Value Problems for Ordinary Differential Equations

Nedialko S. Nedialkov[1] and Kenneth R. Jackson[2]

[1] Department of Computing and Software, McMaster University,
   Hamilton, Ontario, L8S 4L7, Canada.
   nedialk@mcmaster.ca
[2] Department of Computer Science, University of Toronto,
   Toronto, Ontario, M5S 3G4, Canada.
   krj@cs.toronto.edu

Abstract. The problem of reducing the wrapping effect in interval methods for initial value problems for ordinary differential equations has usually been studied from a geometric point of view. We develop a new perspective on this problem by linking the wrapping effect to the stability of the interval method. Thus, reducing the wrapping effect is related to finding a more stable scheme for advancing the solution. This allows us to exploit eigenvalue techniques and to avoid the complicated geometric arguments used previously.

We study the stability of several anti-wrapping schemes, including Lohner's QR-factorization method, which we show can be viewed as a simultaneous iteration for computing the eigenvalues of an associated matrix. Using this connection, we explain how Lohner's method improves the stability of an interval method and show that, for a large class of problems, its global error is not much bigger than that of the corresponding point method.

1   Introduction

We consider interval methods (often called validated methods) for initial value problems (IVPs) for ordinary differential equations (ODEs) of the form

$$y' = f(y), \quad y(t_0) = y_0, \tag{1}$$

where $f : \mathcal{D} \subseteq \mathbb{R}^n \to \mathbb{R}^n$ is sufficiently differentiable. These methods [19,23] compute bounds that are guaranteed to contain the solution of (1) at points $t_j, j = 1, 2, \ldots, m$ in the interval $(t_0, t_m]$, for some $t_m > t_0$. They are usually based on Taylor series or our recent extension of Hermite-Obreschkoff schemes to interval methods [17,18]. Both types of methods use high-order derivatives; thus the requirement that f is sufficiently differentiable is essential. To simplify the notation, we apply our methods to autonomous systems throughout this paper. However, the methods can be extended easily to nonautonomous systems.

To date, only one-step methods have been used as the basis for interval methods for (1). Each step of these methods usually consists of two phases. We refer to them as Algorithm I and Algorithm II [19]. On an integration step from t_{j-1} to t_j,

Algorithm I validates existence and uniqueness of the solution of (1) for all $t \in [t_{j-1}, t_j]$ and computes a priori bounds for this solution for all $t \in [t_{j-1}, t_j]$, [19,20]; and

Algorithm II computes tight bounds for the solution of (1) at t_j.

A major problem in the second phase is the *wrapping effect* [16]. It occurs when a solution set that is not a box in $\mathbb{R}^n$, $n \geq 2$, is enclosed, or wrapped, by a box on each integration step. (By a box we mean a parallelepiped with edges that are parallel to the axes of an orthogonal coordinate system.) As a result of such a wrapping, an overestimation is often introduced on each integration step. These overestimations accumulate as the integration proceeds, and the computed bounds may soon become unacceptably large.

The problem of reducing the wrapping effect has usually been studied from a geometric perspective as finding an enclosing set that introduces as little overestimation of the enclosed set as possible. For example, parallelepipeds [5,13,15,16], ellipsoids [9,10,12,21], convex polygons [24], and zonotopes [14] have been employed in reducing the wrapping effect. Other, rather complicated techniques can be found in [3] and [7]. Recently, Berz and Makino [4] proposed a method for reducing the wrapping effect based on Taylor series expansions not only with respect to time, as in the methods discussed in [19] and [23], but also with respect to the initial conditions.

In this paper, we develop a new point of view on the wrapping effect by viewing it from the perspective of the stability of an interval method. To be more specific, we link the wrapping effect to the stability of an interval Taylor series (ITS) method for IVPs for ODEs and interpret the problem of reducing the wrapping effect as one of finding a more stable scheme for advancing the solution.

To study the stability of ITS methods, and thereby the wrapping effect, we employ eigenvalue techniques, which have proven so useful in the study of the stability of point methods (see for example [2] or [8]). To adopt such techniques, we do not investigate ITS methods for the general problem (1), but consider ITS methods for the simpler, linear, constant-coefficient problem

$$y' = By, \quad y(t_0) = y_0, \tag{2}$$

where $B \in \mathbb{R}^{n \times n}$ and $n \geq 2$. (The wrapping effect does not occur on scalar problems.) We also restrict our considerations to Taylor series methods with a constant number of terms and a constant stepsize.

We study the stability of several schemes for reducing the wrapping effect. In particular, we show that Lohner's QR-factorization method, which is currently the most successful general anti-wrapping scheme, can be viewed as a simultaneous iteration (see for example [25]) for computing the eigenvalues of an associated matrix. Furthermore, using some properties of this iteration, we explain how the QR-factorization method improves the stability of the associated ITS method.

The new perspective on the wrapping effect that we develop also suggests how to search for more stable schemes for advancing the solution using eigenvalue techniques. Hence, we can avoid complicated geometric arguments that have been used to date when constructing methods for reducing the wrapping effect.

This paper is organized as follows. In Sect. 2, we introduce notation and definitions that we use later. In Sect. 3, we present a traditional explanation of the wrapping effect in terms of wrapping a parallelepiped, which is not a box in $\mathbb{R}^n$, by a box. Then, in Sect. 4, we view the wrapping effect as a source of instability in an ITS method and describe how the problem of reducing the wrapping effect can be viewed as the problem of finding a more stable scheme for advancing the solution. In Sect. 5, we outline two methods for reducing the wrapping effect, namely the parallelepiped method and Lohner's QR-factorization method [15]. Section 6 shows that the parallelepiped method can be viewed as the power method for computing the largest eigenvalue of the associated matrix for advancing the solution. This section also explains why the parallelepiped method often fails due to nearly-singular matrices and large errors, while Sect. 7 identifies problems for which this method works well. Section 8 explains how Lohner's QR-factorization method can be interpreted as a simultaneous iteration for computing the eigenvalues of the associated matrix for advancing the solution. Using some properties of this iteration and the QR algorithm [6] for computing eigenvalues, we show that for a wide class of problems, the global error of Lohner's QR-factorization method is not much bigger than an upper bound for the global error of the corresponding point Taylor series (PTS) method, for the same stepsize and order. We also propose a simplified QR-factorization method and show that its global error is not much bigger than a bound on the global error of the corresponding PTS method. Finally, in Sect. 9, we summarize the importance of this work.

2   Preliminaries

In subsection 2.1, we introduce interval arithmetic and present some of its properties that we use later. For a more detailed exposition on interval arithmetic, see for example [1] or [16]. In subsection 2.2, we introduce some "noninterval" notation and definitions.

2.1 Interval Arithmetic

The set of intervals on the real line $\mathbb{R}$ is defined by

$$\mathbb{IR} = \{\, [a] = [\underline{a}, \bar{a}] \mid \underline{a}, \bar{a} \in \mathbb{R}, \ \underline{a} \le \bar{a} \,\}.$$

If $\underline{a} = \bar{a}$ then $[a]$ is a *point* interval; if $\underline{a} \ge 0$ then $[a]$ is *nonnegative* ($[a] \ge 0$); and if $\underline{a} = -\bar{a}$ then $[a]$ is *symmetric*. Two intervals $[a]$ and $[b]$ are equal if $\underline{a} = \underline{b}$ and $\bar{a} = \bar{b}$.

Let $[a]$ and $[b] \in \mathbb{IR}$, and $\circ \in \{+, -, *, /\}$. The interval-arithmetic operations are defined [16] by

$$[a] \circ [b] = \{\, x \circ y \mid x \in [a], \ y \in [b] \,\}, \qquad 0 \notin [b] \text{ when } \circ = /,$$

which can be written in the following equivalent form (we omit $*$ in the notation):

$$[a] + [b] = [\underline{a} + \underline{b}, \bar{a} + \bar{b}], \tag{3}$$

$$[a] - [b] = [\underline{a} - \bar{b}, \bar{a} - \underline{b}], \tag{4}$$

$$[a][b] = [\min\{\underline{a}\underline{b}, \underline{a}\bar{b}, \bar{a}\underline{b}, \bar{a}\bar{b}\}, \ \max\{\underline{a}\underline{b}, \underline{a}\bar{b}, \bar{a}\underline{b}, \bar{a}\bar{b}\}], \tag{5}$$

$$[a]/[b] = [\underline{a}, \bar{a}][1/\bar{b}, 1/\underline{b}], \qquad 0 \notin [b]. \tag{6}$$

We have an inclusion of intervals

$$[a] \subseteq [b] \iff \underline{a} \ge \underline{b} \text{ and } \bar{a} \le \bar{b}.$$

The interval-arithmetic operations are *inclusion monotone*. That is, for real intervals $[a]$, $[a_1]$, $[b]$, and $[b_1]$ such that $[a] \subseteq [a_1]$ and $[b] \subseteq [b_1]$,

$$[a] \circ [b] \subseteq [a_1] \circ [b_1], \qquad \circ \in \{+, -, *, /\}.$$

Although interval addition and multiplication are associative, the distributive law does not hold in general [1]. That is, we can easily find three intervals $[a]$, $[b]$, and $[c]$, for which

$$[a]([b] + [c]) \ne [a][b] + [a][c].$$

However, for any three intervals $[a]$, $[b]$, and $[c]$, the subdistributive law

$$[a]([b] + [c]) \subseteq [a][b] + [a][c],$$

does hold. Moreover, there are important cases in which the distributive law

$$[a]([b] + [c]) = [a][b] + [a][c]$$

does hold. For example, it holds if $[b][c] \geq 0$, if $[a]$ is a point interval, or if $[b]$ and $[c]$ are symmetric. In particular, for $\alpha \in \mathbb{R}$, which can be interpreted as the point interval $[\alpha, \alpha]$, and intervals $[a]$ and $[b]$, we have

$$\alpha([a] + [b]) = \alpha[a] + \alpha[b]. \tag{7}$$

For an interval $[a]$, we define the width and the midpoint of $[a]$ as

$$w([a]) = \bar{a} - \underline{a} \quad \text{and} \tag{8}$$
$$m([a]) = (\bar{a} + \underline{a})/2, \tag{9}$$

respectively [16]. Using (3–5) and (8), one can easily show that

$$w([a] \pm [b]) = w([a]) + w([b]) \quad \text{and} \tag{10}$$
$$w(\alpha[a]) = |\alpha|\, w([a]), \tag{11}$$

for any $\alpha \in \mathbb{R}$.

By an *interval vector* we mean a vector with interval components. We denote the set of n-dimensional real interval vectors by $\mathbb{IR}^n$. For $[a] \in \mathbb{IR}^n$, we denote its ith component by $[a_i]$.

Addition of interval vectors is defined component-wise. That is, for $[a] \in \mathbb{IR}^n$ and $[b] \in \mathbb{IR}^n$, the ith component of $[c] = [a] \pm [b]$ is

$$[c_i] = [a_i] \pm [b_i], \quad \text{for } i = 1, 2, \ldots, n. \tag{12}$$

For a real matrix $A \in \mathbb{R}^{n \times n}$ with components a_{ij} and an interval vector $[b] \in \mathbb{IR}^n$ with components $[b_j]$, the components of $[c] = A[b]$ are given by

$$[c_i] = \sum_{j=1}^{n} a_{ij}[b_j], \quad \text{for } i = 1, 2, \ldots, n. \tag{13}$$

For $A \in \mathbb{R}^{n \times n}$, $[b] \in \mathbb{IR}^n$, and $[c] \in \mathbb{IR}^n$, it follows from (7) and (12–13) that

$$A([b] + [c]) = A[b] + A[c]. \tag{14}$$

We define inclusion, width, and midpoint component-wise for interval vectors:

$$[a] \subseteq [b] \iff [a_i] \subseteq [b_i] \quad \text{(for all } i\text{),} \tag{15}$$
$$w([a]) = \left(w([a_1]), w([a_2]), \ldots, w([a_n])\right)^T, \quad \text{and} \tag{16}$$
$$m([a]) = \left(m([a_1]), m([a_2]), \ldots, m([a_n])\right)^T. \tag{17}$$

Let A be a real matrix. We denote by $|A|$ the matrix with components $|a_{ij}|$, for all i and j. Let $A \in \mathbb{R}^{n \times n}$, $[b] \in \mathbb{IR}^n$, and $[c] \in \mathbb{IR}^n$. We can easily obtain from (7–8), (10–13), and (16) that

$$w([b] \pm [c]) = w([b]) + w([c]), \tag{18}$$
$$w(A[b]) = |A|w([b]), \quad \text{and} \tag{19}$$
$$w(A([b] + [c])) = |A|w([b]) + |A|w([c]). \tag{20}$$

Definition 1. Let $A \in \mathbb{R}^{n \times n}$ and $[x] \in \mathbb{IR}^n$. The wrapping of the parallelepiped

$$\{ Ax \mid x \in [x] \}$$

is the box specified by the interval vector $A[x]$, [11]; see Fig. 1.

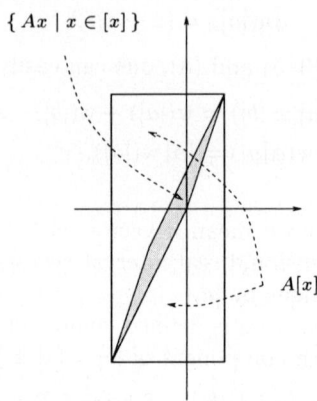

$$\{ Ax \mid x \in [x] \}$$

$A[x]$

Fig. 1. The wrapping of the parallelepiped $\{ Ax \mid x \in [x] \}$ by the box $A[x]$, where $A = \begin{pmatrix} 1 & -2 \\ 3 & -4 \end{pmatrix}$, $[x] = ([0, 1], [0, 1])^T$, and $A[x] = ([-2, 1], [-4, 3])^T$

2.2   Miscellanea

Throughout this paper, we use the infinity norm, which, to reduce the amount of notation, we denote simply by $\| \cdot \|$. By the norm equivalence

$$m\|x\|_Y \leq \|x\| \leq M\|x\|_Y, \tag{21}$$

where $\| \cdot \|_Y$ is any vector norm on $\mathbb{R}^n$, $x \in \mathbb{R}^n$, and m and M are constants that depend on $\| \cdot \|_Y$ but not x, the bounds that we derive in the infinity norm can be translated into any other norm.

Definition 2. For $A \in \mathbb{R}^{n \times n}$, the condition number of A is

$$\text{cond}(A) = \|A\| \, \|A^{-1}\|.$$

Definition 3. The spectral radius of a matrix $A \in \mathbb{R}^{n \times n}$ with eigenvalues $\{ \lambda_i : i = 1, 2, \ldots, n \}$, is

$$\rho(A) = \max_{i=1,2,\ldots,n} \{ |\lambda_i| \} .$$

It is not hard to show that

$$\rho(A) \leq \rho(|A|). \tag{22}$$

Let $\mathrm{Re}(z)$ be the real part of a complex number z.

Definition 4. Let the eigenvalues of B in (2) be $\{\,\lambda_i \,:\, i = 1, 2, \ldots, n\,\}$ and assume $\mathrm{Re}(\lambda_i) < 0$ for all $i = 1, 2, \ldots, n$. An interval method for computing enclosures $[y_j]$ of the true solution of (2) is *asymptotically unstable* if, when applied with constant stepsizes and order to (2),

$$\lim_{j \to \infty} \|\mathrm{w}([y_j])\| = \infty.$$

3   How the Wrapping Effect Arises in Interval Methods for IVPs for ODEs: A Traditional Explanation

In this section, we derive an ITS method for implementing Algorithm II for (2) and show how the wrapping effect arises in this method.

Let

$$T_r(z) = \sum_{i=0}^{r} \frac{z^i}{i!}$$

be the Taylor polynomial of degree r. Since, in this paper, we consider only Taylor series methods with a constant number of terms k and a constant stepsize $h = t_j - t_{j-1}$, we denote

$$T = T_{k-1}(hB) = \sum_{i=0}^{k-1} \frac{(hB)^i}{i!}. \tag{23}$$

If $y(t)$ is the solution of (2), then

$$y(t_j) = Ty(t_{j-1}) + \check{z}_j, \tag{24}$$

where

$$\check{z}_j = \frac{(hB)^k}{k!} \check{y}_{j-1},$$

and the ith component of $\check{y}_{j-1}$ is the ith component of $y(t)$ evaluated at some $\xi_{j-1,i} \in [t_{j-1}, t_j]$.

Suppose that we have computed a tight enclosure $[y_{j-1}]$ on the $(j-1)$st step such that

$$y(t_{j-1}) \in [y_{j-1}]. \tag{25}$$

Suppose also that on the next step, in Algorithm I, we have verified existence and uniqueness of the solution of

$$u' = Bu, \quad u(t_{j-1}) = y_{j-1} \in [y_{j-1}], \tag{26}$$

for all $y_{j-1} \in [y_{j-1}]$ and all $t \in [t_{j-1}, t_j]$, and have computed an a priori enclosure $[\tilde{y}_{j-1}]$ for the solution $u(t)$ of (26) such that

$$u(t) \in [\tilde{y}_{j-1}],$$

for all $t \in [t_{j-1}, t_j]$ and all $y_{j-1} \in [y_{j-1}]$, [19,20]. Then, $y(t) \in [\tilde{y}_{j-1}]$ for all $t \in [t_{j-1}, t_j]$. Hence, $\check{y}_{j-1} \in [\tilde{y}_{j-1}]$, and

$$\check{z}_j = \frac{(hB)^k}{k!}\check{y}_{j-1} \in \frac{(hB)^k}{k!}[\tilde{y}_{j-1}] \equiv [z_j]. \tag{27}$$

Using (24–25) and (27), we derive

$$\begin{aligned}
y(t_j) &= Ty(t_{j-1}) + \check{z}_j \\
&\in \{\, Ty_{j-1} + [z_j] \mid y_{j-1} \in [y_{j-1}]\,\} \tag{28} \\
&\subseteq T[y_{j-1}] + [z_j] \equiv [y_j]. \tag{29}
\end{aligned}$$

We wish to compute the set in (28), but, in practice, we compute instead the more tractable interval vector $[y_j]$ in (29). Thus, we obtain the simple interval Taylor series scheme

$$[y_j] = T[y_{j-1}] + [z_j] \tag{30}$$

for implementing Algorithm II.

We refer to this scheme as the *direct method* [17]. Here, $[z_j]$ can be viewed as an enclosure of the local error of the direct method, and $w([z_j])$ can be interpreted as the size of the bound on this error. We can keep it small by decreasing the stepsize or increasing the order. However, $T[y_{j-1}]$ can be a large overestimation of the set $\{\, Ty_{j-1} \mid y_{j-1} \in [y_{j-1}]\,\}$. The reason for this is that the set $\{\, Ty_{j-1} \mid y_{j-1} \in [y_{j-1}]\,\}$ is not generally a box, but it is enclosed, or wrapped, by the box $T[y_{j-1}]$; see Fig. 1. Therefore, if we compute on each step products of the form $T[y_{j-1}]$, we may incur a wrapping on each step, resulting in unacceptably wide bounds for the solution of (2).

This can also be seen from the following considerations (see also [17] or [23]). Using (14), we derive from (30) that

$$\begin{aligned}
[y_j] &= T[y_{j-1}] + [z_j] = T(T[y_{j-2}] + [z_{j-1}]) + [z_j] \\
&= T(T[y_{j-2}]) + T[z_{j-1}] + [z_j]
\end{aligned}$$

$$\vdots$$

$$\begin{aligned}
&= T^j y_0 + \underbrace{T(\cdots T(T[z_1])\cdots)}_{(j-1)\ T\text{'s}} + \cdots \\
&\quad + T(T[z_{j-2}]) + T[z_{j-1}] + [z_j].
\end{aligned} \tag{31}$$

Consider how $[z_1]$ propagates in the direct method. That is, consider the term $T(\cdots T(T[z_1])\cdots)$ in (31). There is usually one wrapping in computing $T[z_1]$, two wrappings in computing $T(T[z_1])$, and $(j-1)$ wrappings in computing $T\big(\cdots T(T[z_1])\cdots\big)$. Thus, the excess caused by a wrapping on each step accumulates, possibly causing an unacceptably large wrapping effect in propagating $[z_1]$. Figure 2 illustrates the overestimation of the set $\big\{\, A^2x \mid x \in [x] \,\big\}$, when it is wrapped twice in computing $A(A[x])$, where A and $[x]$ are given in Fig. 1.

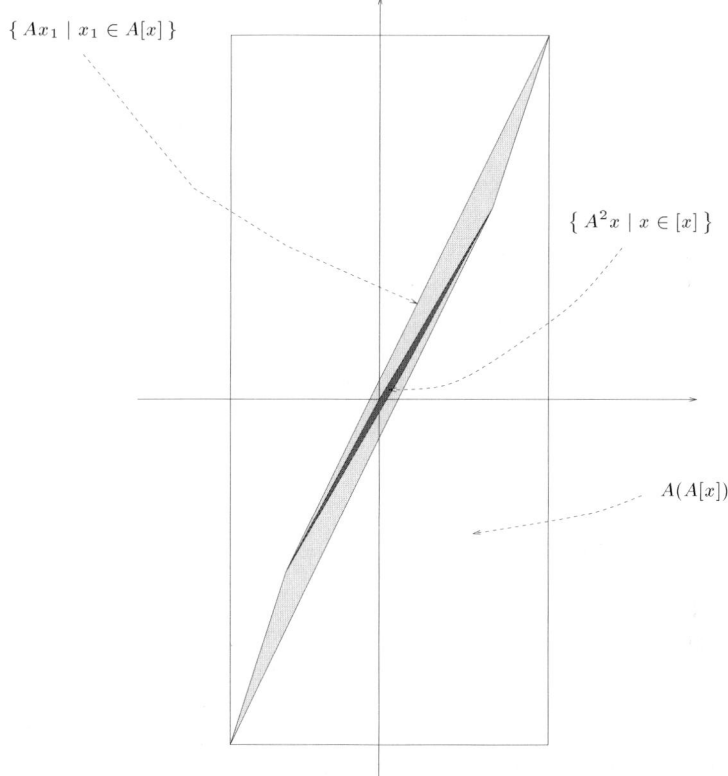

Fig. 2. The box $A[x]$ from Fig. 1 is mapped into $\big\{\, Ax_1 \mid x_1 \in A[x] \,\big\}$, which is wrapped by $A(A[x])$

Below, we present further insights into the wrapping effect in the direct method. From (24) and (27),

$$
\begin{aligned}
y(t_j) &= Ty(t_{j-1}) + \check{z}_j \\
&= T^j y_0 + T^{j-1}\check{z}_1 + \cdots + T\check{z}_{j-1} + \check{z}_j \\
&\in T^j y_0 + T^{j-1}[z_1] + \cdots + T[z_{j-1}] + [z_j] \equiv [v_j].
\end{aligned}
\tag{32}
$$

Note that we are interested in computing an easily-representable enclosure of the set

$$\{ T^{j-1} z_1 \mid z_1 \in [z_1] \} \tag{33}$$

such that this enclosure contains as little overestimation of (33) as possible. If we enclose (33) by $T^{j-1}[z_1]$, we introduce at most one wrapping, while in (31), we can have $(j-1)$ wrappings when enclosing (33) by $T(\cdots T(T[z_1])\cdots)$. Note also that if $|T|^{j-1} \gg |T^{j-1}|$, as often happens, then typically

$$w\Big(T(\cdots T(T[z_1])\cdots)\Big) = |T|^{j-1} w([z_1])$$
$$\gg |T^{j-1}| w([z_1]) = w\big(T^{j-1}[z_1]\big).$$

Since in this paper we consider only problems with point initial conditions, we can interpret $w([y_j])$ as the size of the bound on the global error of the direct method on the jth step. Taking widths in (31) and (32) and using (18–19), we derive

$$w([y_j]) = |T| w([y_{j-1}]) + w([z_j])$$
$$= \sum_{i=1}^{j} |T|^{j-i} w([z_i]), \tag{34}$$

while

$$w([v_j]) = \sum_{i=1}^{j} |T^{j-i}| w([z_i]). \tag{35}$$

Obviously, if $|T|^{j-i} \gg |T^{j-i}|$ for some $i \in \{1, \ldots, j-1\}$, then $w([y_j])$ may be much larger than $w([v_j])$. Therefore, because of the wrapping effect, or $|T|^{j-i} \gg |T^{j-i}|$, the global error $w([y_j])$ of the direct method often becomes unacceptably large.

The reason why we do not use a scheme of the form (32) in practice is that it can be significantly more expensive than (30), because we have to store the necessary powers of T or recompute them on each step. Moreover, if the ITS method uses a stepsize control, it follows from (23) that the $T_j = T_{k-1}(h_j B)$ are normally not all equal. Hence, we have to store all matrix products of the form $T_1, T_1 T_2, \ldots, T_1 T_2 \cdots T_j$.

4 The Wrapping Effect as a Source of Instability in Interval Methods for IVPs for ODEs

Consider now the point Taylor series (PTS) method given by

$$u_j = T u_{j-1}, \quad u_0 = y_0. \tag{36}$$

Denote by

$$\delta_j = y(t_j) - u_j \tag{37}$$

the global error of this method at t_j, $j = 1, 2, \ldots$. Then, from (24) and (36–37), we can express the global error of the PTS method at t_j using the local truncation errors $\check{z}_i$, $i = 1, 2, \ldots, j$, as

$$\delta_j = T\delta_{j-1} + \check{z}_j$$
$$= \sum_{i=1}^{j} T^{j-i} \check{z}_i. \tag{38}$$

In this section, we investigate how $\|w([y_j])\|$ compares to $\|\delta_j\|$, where $w([y_j])$ is given in (34). An important observation is that the global error in the PTS method propagates with T, while the global error in the direct method propagates with $|T|$. Assuming that T and $|T|$ are diagonalizable (to simplify the analysis), we show that

- if $\rho(T) = \rho(|T|)$, then the bounds on $\|w([y_j])\|$ and $\|\delta_j\|$ may not be much different, but
- if $\rho(T) < \rho(|T|)$, then $\|w([y_j])\|$ may be much larger than $\|\delta_j\|$.

Recall that $\rho(T) \le \rho(|T|)$; see subsection 2.2.

Let V_T and $V_{|T|}$ be the matrices that diagonalize T and $|T|$, respectively. Let D_T and $D_{|T|}$ be the associated diagonal matrices. Then

$$T = V_T^{-1} D_T V_T \quad \text{and} \tag{39}$$
$$|T| = V_{|T|}^{-1} D_{|T|} V_{|T|}. \tag{40}$$

Assume also that

$$\|\check{z}_i\| \le \delta \quad \text{and} \tag{41}$$
$$\|w([z_i])\| \le \beta, \quad i = 1, 2, \ldots, j, \tag{42}$$

for some constants δ and β, respectively.

We derive from (38–39) and (41) that

$$\|\delta_j\| \le \delta \left(\text{cond}(V_T) \sum_{i=1}^{j-1} (\rho(T))^i + 1 \right), \tag{43}$$

and we derive from (34), (40), and (42) that

$$\|w([y_j])\| \le \beta \left(\text{cond}(V_{|T|}) \sum_{i=1}^{j-1} (\rho(|T|))^i + 1 \right). \tag{44}$$

Obviously, if $\rho(T) = \rho(|T|)$, the upper bound for $\|w([y_j])\|$ is approximately a constant times the upper bound for $\|\delta_j\|$.

Example 1. Consider

$$y' = By = \begin{pmatrix} -2 & 1 \\ 1 & -4 \end{pmatrix} y, \quad y(0) = (1,1)^T. \tag{45}$$

Since this is a quasi-monotone problem, the wrapping effect should not cause a significant overestimation in propagating the error in the direct method on this problem [22].

In Figs. 3(a) and 3(b), we plot $\rho(T_{k-1}(hB))$ and $\rho(|T_{k-1}(hB)|)$ versus h for $k = 10$ and $k = 30$.[1] For the ranges of stepsize considered,

$$\rho(T_{k-1}(hB)) \approx \rho(|T_{k-1}(hB)|). \tag{46}$$

Note that the eigenvalues of B are (approximately) -1.5858 and -4.4142.

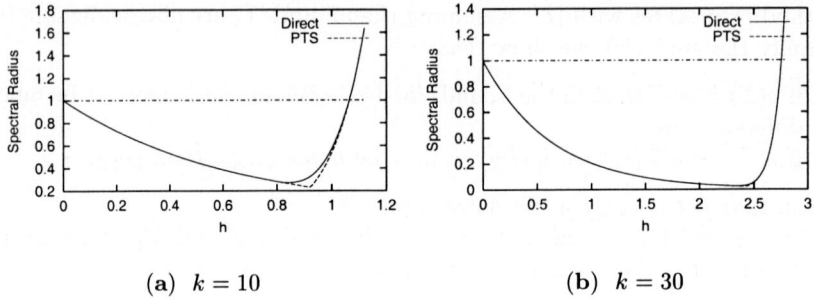

(a) $k = 10$ (b) $k = 30$

Fig. 3. Graphs of Direct $\equiv \rho(|T_{k-1}(hB)|)$ and PTS $\equiv \rho(T_{k-1}(hB))$ versus h for $k = 10$ and $k = 30$ for Example 1

Using (46) together with (43) and (44), we see that the bound on $\|\mathrm{w}([y_j])\|$ cannot be much bigger than the bound on $\|\delta_j\|$. Therefore, if these bounds are fairly tight, the wrapping effect does not cause a significant overestimation of the global error in the direct method on this problem.

To support this claim, we integrated (45) for $t \in [0, 20]$ with the PTS method and the direct method.[2] In both cases, we used $h = 0.1$, and $k = 30$. In Fig. 4, we plot the base-10 logarithm of the global error in these two methods versus $t_j = jh = 0.1j$.

Now, we show that, if $\rho(T) < \rho(|T|)$, $\|\mathrm{w}([y_j])\|$ may be much bigger than $\|\delta_j\|$. Denote the eigenvalues of T by $\gamma_1, \gamma_2, \ldots, \gamma_n$ and assume

$$|\gamma_1| > |\gamma_2| \geq |\gamma_3| \geq \cdots \geq |\gamma_{n-1}| \geq |\gamma_n|. \tag{47}$$

---

[1] The computations are performed in Matlab. The plots in this paper are produced with Gnuplot.

[2] The PTS method is implemented in Matlab. The interval methods in this paper are implemented in the VNODE [17] package.

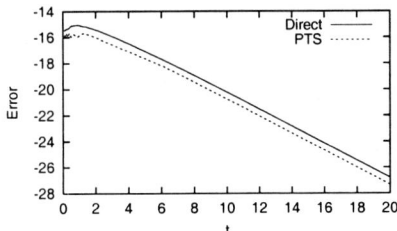

Fig. 4. $\log_{10} \|w([y_j])\|$ and $\log_{10} \|\delta_j\|$ versus t_j for Example 1

Assume also for simplicity that T has a full set of linearly independent eigenvectors $v_1, v_2, \ldots, v_n$, where $Tv_i = \gamma_i v_i$ for $i = 1, 2, \ldots, n$. We can express $\check{z}_1$ as a linear combination of the eigenvectors of T,

$$\check{z}_1 = \sum_{i=1}^{n} c_i v_i,$$

where $c_i \in \mathbb{R}$, $i = 1, 2, \ldots, n$. If $c_1 \neq 0$, then

$$T^{j-1}\check{z}_1 = \gamma_1^{j-1}\left(c_1 v_1 + \left(\frac{\gamma_2}{\gamma_1}\right)^{j-1} c_2 v_2 + \cdots + \left(\frac{\gamma_n}{\gamma_1}\right)^{j-1} c_n v_n\right)$$

$$\rightarrow \gamma_1^{j-1} c_1 v_1, \quad \text{as } j \rightarrow \infty.$$

Thus, for sufficiently large j,

$$T^{j-1}\check{z}_1 \approx \left(\rho(T)\right)^{j-1} c_1 v_1. \tag{48}$$

(In practice, $T^{j-1}\check{z}_1$ is computed in floating-point arithmetic with an iteration of the form $p_l = Tp_{l-1}$, $p_1 = \check{z}_1$, $l = 2, 3, \ldots, j$. In general, if $c_1 = 0$, then roundoff errors in this iteration eventually introduce a nonzero component in the direction of v_1.)

Similarly, denote the eigenvalues of $|T|$ by $\nu_1, \nu_2, \ldots, \nu_n$ and assume

$$|\nu_1| > |\nu_2| \geq |\nu_3| \geq \cdots \geq |\nu_{n-1}| \geq |\nu_n|.$$

Assume also for simplicity that $|T|$ has a full set of linearly independent eigenvectors $w_1, w_2, \ldots, w_n$, where $|T|w_i = \nu_i w_i$ for $i = 1, 2, \ldots, n$. We can express $w([z_1])$ as a linear combination of the eigenvectors of $|T|$,

$$w([z_1]) = \sum_{i=1}^{n} d_i w_i,$$

where $d_i \in \mathbb{R}$, $i = 1, 2, \ldots, n$. If $d_1 \neq 0$,

$$T^{j-1}w([z_1]) = \nu_1^{j-1}\left(d_1 w_1 + \left(\frac{\nu_2}{\nu_1}\right)^{j-1} d_2 w_2 + \cdots + \left(\frac{\nu_n}{\nu_1}\right)^{j-1} d_n w_n\right)$$

$$\rightarrow \nu_1^{j-1} d_1 w_1, \quad \text{as } j \rightarrow \infty.$$

Thus, for sufficiently large j,

$$|T|^{j-1}\mathrm{w}([z_1]) \approx \left(\rho(|T|)\right)^{j-1} d_1 w_1. \tag{49}$$

From (48) and (49), we see that, for sufficiently large j,

$$\frac{\||T|^{j-1}\mathrm{w}([z_1])\|}{\|T^{j-1}\check{z}_1\|} \approx \left(\frac{\rho(|T|)}{\rho(T)}\right)^{j-1} \frac{\|d_1 w_1\|}{\|c_1 v_1\|}. \tag{50}$$

Hence, if $\rho(T) < \rho(|T|)$, $\||T|^{j-1}\mathrm{w}([z_1])\|$ may be much bigger than $\|T^{j-1}z_1\|$. This implies that the global error bound for the direct method may be significantly larger than the global error for the PTS method, as j becomes sufficiently large.

For example, assume $\rho(T) < 1$. Then the PTS method computes approximations for which $\lim_{j\to\infty} \|u_j\| = 0$. However, $\rho(T) < 1$ does not imply $\rho(|T|) < 1$. Therefore, it is possible to have $\rho(T) < 1$, but $\rho(|T|) > 1$. Then from (49),

$$\||T|^{j-1}\mathrm{w}([z_1])\| \approx \left(\rho(|T|)\right)^{j-1}\|d_1 v_1\| \to \infty, \quad \text{as } j \to \infty,$$

and our bound on $\|\mathrm{w}([y_j])\| \to \infty$, as $j \to \infty$. That is, the direct method may be asymptotically unstable.

Example 2. The problem

$$y' = By = \begin{pmatrix} 1 & -2 \\ 3 & -4 \end{pmatrix} y, \quad y(0) = (1, -1)^T \tag{51}$$

is an example for which the true solution components converge rapidly to zero, but the wrapping effect can cause significant overestimation of the error [15,17]. Here, the eigenvalues of B are -1 and -2.

Let us look at how our analysis of the direct method applies to (51). For this purpose, we plot in Figs. 5(a) and 5(b) $\rho\left(T_{k-1}(hB)\right)$ and $\rho\left(|T_{k-1}(hB)|\right)$ versus h for $k = 10$ and $k = 30$.

Consider the case $k = 30$. For $h \lesssim 6.09$, $\rho(T_{29}(hB)) < 1$, and the PTS method computes approximations for which $\lim_{j\to\infty} \|u_j\| = 0$. However, for $h \lesssim 1.32$ and $h \gtrsim 5.7$, $\rho\left(|T_{29}(hB)|\right) > 1$, and the bounds for the direct method blow-up, suggesting that $\lim_{j\to\infty} \|\mathrm{w}([y_j])\| = \infty$. To support this conjecture, we used the direct method with $h = 0.2$ and $k = 30$ to compute $\|\mathrm{w}([y_j])\| \approx 5.2 \times 10^5$ at $t = 40$.

It is interesting to note that for $1.32 \lesssim h \lesssim 5.7$, the direct method may not suffer significantly from the wrapping effect. For example, with $h = 1.5$, we computed with the direct method $\|\mathrm{w}([y_j])\| \approx 2.6 \times 10^{-14}$ at $t = 300$. Therefore, even if the direct method suffers from the wrapping effect for a particular problem and range of stepsizes, there might be another range of stepsizes for which the wrapping effect is not a significant difficulty for this method.

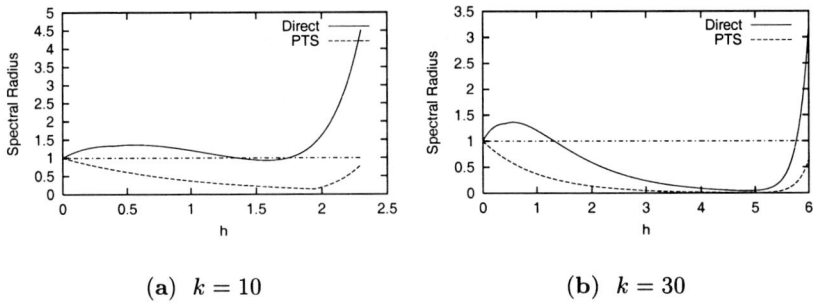

(a) $k = 10$ (b) $k = 30$

Fig. 5. Graphs of Direct $\equiv \rho\big(|T_{k-1}(hB)|\big)$ and PTS $\equiv \rho\big(T_{k-1}(hB)\big)$ versus h for $k = 10$ and $k = 30$ for Example 2

Intuitively, we have associated the wrapping effect in an ITS method with its global error being significantly larger than the global error of the corresponding PTS method. We showed that when $\rho(T) = \rho(|T|)$, the wrapping effect is not a serious difficulty for the direct method. However, when $\rho(T) < \rho(|T|)$, the direct method may suffer significantly from the wrapping effect. In particular, if $\rho(T) < 1$ and $\rho(|T|) > 1$, the PTS method is stable, but the associated ITS method is likely asymptotically unstable. To reduce the wrapping effect, or improve the stability of an ITS method, we have to find a more stable scheme for advancing the solution.

5 The Parallelepiped and Lohner's QR-Factorization Methods

In this section, we describe and briefly discuss two methods, namely the parallelepiped and Lohner's QR-factorization methods, for reducing the wrapping effect.

Let

$$A_0 = I,$$
$$\hat{y}_0 = y_0,$$
$$s_j = \mathrm{m}([z_j]), \quad \text{and}$$
$$\hat{y}_j = T\hat{y}_{j-1} + s_j \quad (j \geq 1),$$

where I is the identity matrix, $[z_j]$ is defined in (27), and T is defined in (23). Let also $[r_0]$ be an interval vector with each component equal to $[0,0]$. We compute (for $j \geq 1$)

$$[y_j] = T\hat{y}_{j-1} + (T A_{j-1})\,[r_{j-1}] + [z_j] \tag{52}$$

and propagate for the next step

$$[r_j] = \left(A_j^{-1} T A_{j-1}\right)[r_{j-1}] + A_j^{-1}\left([z_j] - s_j\right), \tag{53}$$

where $A_j \in \mathbb{R}^{n \times n}$ is nonsingular. We discuss in the next two subsections how A_j may be chosen.

5.1   The Parallelepiped Method

If $A_j = T A_{j-1}$, we obtain the parallelepiped method [15]. This choice for A_j has also been suggested in [5,13,16]. As Lohner points out [15], the parallelepiped method works particularly well for pure rotations, but breaks down when B (in (2)) has two eigenvalues with different real components. In this case, after a sufficiently long integration, A_j becomes nearly singular.

Using eigenvalue techniques, we show in Sect. 6 that the matrices A_j often become nearly singular and that the global error of this method is often much larger than the global error of the corresponding PTS method. In Sect. 7, we show that the parallelepiped method works well when all the eigenvalues of T are equal in magnitude, as observed by Lohner. In this case, the global error associated with the parallelepiped method is not much bigger than the bound for the global error of the corresponding PTS method.

5.2   Lohner's QR-Factorization Method

Lohner proposes [15] the following two schemes for selecting the transformation matrices A_j, $j \geq 1$ (in both schemes $A_0 = Q_0 = I$):

$$\begin{aligned} T A_{j-1} &= Q_j R_j, \\ A_j &= Q_j \end{aligned} \tag{54}$$

and

$$\begin{aligned} T A_{j-1} P_j &= Q_j R_j, \\ A_j &= Q_j. \end{aligned} \tag{55}$$

Here, Q_j is orthogonal, R_j is upper triangular, and P_j is a permutation matrix chosen as follows.

Let $l_{j-1,i}$ be the length (in the Euclidean norm) of the ith column of $T A_{j-1}$ in (55) and $\left(\mathrm{w}([r_{j-1}])\right)_i$ be the ith component of $\mathrm{w}([r_{j-1}])$. Then P_j is such that the components of

$$\left(l_{j-1,1}\left(\mathrm{w}([r_{j-1}])\right)_1,\ l_{j-1,2}\left(\mathrm{w}([r_{j-1}])\right)_2,\ \ldots,\ l_{j-1,n}\left(\mathrm{w}([r_{j-1}])\right)_n\right) P_j$$

are in non-increasing order.

We refer to (52–53) with A_j's chosen from either (54) or (55) as Lohner's *QR-factorization method*. However, to distinguish between the methods, we also refer to (52–53) with A_j's from (54) as the *QR method* and to (52–53) with A_j's from (55) as the *QR-P method*.

In this section, we outline a geometric explanation of why the QR-factorization method is successful at reducing the wrapping effect. For more details see [15] or [19]. In Sect. 8, we use eigenvalue techniques to show how the QR method improves the stability of the direct and parallelepiped methods.

The interval vector $[r_j]$ in (53) can be interpreted as an enclosure of the global error (at t_j) that is propagated to the next step. Obviously, we are interested in keeping the overestimation in $[r_j]$ as small as possible.

Since $A_j = Q_j$ is orthogonal, $A_j^{-1} = Q_j^T$. Thus, we avoid the problem of having to compute the inverse of a nearly singular matrix. In addition, since Q_j is orthogonal,

$$|A_j^{-1}|\mathrm{w}([z_j]) = |Q_j^T|\mathrm{w}([z_j])$$

is not much bigger than $\mathrm{w}([z_j])$, which we keep sufficiently small (by reducing the stepsize or increasing the order).

The set

$$\{\,(TA_{j-1})r_{j-1} = (TQ_{j-1})r_{j-1} \mid r_{j-1} \in [r_j]\,\} \tag{56}$$

is generally a parallelepiped in $\mathbb{R}^n$. The set

$$\{\,(A_j^{-1}TA_{j-1})r_{j-1} = (Q_j^T TQ_{j-1})r_{j-1} \mid r_{j-1} \in [r_j]\,\} \tag{57}$$

is the parallelepiped in (56) rotated such that one of its edges is parallel to an axis in the standard coordinate system for $\mathbb{R}^n$. Hence, the box $(A_j^{-1}TA_{j-1})[r_{j-1}]$ that encloses the parallelepiped (57) is such that one of the edges of $(A_j^{-1}TA_{j-1})[r_{j-1}]$ is parallel to an edge of (57). Intuitively, if we wrap a parallelepiped by a box that matches one of the edges of this parallelepiped, we should normally introduce a smaller overestimation than if we do not match any of its edges. Furthermore, if we match the longest edge of the enclosed parallelepiped, as we do in the QR-P method, we should have a smaller overestimation than if we match an edge that is not the longest one.

6   Why the Parallelepiped Method Often Fails

Since $A_0 = I$ and $A_j = TA_{j-1}$ for $j \geq 1$, $A_j = T^j$. Substituting T^j for A_j in (52) and (53), we write the parallelepiped method as

$$[y_j] = T\hat{y}_{j-1} + T^j[r_{j-1}] + [z_j] \quad \text{and} \tag{58}$$

$$[r_j] = [r_{j-1}] + T^{-j}\,([z_j] - s_j)\,. \tag{59}$$

Obviously, we require that T is nonsingular, which is the case when the stepsize, h, is sufficiently small.

In the next two subsections, we show that, as we integrate with this method,

- the matrices $A_j = T^j$ often become nearly singular, and
- $w([y_j])$ often grows substantially, even when $\rho(T) < 1$. In this case, the PTS method computes approximations for which $\lim_{j \to \infty} \|u_j\| = 0$.

6.1   Nearly Singular Matrices

As in Sect. 4, we denote the eigenvalues of T by $\gamma_1, \gamma_2, \ldots, \gamma_n$ and assume again (to simplify the analysis) that T has a set of linearly independent eigenvectors $v_1, v_2, \ldots, v_n$, where $Tv_i = \gamma_i v_i$. Assume also that

$$|\gamma_1| > |\gamma_2| \geq |\gamma_3| \geq \cdots \geq |\gamma_{n-1}| \geq |\gamma_n| > 0, \tag{60}$$

where the last inequality, $|\gamma_n| > 0$, follows from our assumption that T is nonsingular.

We can view the ith column of T^j as the vector obtained from the power method applied to the unit vector e_i (for which the ith component is one and the rest are zero). For each $i = 1, 2, \ldots, n$, we can express e_i as a linear combination of the eigenvectors of T,

$$e_i = \sum_{j=1}^{n} c_{i,j} v_j, \quad c_{i,j} \in \mathbb{R}. \tag{61}$$

If $c_{i,1} \neq 0$ for all i's, then

$$T^j e_i = \gamma_1^j \left(c_{i,1} v_1 + \left(\frac{\gamma_2}{\gamma_1}\right)^j c_{i,2} v_2 + \cdots + \left(\frac{\gamma_n}{\gamma_1}\right)^j c_{i,n} v_n \right)$$
$$\to \gamma_1^j c_{i,1} v_1, \quad \text{as } j \to \infty. \tag{62}$$

That is, for sufficiently large j, each column of T^j becomes nearly parallel to the eigenvector v_1.

In practice, we compute T^j in floating-point arithmetic using the iteration $A_j = TA_{j-1}$, $A_0 = I$. If $c_{i,1} = 0$ for some $i = 1, 2, \ldots, n$, the ith column of the computed T^j (in floating-point arithmetic) usually becomes nearly parallel to v_1 since roundoff errors eventually introduce a nonzero component in the direction of v_1. Note, though, that this is not always the case. For example, if T is diagonal, then roundoff errors do not introduce nonzero components in the direction of v_1 for the columns $T^j e_i$, where $i = 2, 3, \ldots, n$.

Since the computed T^j matrices become closer and closer to singular, as j increases, computing tight enclosures for T^{-j} becomes impossible, and the parallelepiped method breaks down. In addition, in floating-point arithmetic, the columns of T^j may overflow, if $|\gamma_1| > 1$, or underflow to zero, if $|\gamma_1| < 1$.

6.2   Large Errors

In this subsection, we assume, in addition to (60), that

$$|\gamma_{n-1}| > |\gamma_n|$$

holds. Let

$$G = \frac{|hB|^k}{k!}.$$

Since $[\tilde{y}_{j-1}]$ must contain $[y_{j-1}]$,

$$[z_j] = \frac{(hB)^k}{k!}[\tilde{y}_{j-1}] \supseteq \frac{(hB)^k}{k!}[y_{j-1}],$$

and

$$w([z_j]) = \frac{|hB|^k}{k!}w([\tilde{y}_{j-1}]) \geq \frac{|hB|^k}{k!}w([y_{j-1}]) = G\,w([y_{j-1}]). \qquad (63)$$

Taking widths in (58–59) and using (18–19), we obtain[3] that

$$w([y_j]) = |T^j|w([r_{j-1}]) + w([z_j]) \geq |T^j|w([r_{j-1}]) \quad \text{and} \qquad (64)$$
$$w([r_j]) = w([r_{j-1}]) + |T^{-j}|w([z_j]) \geq |T^{-j}|w([z_j]). \qquad (65)$$

Let j be an odd number. Using (63–65), we derive

$$w([y_j]) \geq |T^j|w([r_{j-1}]) \geq |T^j||T^{-(j-1)}|\,w([z_{j-1}])$$
$$\geq |T^j||T^{-(j-1)}|\,G\,w([y_{j-2}])$$
$$\vdots \qquad\qquad\qquad (66)$$
$$\geq \left(|T^j||T^{-(j-1)}|\,G\right)\left(|T^{j-2}||T^{-(j-3)}|\,G\right)\cdots$$
$$\left(|T^3||T^{-2}|\,G\right)w([z_1]).$$

Denote by V_1 the $n \times n$ matrix with columns v_1 and by V_n the $n \times n$ matrix with columns v_n. Let

$$C_1 = \text{diag}(c_{1,1}, c_{2,1}, \cdots, c_{n,1}) \quad \text{and} \quad C_n = \text{diag}(c_{1,n}, c_{2,n}, \cdots, c_{n,n}),$$

where $c_{i,1}$ and $c_{i,n}$, $i = 1, 2, \ldots, n$, are the coefficients from (61). We now assume not only that $c_{i,1} \neq 0$ for all $i = 1, 2, \ldots, n$ (as in the previous subsection) but also that $c_{i,n} \neq 0$ for all $i = 1, 2, \ldots, n$. Now let

$$F = |\gamma_n||V_1 C_1||V_n C_n|. \qquad (67)$$

We show that $|T^j||T^{-(j-1)}|$ in (66) becomes closer and closer to $|\gamma_1/\gamma_n|^j\,F$, as we integrate with the parallelepiped method.

For sufficiently large j, it follows from (62) that

$$|T^j| \approx |\gamma_1^j||V_1 C_1|. \qquad (68)$$

---

[3] Throughout the rest of this paper, we shall not refer to properties from Sect. 2 when we use them.

Similarly,

$$|T^{-(j-1)}| \approx \frac{1}{|\gamma_n^{j-1}|}|V_n C_n|. \tag{69}$$

From (67–69),

$$|T^j||T^{-(j-1)}| \approx \left|\frac{\gamma_1}{\gamma_n}\right|^j |\gamma_n||V_1 C_1||V_n C_n| = \left|\frac{\gamma_1}{\gamma_n}\right|^j F. \tag{70}$$

From (66) and (70), we obtain for sufficiently large j that

$$\mathrm{w}([y_j]) \geq |T^j||T^{-(j-1)}|\, G\, \mathrm{w}([y_{j-2}])$$

$$\approx \left|\frac{\gamma_1}{\gamma_n}\right|^j F\, G\, \mathrm{w}([y_{j-2}]). \tag{71}$$

Since we assumed $|\gamma_1| > |\gamma_n|$, $|\gamma_1/\gamma_n|^j \gg 1$ for j sufficiently large. Therefore, $\mathrm{w}([y_j])$ can be much larger than $\mathrm{w}([y_{j-2}])$. Hence, as we integrate, we obtain enclosures with increasing widths. Thus, when $|\gamma_1| > |\gamma_2| \geq \cdots \geq |\gamma_{n-1}| > |\gamma_n| > 0$, the parallelepiped method is likely asymptotically unstable. Note that this happens even when $\rho(T) = |\gamma_1| < 1$, in which case the PTS method is stable.

Example 3. Consider

$$y' = By = \begin{pmatrix} 1 & -2 \\ 3 & -4 \end{pmatrix} y, \quad y(0) = (1, -1)^T. \tag{72}$$

As noted in Example 2, the eigenvalues of B are -1 and -2. Let $h = 0.1$ and $k = 16$. The eigenvalues of $T = T_{15}(0.1B)$ are $\gamma_1 \approx 0.9048$ and $\gamma_2 \approx 0.8187$. Since $\gamma_1 > \gamma_2 > 0$, from the analysis in this subsection, we expect, for a sufficiently long integration of this problem, the parallelepiped to produce a large error and to break down.

We integrated (72) with the parallelepiped method with $h = 0.1$ and $k = 16$. After computing $[y_{134}]$, our solver could not continue since it could not enclose the inverse of $A_{134} = T^{134}$. Indeed, for $j = 134$,

$$T^{134} \approx 10^{-6} \begin{pmatrix} 4.545428 & -3.030284 \\ 4.545425 & -3.030281 \end{pmatrix}$$

is close to singular with $\mathrm{cond}(T^{134}) \approx 1.98 \times 10^7$.

Furthermore, the bound on the global error of the parallelepiped method at $t = 13.4$ is $\|\mathrm{w}([y_{134}])\| \approx 1.98 \times 10^2$, while the bound on the global error of the PTS method at $t = 13.4$ is $\|\delta_{134}\| \approx 1.8 \times 10^{-18}$. Note that, for the exact solution, $\|y(13.4)\| \approx 7.6 \times 10^{-6}$. To illustrate the growth in the error bounds in the parallelepiped method on this problem, we plot in Fig. 6 $\log_{10} \|\mathrm{w}([y_j])\|$ versus $t_j = jh = 0.1j$.

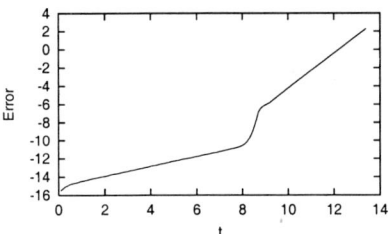

Fig. 6. Plot of $\log_{10} \|\mathrm{w}([y_j])\|$ versus t_j for the parallelepiped method on Example 3

7   When Does the Parallelepiped Method Work Well

Assume that the eigenvalues of T satisfy

$$|\gamma_1| \geq |\gamma_2| \geq |\gamma_3| \geq \cdots \geq |\gamma_{n-1}| \geq |\gamma_n| > 0$$

and that T can be diagonalized. Let V_T be the matrix that diagonalizes T and $D_T = \mathrm{diag}(\gamma_1, \gamma_2, \gamma_3, \ldots, \gamma_n)$. We derive an upper bound for the global error of the parallelepiped method and show that, if $|\gamma_1| = |\gamma_n|$, this bound is not much bigger than the bound in (43) for the global error of the corresponding PTS method.

Using (59) and then (58), we derive

$$\|\mathrm{w}([r_{j-1}])\| \leq \|\mathrm{w}([r_{j-2}])\| + \||T^{-(j-1)}|\| \|\mathrm{w}([z_{j-1}])\|$$
$$\leq \sum_{i=1}^{j-1} \||T^{-i}|\| \|\mathrm{w}([z_i])\| \tag{73}$$

and

$$\|\mathrm{w}([y_j])\| \leq \||T^j|\| \|\mathrm{w}([r_{j-1}])\| + \|\mathrm{w}([z_j])\|$$
$$\leq \||T^j|\| \sum_{i=1}^{j-1} \||T^{-i}|\| \|\mathrm{w}([z_i])\| + \|\mathrm{w}([z_j])\|. \tag{74}$$

Substituting

$$\||T^j|\| = \|T^j\| \leq \mathrm{cond}(V_T) |\gamma_1|^j \quad \text{and}$$
$$\||T^{-i}|\| = \|T^{-i}\| \leq \mathrm{cond}(V_T) \frac{1}{|\gamma_n|^i}$$

into (74) and assuming that

$$\|\mathrm{w}([z_i])\| \leq \beta, \quad \text{for } i = 1, 2, \ldots, j,$$

we obtain that

$$\|\mathrm{w}([y_j])\| \leq \beta \left((\mathrm{cond}(V_T))^2 \sum_{i=1}^{j-1} \frac{|\gamma_1|^j}{|\gamma_n|^i} + 1 \right).$$

If $\rho(T) = |\gamma_1| = |\gamma_n|$, then

$$\|w([y_j])\| \leq \beta \left((\text{cond}(V_T))^2 \sum_{i=1}^{j-1} (\rho(T))^i + 1 \right). \tag{75}$$

Comparing the bounds in (75) and (43), we see that, if $|\gamma_1| = |\gamma_n|$, the global error of the parallelepiped method is not much bigger than the bound for the global error of the corresponding point method. Therefore, the parallelepiped method works well for problems for which the eigenvalues of T are of the same magnitude. Such problems arise, for example, when the matrix B is skew symmetric, $B = -B^T$. Then, for a given h,

$$e^{hB} \left(e^{hB} \right)^T = e^{hB} e^{hB^T} = e^{hB} e^{-hB} = I,$$

which implies that e^{hB} is orthogonal. Hence, each eigenvalue of e^{hB} is of magnitude one. Since $T = T_{k-1}(hB)$ is normally a good approximation to e^{hB}, the eigenvalues of T are approximately of magnitude one.

Example 4. Consider the IVP problem

$$y' = By = \begin{pmatrix} 0 & 1 \\ -1 & 0 \end{pmatrix} y, \quad y(0) = (1,0)^T.$$

Moore [16] used this problem to illustrate the wrapping effect. Lohner [15] reports that the parallelepiped method with $h = 0.32$ and $k = 10$ produces tight bounds for this problem at $t_m = 10000$.

Since B is skew symmetric, it is easy to see from our analysis why the parallelepiped method produces tight bounds for this problem: the eigenvalues of $T_9(0.32B)$ are $\approx 0.9492 \pm 0.3146i$, which are within 5.0×10^{-5} of magnitude one.

However, the eigenvalues of $|T|$ are ≈ 1.2638 and ≈ 0.6347. It follows from (49) that $|T|^{j-1} w([z_1])$ grows like $(1.2638)^{j-1}$. Therefore, the global error in the direct method on this problem blows-up quickly.

8   How the QR Method Improves Stability

A key observation in the derivations of the results that follow is that Lohner's QR method, when applied with a constant stepsize and order to problem (2), is closely related to Francis' QR algorithm [6] for finding the eigenvalues of T. To see this, recall that in Lohner's QR method,

$$T A_{j-1} = Q_j R_j \tag{76}$$

$$A_j = Q_j, \tag{77}$$

where $A_0 = Q_0 = I$, Q_j is orthogonal, and R_j is upper triangular. Let

$$S_j = Q_j^T T Q_j \quad \text{and} \tag{78}$$

$$\widehat{Q}_j = Q_{j-1}^T Q_j \quad (j \geq 1). \tag{79}$$

Note that $S_0 = T$ since $Q_0 = I$. It follows from (78–79) that

$$S_{j-1} = \widehat{Q}_j R_j \quad \text{and} \tag{80}$$

$$S_j = R_j \widehat{Q}_j. \tag{81}$$

Now observe that the iteration (80–81) is just the *unshifted QR algorithm* for finding the eigenvalues of T. Observe also that (76–77) is the *simultaneous iteration* (see for example [25]) for computing the eigenvalues of T.

Using some properties of the QR algorithm, we show how Lohner's QR method improves the stability of an interval method for IVPs for ODEs. We consider first the case that T has eigenvalues of different magnitudes and then the case that T has eigenvalues of the same magnitude.

8.1   Eigenvalues of Different Magnitudes

Assuming that T is nonsingular with eigenvalues of different magnitudes, we derive an upper bound for the global error of Lohner's QR method. Then, we show that this bound is not much bigger than the bound (43) for the global error of the PTS method (36).

Based on our analysis of the QR method, we propose a simplified QR method, which we refer to as the *QR-S method*. We derive a simpler bound for its global error than the bound for the global error of the QR method. We also show that the global error of the QR-S method is not much bigger than the bound for the global error of the corresponding PTS method. Then, we compare empirically the global errors in the QR, QR-P, and QR-S methods to the global error in the PTS method.

Using (76–77), we write (52–53) in the following equivalent form:

$$[y_j] = T\hat{y}_{j-1} + (Q_j R_j)[r_{j-1}] + [z_j] \tag{82}$$

$$[r_j] = R_j[r_{j-1}] + Q_j^T ([z_j] - s_j). \tag{83}$$

Obviously, this method does not require that we enclose the inverse of a nearly singular matrix. Therefore, unlike the parallelepiped method, the QR method cannot break down for that reason. (In practice, we have to enclose the inverse of a floating-point approximation to an orthogonal matrix.)

We can view $[r_j]$, $j = 1, 2, \ldots$, as an enclosure of the global error at t_j that we propagate to the next step. Since

$$w([r_j]) = |R_j| w([r_{j-1}]) + |Q_j^T| w([z_j]),$$

we can consider $|R_j|$ as the matrix for propagating the global error in the QR method. As we show in this section, the nature of the R_j matrices ensures better stability of this method, compared to the previous two methods.

From (82) and (83), we have for the width of $[y_j]$,

$$
\begin{aligned}
\mathrm{w}([y_j]) &= |Q_j R_j| \mathrm{w}([r_{j-1}]) + \mathrm{w}([z_j]) \\
&= |Q_j R_j||R_{j-1}| \cdots |R_2||Q_1^T| \mathrm{w}([z_1]) \\
&\quad + |Q_j R_j||R_{j-1}| \cdots |R_3||Q_2^T| \mathrm{w}([z_2]) \\
&\quad \vdots \\
&\quad + |Q_j R_j||Q_{j-1}^T| \mathrm{w}([z_{j-1}]) \\
&\quad + \mathrm{w}([z_j]).
\end{aligned}
\tag{84}
$$

To bound $\|\mathrm{w}([y_j])\|$, we shall bound the norm of each of the terms (after the second equal sign) in (84).

For any orthogonal $Q \in \mathbb{R}^{n \times n}$,

$$
\|Q\| = \||Q|\| \leq \sqrt{n}.
\tag{85}
$$

Using (42) and (85) and assuming again that $\|\mathrm{w}([z_{i-1}])\| \leq \beta$ for all i, we bound the norm of

$$
|Q_j R_j||R_{j-1}| \cdots |R_i||Q_{i-1}^T| \mathrm{w}([z_{i-1}]) \qquad (i = 2, 3, \ldots, j)
$$

in (84) as

$$
\left\| |Q_j R_j||R_{j-1}| \cdots |R_i||Q_{i-1}^T| \mathrm{w}([z_{i-1}]) \right\| \leq \beta\, n\, \left\| |R_j||R_{j-1}| \cdots |R_i| \right\|.
\tag{86}
$$

Below, we derive a bound for $\left\| |R_j||R_{j-1}| \cdots |R_i| \right\|$.

We show first that

$$
\lim_{j \to \infty} |R_j| = R
\tag{87}
$$

for some upper-triangular matrix R with eigenvalues $\{ |\gamma_i| : i = 1, 2, \ldots, n \}$, where $\{ \gamma_i : i = 1, 2, \ldots, n \}$ are the eigenvalues of T. Thus,

$$
\lim_{j \to \infty} \rho(|R_j|) = \rho(R) = \rho(T).
\tag{88}
$$

From Theorem 3 in [6], under the QR iteration (80–81), the elements below the principal diagonal of S_j tend to zero, the moduli of those above the diagonal tend to fixed values, and the elements on the principal diagonal tend to the eigenvalues of T. Therefore,

$$
\lim_{j \to \infty} |S_j| = R,
$$

where R is upper-triangular with eigenvalues $\{\,|\gamma_i| : i = 1, 2, \ldots, n\,\}$. Since $S_{j-1} = \widehat{Q}_j R_j$ approaches an upper-triangular form, $\widehat{Q}_j$ approaches a diagonal form, where the magnitude of each diagonal entry tends to one. Hence,

$$\lim_{j \to \infty} |R_j| = \lim_{j \to \infty} |S_j \widehat{Q}_j^T| = \lim_{j \to \infty} |S_j| = R. \tag{89}$$

Let $E_l = |R_l| - R \in \mathbb{R}^{n \times n}$, whence

$$|R_l| = R + E_l. \tag{90}$$

From (89) and (90), for any $\epsilon > 0$, there exists $d \in \mathbb{N}$, $d > 0$, and $E \in \mathbb{R}^{n \times n}$ with $\|E\| \leq \epsilon$, such that

$$|E_l| \leq E, \qquad \text{for } l \geq d. \tag{91}$$

Let $i \geq d$ in (86). By (90) and (91),

$$\begin{aligned}
\big\||R_j||R_{j-1}| \cdots |R_i|\big\| &= \big\||R + E_j||R + E_{j-1}| \cdots |R + E_i|\big\| \\
&\leq \big\|(R + E)(R + E) \cdots (R + E)\big\| \\
&= \big\|(R + E)^{j-i+1}\big\|.
\end{aligned} \tag{92}$$

Since R is nonsingular with distinct eigenvalues, there exists a matrix V_R that diagonalizes R. Then, from (92), Lemma 1, and (88),

$$\begin{aligned}
\big\||R_j||R_{j-1}| \cdots |R_i|\big\| &\leq \mathrm{cond}(V_R)\big(\rho(R) + \mathrm{cond}(V_R)\,\epsilon\big)^{j-i+1} \\
&= \mathrm{cond}(V_R)\big(\rho(T) + \mathrm{cond}(V_R)\,\epsilon\big)^{j-i+1}.
\end{aligned} \tag{93}$$

Substituting (93) into (86), we obtain for $i \geq d$ that

$$\begin{aligned}
\big\|Q_j R_j &|R_{j-1}| \cdots |R_i| Q_{i-1}^T |\mathrm{w}([z_{i-1}])|\big\| \\
&\leq \beta\,n\,\mathrm{cond}(V_R)\big(\rho(T) + \mathrm{cond}(V_R)\,\epsilon\big)^{j-i+1}.
\end{aligned} \tag{94}$$

Let $r \in \mathbb{R}$ be such that

$$\big\||R_l|\big\| \leq r \quad \text{for } l = 2, 3, \ldots, (d-1), \quad d \geq 3. \tag{95}$$

Such an r always exists. For example, we can chose $r = n\big\|\,|T|\,\big\| = n\|T\|$, because

$$\big\||R_l|\big\| = \big\||Q_l^T T Q_{l-1}|\big\| \leq n\big\|\,|T|\,\big\| = n\|T\|.$$

If $i < d$, then from (93) with $i = d$ and (95), we obtain

$$\begin{aligned}
\big\|Q_j R_j &|R_{j-1}| \cdots |R_i| Q_{i-1}^T |\mathrm{w}([z_{i-1}])|\big\| \\
&\leq \beta\,n\,\mathrm{cond}(V_R)\big(\rho(T) + \mathrm{cond}(V_R)\,\epsilon\big)^{j-d+1} r^{d-i}.
\end{aligned} \tag{96}$$

Taking norms on both sides of (84) and using (94) and (96), we finally derive

$$\|w([y_j])\| \le \beta\, n\, \text{cond}(V_R) \left((\rho(T) + \text{cond}(V_R)\, \epsilon)^{j-d+1} \sum_{i=1}^{d-2} r^i \right.$$
$$\left. + \sum_{i=1}^{j-d+1} (\rho(T) + \text{cond}(V_R)\, \epsilon)^i \right) + \beta, \tag{97}$$

which gives an upper bound for the global error of the QR method at t_j. Recall that we used the infinity norm to derive (97). However, by the equivalence of vector norms (21), we have essentially the same result for any other norm.

From (43),

$$\|\delta_j\| \le \delta \left(\text{cond}(V_T) \sum_{i=1}^{j-1} (\rho(T))^i + 1 \right)$$
$$= \delta\, \text{cond}(V_T) \left((\rho(T))^{j-d+1} \sum_{i=1}^{d-2} (\rho(T))^i + \sum_{i=1}^{j-d+1} (\rho(T))^i \right) + \delta. \tag{98}$$

Let us compare the bounds in (97) and (98).

- We can assume that $\epsilon \ll 1$. Then,

$$(\rho(T) + \text{cond}(V_R)\, \epsilon)^{j-d+1} \sum_{i=1}^{d-2} r^i \approx (\rho(T))^{j-d+1} \sum_{i=1}^{d-2} r^i$$
$$= \frac{\sum_{i=1}^{d-2} r^i}{\sum_{i=1}^{d-2} (\rho(T))^i} \left((\rho(T))^{j-d+1} \sum_{i=1}^{d-2} (\rho(T))^i \right).$$

That is, the first term in (97) is approximately a constant times the first term in (98).
- The second sums in (97) and (98) are almost the same. That is,

$$\sum_{i=1}^{j-d+1} (\rho(T) + \text{cond}(V_R)\, \epsilon)^i \approx \sum_{i=1}^{j-d+1} (\rho(T))^i.$$

- Obviously, $\beta\, n\, \text{cond}(V_R)$ from (97) is equal to some constant times $\delta\, \text{cond}(V_T)$ from (98).

Thus, the bound for the global error of the QR method is not much bigger than the bound for the global error of the corresponding PTS method.

A Simplified QR Method Let Q be an $n \times n$ orthogonal matrix such that

$$S = Q^T T Q \tag{99}$$

is upper triangular. As we have assumed throughout this subsection that T has eigenvalues of different magnitudes, such a Q always exists, since (99) is the real Schur decomposition of T.

Consider now the QR method, but instead of $A_0 = Q_0 = I$, we set $A_0 = Q_0 = Q$. Then, we can choose

$$A_j = Q_j = Q, \quad \text{for } j \geq 1$$

in (52–53). With this choice for the transformation matrices, (52–53) becomes

$$[y_j] = T\hat{y}_{j-1} + (QS)[r_{j-1}] + [z_j] \quad \text{and} \tag{100}$$
$$[r_j] = S[r_{j-1}] + Q^T([z_j] - s_j), \tag{101}$$

for $j \geq 1$. We refer to (100–101) as the *QR-S method*.

From (100–101),

$$\mathrm{w}([y_j]) = |QS|\mathrm{w}([r_{j-1}]) + \mathrm{w}([z_j])$$

$$= |QS| \sum_{i=1}^{j-1} |S|^{j-i-1} |Q^T| \mathrm{w}([z_i]) + \mathrm{w}([z_j])$$

$$\leq |Q| \sum_{i=1}^{j-1} |S|^{j-i} |Q^T| \mathrm{w}([z_i]) + \mathrm{w}([z_j]). \tag{102}$$

Taking norms on both sides of (102) and employing (42) and (85), we derive that

$$\|\mathrm{w}([y_j])\| \leq \||Q|\| \sum_{i=1}^{j-1} \||S|^{j-i}\| \, \||Q^T|\|\beta + \beta$$

$$\leq \beta \left(n \sum_{i=1}^{j-1} \||S|^i\| + 1 \right). \tag{103}$$

Since T has eigenvalues of distinct magnitudes, and S is upper triangular with the same eigenvalues as T, $|S|$ has distinct eigenvalues. Therefore, it can be diagonalized. Let $|S| = V_{|S|}^{-1} |D_S| V_{|S|}$, where D_S is a diagonal matrix with the eigenvalues of T on its diagonal. Then,

$$\||S|^i\| = \|V_{|S|}^{-1} |D_S|^i V_{|S|}\| \leq \mathrm{cond}(V_{|S|}) (\rho(T))^i. \tag{104}$$

Substituting (104) into (103), we derive that

$$\|\mathrm{w}([y_j])\| \leq \beta \left(n \, \mathrm{cond}(V_{|S|}) \sum_{i=1}^{j-1} (\rho(T))^i + 1 \right). \tag{105}$$

Recall that for the PTS method,

$$\|\delta_j\| \le \delta \left(\text{cond}(V_T) \sum_{i=1}^{j-1} (\rho(T))^i + 1 \right). \tag{106}$$

Comparing (105) and (106), we make the following conclusions.

- The global error in the QR-S method cannot be much bigger than $\approx cn$ times the upper bound for the global error of the PTS method, where c is some constant.
- If T is a normal matrix, then S is diagonal. In this case, $V_{|S|} = I$, and V_T is orthogonal. Therefore,

$$\|\text{w}([y_j])\| \le \beta \left(n \sum_{i=1}^{j-1} (\rho(T))^i + 1 \right) \quad \text{and}$$

$$\|\delta_j\| \le \delta \left(n \sum_{i=1}^{j-1} (\rho(T))^i + 1 \right).$$

That is, the bounds are almost the same.

Numerical Experiments We use the numerical experiments described below to compare

- the error propagation in the QR, QR-S, and QR-P methods to the error propagation in the PTS method, and
- the global errors in these methods,

assuming T is nonsingular with real eigenvalues of distinct magnitudes.
We consider the cases in which B is

1. symmetric,
2. non-normal, and
3. highly non-normal.

For each of these cases, we consider the subcases in which B has

1. negative eigenvalues,
2. positive eigenvalues, and
3. positive and negative eigenvalues.

In each of the examples below, we construct a 5×5 matrix B as follows. We compute an orthogonal matrix Q from the QR-factorization of a 5×5 Hilbert matrix, choose an upper-triangular or diagonal matrix R, as described in each example, and set $B = Q^T R Q$. Then we choose h, which can be different in the various examples, set $k = 30$, and compute $T = T_{29}(hB)$. For each problem, we use an initial condition

$$y(0) = (1, 1, 1, 1, 1)^T$$

and choose an appropriate t_m.

To simplify our study of the error propagation in the interval methods, we investigate only the propagation of the first error in these methods. Consider the QR method and the PTS method. Since $\mathrm{w}([z_1])$ in the QR method propagates to

$$|Q_j R_j||R_{j-1}| \cdots |R_2||Q_1^T|\mathrm{w}([z_1])$$

on the jth step (see (84)), and $\check{z}_1$ in the PTS method propagates to $T^{j-1}\check{z}_1$ on the jth step (see (38)), the ratio

$$\kappa_1(j) = \frac{\||Q_j R_j||R_{j-1}| \cdots |R_2||Q_1^T|\|}{\|T^{j-1}\|} \tag{107}$$

($j \geq 2$) is a reasonable measure of the overestimation in propagating the errors in the QR method compared to the error propagation in the PTS method.

Similarly, we write

$$\kappa_2(j) = \frac{\||Q_j^{(s)} R_j^{(s)}||R_{j-1}^{(s)}| \cdots |R_2^{(s)}||Q_1^{(s)T}|\|}{\|T^{j-1}\|} \tag{108}$$

for the QR-S method, where

$$Q_1^{(s)} = Q \quad \text{from a Schur form } S = Q^T T Q \quad \text{and} \tag{109}$$

$$TQ_{j-1}^{(s)} = Q_j^{(s)} R_j^{(s)}, \quad \text{for } j \geq 2. \tag{110}$$

Note that, if the eigenvalues of T are not in decreasing order of magnitudes on the diagonal of S, then this iteration will normally arrange them in such an order.

For the QR-P method, we write

$$\kappa_3(j) = \frac{\||Q_j^{(p)} R_j^{(p)} P^{(p)}{}_j^T||R_{j-1}^{(p)} P^{(p)}{}_{j-1}^T| \cdots |R_2^{(p)} P^{(p)}{}_2^T||Q_1^{(p)T}|\|}{\|T^{j-1}\|}, \tag{111}$$

where $Q_l^{(p)}$ and $R_l^{(p)}$ are determined from the iteration

$$T = Q_1^{(p)} R_1^{(p)}, \tag{112}$$

$$TQ_{j-1}^{(p)} P_j^{(p)} = Q_j^{(p)} R_j^{(p)}, \quad \text{for } j \geq 2. \tag{113}$$

We compute $P_j^{(p)}$ as described in subsection 5.2, except that we assume that we have introduced only one local error $w([z_1])$ with all components of the same size, and all subsequent local errors are zero in this method.

In Figs. 7–15, we plot $\kappa_1(j)$, $\kappa_2(j)$, and $\kappa_3(j)$ versus $t_j = jh$ as well as the base-10 logarithm of the global errors, $\log_{10} \|w([y_j])\|$, of the QR, QR-S, and QR-P methods versus $t_j = jh$, and the base-10 logarithm of the global error of the PTS method, $\log_{10} \|\delta_j\|$, versus $t_j = jh$.

The computations of $\kappa_1(j)$, $\kappa_2(j)$, $\kappa_3(j)$, and $\|\delta_j\|$ are performed in Matlab, and the computations of $\|w([y_j])\|$ are performed with the VNODE package [17]. The plots are produced with Gnuplot.

Example 5. B is symmetric with negative eigenvalues; see Fig. 7.

$$R = \mathrm{diag}(-1, -2, -3, -4, -5),$$

$h = 0.1$, and the eigenvalues of T are

$$\gamma(T) \approx \{\, 0.9048,\, 0.8187,\, 0.7408,\, 0.6703,\, 0.6065 \,\}.$$

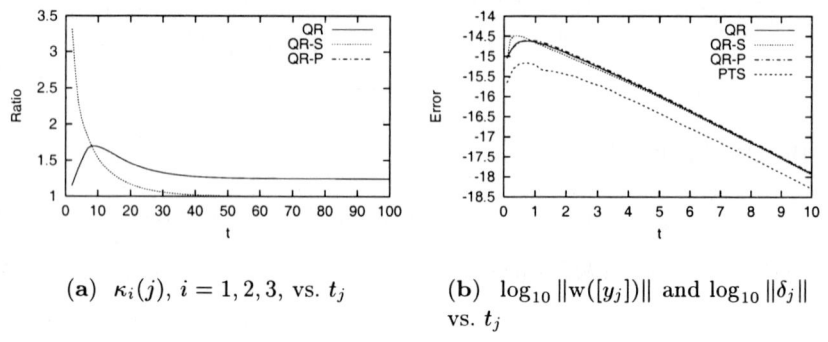

(a) $\kappa_i(j)$, $i = 1, 2, 3$, vs. t_j

(b) $\log_{10} \|w([y_j])\|$ and $\log_{10} \|\delta_j\|$ vs. t_j

Fig. 7. B is symmetric with negative eigenvalues

Example 6. B is symmetric with positive eigenvalues; see Fig. 8.

$$R = \mathrm{diag}(1, 2, 3, 4, 5),$$

$h = 0.1$, and

$$\gamma(T) \approx \{\, 1.1052,\, 1.2214,\, 1.3499,\, 1.4918,\, 1.6487 \,\}.$$

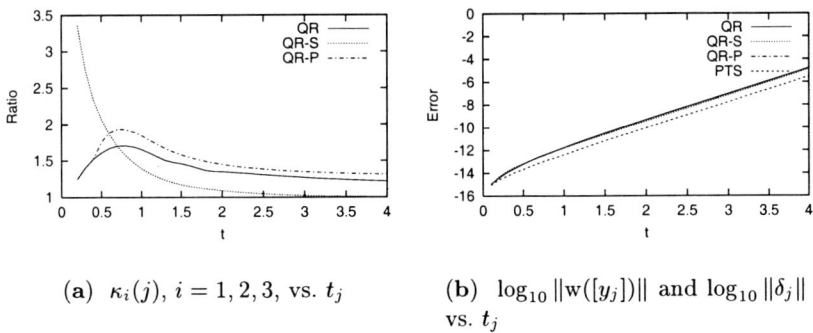

(a) $\kappa_i(j)$, $i = 1, 2, 3$, vs. t_j **(b)** $\log_{10}\|\mathrm{w}([y_j])\|$ and $\log_{10}\|\delta_j\|$ vs. t_j

Fig. 8. B is symmetric with positive eigenvalues

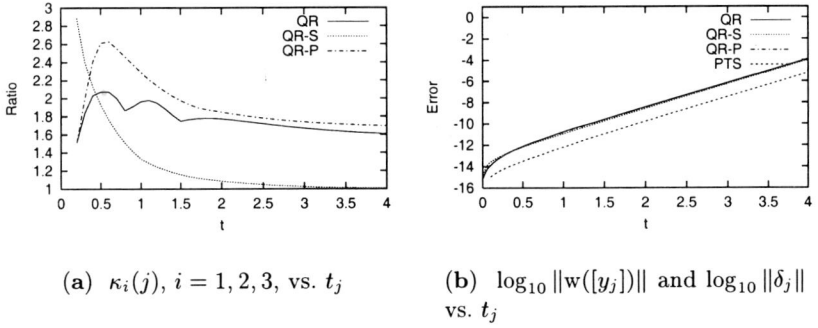

(a) $\kappa_i(j)$, $i = 1, 2, 3$, vs. t_j **(b)** $\log_{10}\|\mathrm{w}([y_j])\|$ and $\log_{10}\|\delta_j\|$ vs. t_j

Fig. 9. B is symmetric with positive and negative eigenvalues

Example 7. B is symmetric with positive and negative eigenvalues; see Fig. 9.

$$R = \mathrm{diag}(-1, -2, 3, 4, 5),$$

$h = 0.1$, and

$$\gamma(T) \approx \{\, 0.8187, 0.9048, 1.3499, 1.6487, 1.4918 \,\}.$$

Example 8. B is non-normal with negative eigenvalues; see Fig. 10.

$$R = \begin{pmatrix} -1 & 6 & 0 & 0 & 0 \\ 0 & -3 & 6 & 0 & 0 \\ 0 & 0 & -5 & 6 & 0 \\ 0 & 0 & 0 & -7 & 6 \\ 0 & 0 & 0 & 0 & -9 \end{pmatrix},$$

$h = 0.1$, and

$$\gamma(T) \approx \{\, 0.9048,\ 0.7408,\ 0.6065,\ 0.4966,\ 0.4066 \,\}.$$

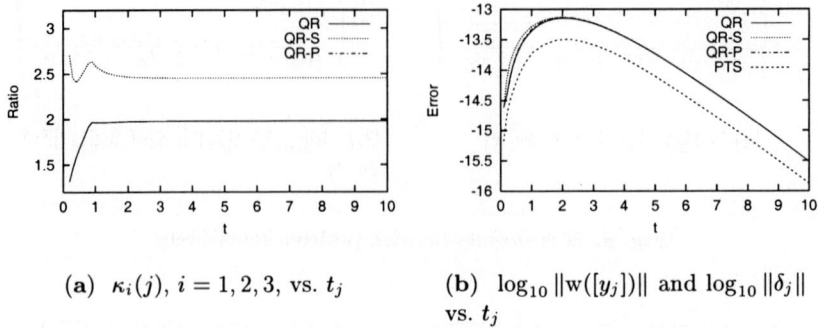

(a) $\kappa_i(j)$, $i = 1, 2, 3$, vs. t_j (b) $\log_{10} \| \mathrm{w}([y_j]) \|$ and $\log_{10} \| \delta_j \|$ vs. t_j

Fig. 10. B is non-normal with negative eigenvalues

Example 9. B is non-normal with positive eigenvalues; see Fig. 11.

$$R = \begin{pmatrix} 1 & 6 & 0 & 0 & 0 \\ 0 & 3 & 6 & 0 & 0 \\ 0 & 0 & 5 & 6 & 0 \\ 0 & 0 & 0 & 7 & 6 \\ 0 & 0 & 0 & 0 & 9 \end{pmatrix},$$

$h = 0.01$, and

$$\gamma(T) \approx \{\, 1.0101,\ 1.0305\ 1.0513,\ 1.0725,\ 1.0942 \,\}.$$

Example 10. B is non-normal with positive and negative eigenvalues; see Fig. 12.

$$R = \begin{pmatrix} -1 & 6 & 0 & 0 & 0 \\ 0 & -3 & 6 & 0 & 0 \\ 0 & 0 & 5 & 6 & 0 \\ 0 & 0 & 0 & 7 & 6 \\ 0 & 0 & 0 & 0 & 9 \end{pmatrix},$$

$h = 0.01$, and

$$\gamma(T) \approx \{\, 0.9704,\ 0.9900,\ 1.0513,\ 1.0725,\ 1.0942 \,\}.$$

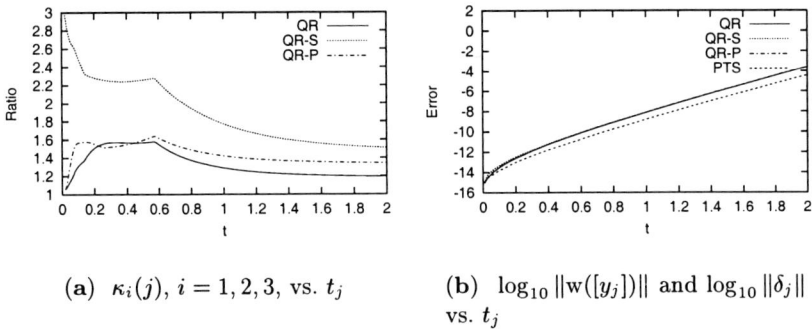

(a) $\kappa_i(j)$, $i = 1, 2, 3$, vs. t_j

(b) $\log_{10} \|w([y_j])\|$ and $\log_{10} \|\delta_j\|$ vs. t_j

Fig. 11. B is non-normal with positive eigenvalues

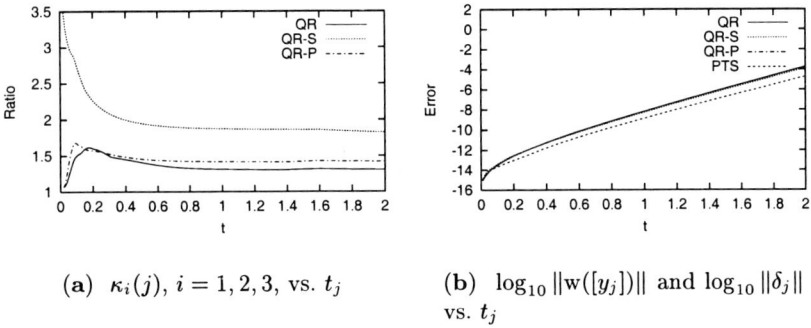

(a) $\kappa_i(j)$, $i = 1, 2, 3$, vs. t_j

(b) $\log_{10} \|w([y_j])\|$ and $\log_{10} \|\delta_j\|$ vs. t_j

Fig. 12. B is non-normal with positive and negative eigenvalues

Example 11. B is highly non-normal with negative eigenvalues; see Fig. 13.

$$
R = \begin{pmatrix}
-1 & 50 & 0 & 0 & 0 \\
0 & -3 & 50 & 0 & 0 \\
0 & 0 & -5 & 50 & 0 \\
0 & 0 & 0 & -7 & 50 \\
0 & 0 & 0 & 0 & -9
\end{pmatrix},
$$

$h = 0.05$, and

$$
\gamma(T) \approx \{\, 0.9512, 0.8607, 0.7788, 0.7047, 0.6376 \,\}.
$$

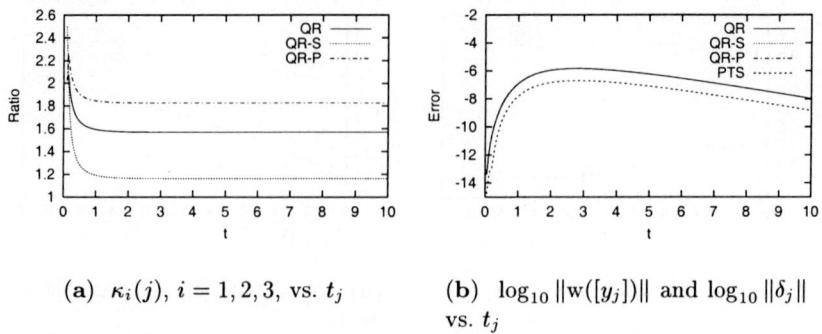

(a) $\kappa_i(j)$, $i = 1, 2, 3$, vs. t_j

(b) $\log_{10} \|\mathrm{w}([y_j])\|$ and $\log_{10} \|\delta_j\|$ vs. t_j

Fig. 13. B is highly non-normal with negative eigenvalues

Example 12. B is highly non-normal with positive eigenvalues; see Fig. 14.

$$R = \begin{pmatrix} 1 & 50 & 0 & 0 & 0 \\ 0 & 3 & 50 & 0 & 0 \\ 0 & 0 & 5 & 50 & 0 \\ 0 & 0 & 0 & 7 & 50 \\ 0 & 0 & 0 & 0 & 9 \end{pmatrix},$$

$h = 0.01$, and

$$\gamma(T) \approx \{\, 1.0101,\ 1.0305,\ 1.0513,\ 1.0725,\ 1.0942 \,\}.$$

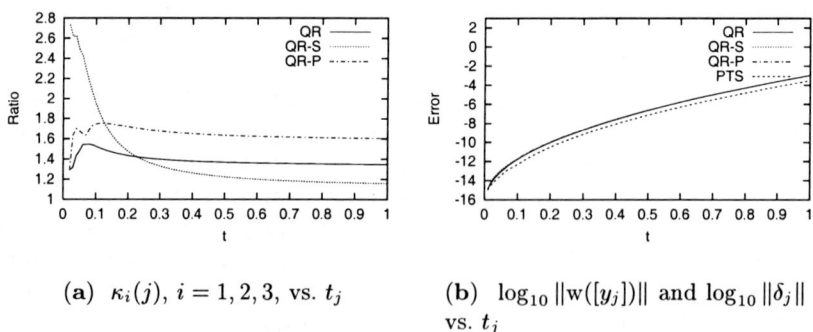

(a) $\kappa_i(j)$, $i = 1, 2, 3$, vs. t_j

(b) $\log_{10} \|\mathrm{w}([y_j])\|$ and $\log_{10} \|\delta_j\|$ vs. t_j

Fig. 14. B is highly non-normal with positive eigenvalues

Example 13. B is highly non-normal with positive and negative eigenvalues; see Fig. 15.

$$R = \begin{pmatrix} -1 & 50 & 0 & 0 & 0 \\ 0 & -3 & 50 & 0 & 0 \\ 0 & 0 & 5 & 50 & 0 \\ 0 & 0 & 0 & 7 & 50 \\ 0 & 0 & 0 & 0 & 9 \end{pmatrix},$$

$h = 0.05$, and

$$\gamma(T) \approx \{\, 0.8607,\, 0.9512,\, 1.2840,\, 1.4191,\, 1.5683 \,\}.$$

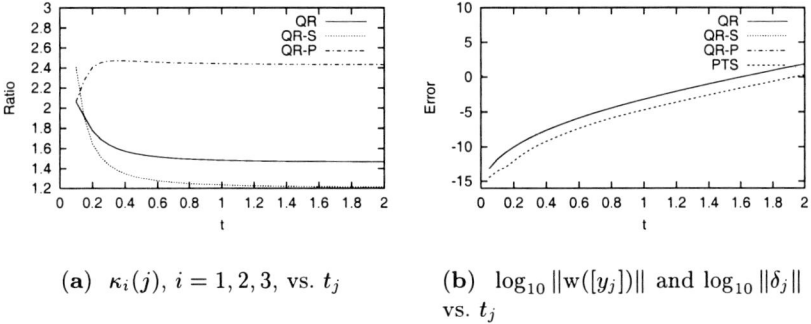

(a) $\kappa_i(j)$, $i = 1, 2, 3$, vs. t_j (b) $\log_{10} \| w([y_j]) \|$ and $\log_{10} \| \delta_j \|$ vs. t_j

Fig. 15. B is highly non-normal with positive and negative eigenvalues

We have performed significantly more experiments than the ones reported in this subsection. Here, we summarize our observations.

As j becomes sufficiently large, the ratios $\kappa_i(j)$, $i = 1, 2, 3$, are usually not much bigger than one. Often, but not always, $\kappa_2(j)$ becomes smaller than $\kappa_1(j)$ and $\kappa_3(j)$, and $\kappa_3(j)$ generally becomes bigger than $\kappa_1(j)$.

This suggests that the propagated first error in the interval methods (QR, QR-S, and QR-P) should not be much larger than the propagated first error in the PTS method. This also suggests that the QR-S method should generally produce tighter bounds than the QR and QR-P methods, while the QR-P should generally produce wider bounds than the QR method.

However, from the (b) parts of Figs. 7–15, the global errors in the interval methods at each t_j are nearly the same and not much bigger than the corresponding global error of the PTS method. Moreover, no interval method produces a substantially smaller global error than the other two at any t_j.

8.2   Eigenvalues of Equal Magnitude

If T is nonsingular with p eigenvalues of equal magnitude, then the matrices S_j defined in (78) tend to a block upper-triangular form with a $p \times p$ block on its main diagonal [26, pp. 520–521]. The eigenvalues of this block tend to these p eigenvalues.

In this subsection, we consider the case that T is nonsingular with at least one complex conjugate pair of eigenvalues and at most two eigenvalues of the same magnitude. In this case, S_j tends to a block upper-triangular form with 1×1 and 2×2 blocks on its diagonal. The eigenvalues of these blocks tend to the eigenvalues of T. Note that S_j does not converge to a fixed matrix, but the eigenvalues of its diagonal blocks converge to the eigenvalues of T.

When $S_{j-1} = \widehat{Q}_j R_j$ approaches such a block upper-triangular form, the orthogonal matrices $\widehat{Q}_j$ approach a block-diagonal form with 1×1 and 2×2 blocks on its main diagonal, where the 2×2 blocks are at the same locations as the 2×2 blocks of S_{j-1}.

For simplicity, assume that each sub-diagonal entry ("diagonal" includes 2×2 blocks) of S_{j-1} and each off-diagonal entry of $\widehat{Q}_j$ is zero, while in practice, they converge to zero. Consider a 2×2 block of S_{j-1},

$$\begin{pmatrix} (S_{j-1})_{l,l} & (S_{j-1})_{l,l+1} \\ (S_{j-1})_{l+1,l} & (S_{j-1})_{l+1,l+1} \end{pmatrix},$$

and denote its complex eigenvalues by γ_l and $\gamma_{l+1} = \overline{\gamma}_l$, which are also eigenvalues of T. From Lemma 2, the components of R_j that correspond to these eigenvalues are

$$(R_j)_{l,l} = s_1 \alpha_{l,j-1} \quad \text{and}$$
$$(R_j)_{l+1,l+1} = s_2 |\gamma_l|^2 / \alpha_{l,j-1},$$

where $s_1, s_2 = \pm 1$ and $\alpha_{l,j-1} = \sqrt{(S_{j-1})_{l,l}^2 + (S_{j-1})_{l+1,l}^2}$.

Hence, as $j \to \infty$,

$$|(R_j)_{l,l}| = \alpha_{l,j-1} \quad \text{and}$$
$$|(R_j)_{l+1,l+1}| = |\gamma_l|^2 / \alpha_{l,j-1}$$

oscillate around $|\gamma_l|$.

If $\rho(T) = |\gamma_l|$, then

$$\rho(|R_j|) \geq \rho(T), \quad \text{as } j \to \infty.$$

In fact, if $\alpha_{l,j-1} \ll |\gamma_l|$ or $\alpha_{l,j-1} \gg |\gamma_l|$, then $\rho(|R_j|)$ may be much larger than $\rho(T)$. However, we have not observed this in practice. The examples reported below are typical of our much larger class of test problems in that $\rho(|R_j|)$ is normally either equal to or just a little larger than $\rho(T)$.

On the other hand, if T has a dominant real eigenvalue, then we should generally expect that

$$\rho(|R_j|) \to \rho(T), \quad \text{as } j \to \infty.$$

Since the matrices $|R_j|$ do not approach a fixed matrix, as in the case with eigenvalues of distinct magnitudes, deriving a bound for the global error in this case is more difficult. However, we show in four examples, in which T has some eigenvalues of equal magnitude, that the global errors in the QR, QR-P, and QR-S methods are not much bigger than the global error of the PTS method.

Numerical Experiments We investigate the behavior of the QR, QR-S, and QR-P methods assuming T has at least one complex conjugate pair of eigenvalues. We consider the following four cases:

1. B has a dominant complex conjugate pair;
2. B has a dominant real eigenvalue;
3. B is normal, but neither symmetric nor skew symmetric; and
4. B is skew symmetric.

For each of the four examples, we describe below how B and the corresponding T matrices are constructed. Moreover, for each example, we plot

- $\rho(|R_j|)$, $\rho(|R_j^{(s)}|)$, $\rho(|R_j^{(p)} P_j^T|)$, and $\rho(T)$ versus $t_j = jh$, where $R_j^{(s)}$, $R_j^{(p)}$, and P_j are computed as described in the numerical experiments part of subsection 8.1;
- the ratios $\kappa_1(j)$, $\kappa_2(j)$, and $\kappa_3(j)$ defined in (107), (108), and (111) versus t_j;
- the norms

$$\eta_1(j) = \left\||Q_j R_j||R_{j-1}|\cdots|R_2||Q_1^T|\right\|,$$
$$\eta_2(j) = \left\||Q_j^{(s)} R_j^{(s)}||R_{j-1}^{(s)}|\cdots|R_2^{(s)}||Q_1^{(s)T}|\right\|,$$
$$\eta_3(j) = \left\||Q_j^{(p)} R_j^{(p)} P_j^T||R_{j-1}^{(p)} P_{j-1}^T|\cdots|R_2^{(p)} P_2^T||Q_1^{(p)T}|\right\|, \quad \text{and}$$
$$\eta_4(j) = \left\|T^{j-1}\right\|,$$

which correspond to the QR, QR-S, QR-P, and PTS methods, versus t_j; and
- the base-10 logarithm of the global errors in the QR, QR-S, QR-P, and PTS methods versus t_j. The initial condition vector in each example has all components equal to one.

The computations of the spectral radii, $\kappa_i(j)$, $i = 1, 2, 3$, $\eta_i(j)$, $i = 1, 2, 3, 4$, and $\|\delta_j\|$ are performed in Matlab, and the computations of $\|w([y_j])\|$ are performed with the VNODE package [17]. The plots are produced with Gnuplot.

Example 14. *B* has a dominant complex conjugate pair.
    Let

$$B = \begin{pmatrix} -0.5 & 0.5 & 0 \\ -2.0 & -0.5 & 1 \\ 0 & 1 & -2 \end{pmatrix}.$$

For $k = 30$ and $h = 0.1$,

$$\gamma(T) \approx \{\, 0.9674 \pm 0.0836i, \, 0.7858 \,\}.$$

Here, $|\gamma_1| = |\gamma_2| \approx 0.9710 > |\gamma_3| \approx 0.7858$. That is, the complex conjugate pair of eigenvalues dominates.

In Fig. 16(a), we plot $\rho(|R_j|)$, $\rho(|R_j^{(s)}|)$, $\rho(|R_j^{(p)} P_j^T|)$, and $\rho(T)$ versus t_j. The oscillations in $\rho(|R_j|)$ are caused by the changes in the diagonal elements $(R_j)_{1,1}$ and $(R_j)_{2,2}$, which correspond to the complex eigenvalues $\gamma_{1,2} \approx 0.9674 \pm 0.0836i$. Since the QR and QR-S method differ in the starting matrix for the simultaneous iteration (identity in the QR method, and Q from a Schur form $S = Q^T T Q$ in the QR-S method), the oscillations in $\rho(|R_j^{(s)}|)$ can be explained by the same arguments as the oscillations in $\rho(|R_j|)$. However, we do not propose an explanation of why $\rho(|R_j^{(p)} P_j^T|)$ behaves similarly to $\rho(|R_j^{(s)}|)$ and $\rho(|R_j|)$.

In Fig. 16(b), we plot the ratios $\kappa_1(j)$, $\kappa_2(j)$, and $\kappa_3(j)$. Although $\kappa_2(j)$ can become much larger than one, the first error in the QR-S method is still damped to zero. This can be seen from Fig. 16(c), where we plot $\eta_1(j)$, $\eta_2(j)$, $\eta_3(j)$, and $\eta_4(j)$ versus t_j. In this figure, the above norms approach zero, as j increases, suggesting that $\mathrm{w}([z_1])$ is damped to zero, as j increases. Note that this happens even though the corresponding spectral radii in the interval methods are not always smaller than one.

From Fig. 16(d), we see that at each t_j, the global errors in the interval methods are not much larger than the corresponding global error in the PTS method. In addition, all these errors decrease in a similar manner, as j becomes sufficiently large.

Example 15. *B* has a dominant real eigenvalue.
    In this example, we choose

$$B = \begin{pmatrix} -2 & 2 & -3 \\ -1 & -3 & 1 \\ 1 & 1 & -1 \end{pmatrix},$$

$k = 30$, and $h = 0.4$. The eigenvalues of $T = T_{29}(0.4B)$ are

$$\gamma(T) \approx \{\, 0.6703, \, 0.2629 \pm 0.2573i \,\}.$$

Here, $|\gamma_1| \approx 0.6703 > |\gamma_2| = |\gamma_3| \approx 0.3679$. That is, the real eigenvalue dominates.

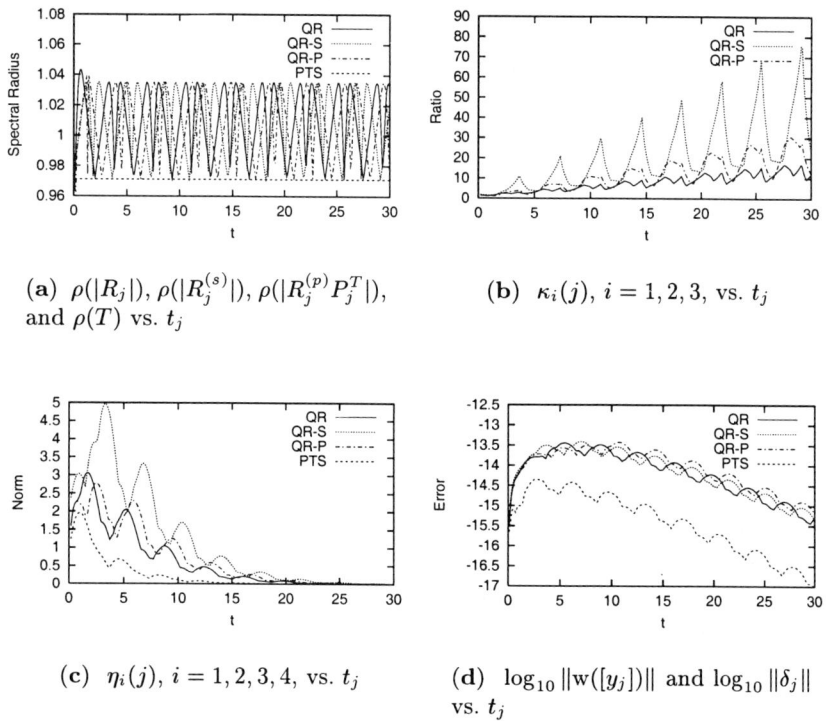

(a) $\rho(|R_j|)$, $\rho(|R_j^{(s)}|)$, $\rho(|R_j^{(p)}P_j^T|)$, and $\rho(T)$ vs. t_j

(b) $\kappa_i(j)$, $i = 1, 2, 3$, vs. t_j

(c) $\eta_i(j)$, $i = 1, 2, 3, 4$, vs. t_j

(d) $\log_{10}\|w([y_j])\|$ and $\log_{10}\|\delta_j\|$ vs. t_j

Fig. 16. B has a dominant complex conjugate pair

In Fig. 17(a), $\rho(|R_j|)$, $\rho(|R_j^{(s)}|)$, and $\rho(|R_j^{(p)}P_j^T|)$ settle to $\rho(T)$, but then $\rho(|R_j^{(s)}|)$ oscillates and settles again to $\rho(T)$. Thus, when the dominant eigenvalue is real, the methods behave like they do with real distinct eigenvalues. To support this point, compare Fig. 17(d) with the (b) parts of the figures in subsection 8.1. The reason for the "peculiar" oscillations in $\rho(|R_j^{(s)}|)$ is that the block upper-triangular matrix returned by the Matlab function schur (which we use to compute Q and S such that $S = Q^T T Q$ is block upper triangular) contains the dominant real eigenvalue in position $(3, 3)$ (that is, $S_{3,3} \approx 0.6703$), and the simultaneous iteration (110) moves this eigenvalue to position $(1, 1)$ in $R_j^{(s)}$, as j becomes sufficiently large.

Although the ratios $\kappa_1(j)$, $\kappa_2(j)$, and $\kappa_3(j)$ are bigger than one, the propagated error $w([z_1])$ in the QR, QR-S, and QR-P is damped in these methods; see the plots in Figs. 17(c) and (d).

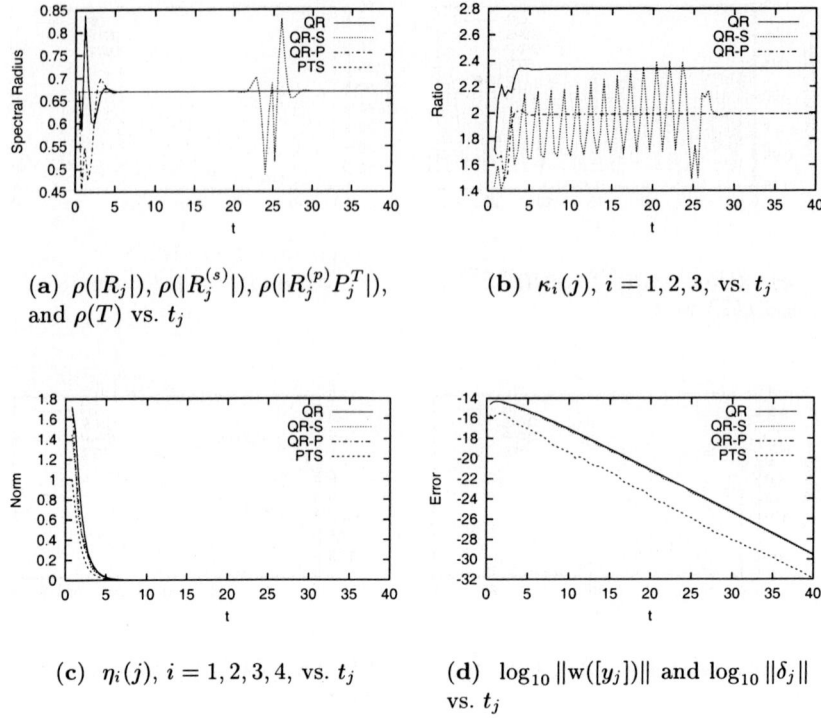

(a) $\rho(|R_j|)$, $\rho(|R_j^{(s)}|)$, $\rho(|R_j^{(p)}P_j^T|)$, and $\rho(T)$ vs. t_j

(b) $\kappa_i(j)$, $i = 1, 2, 3$, vs. t_j

(c) $\eta_i(j)$, $i = 1, 2, 3, 4$, vs. t_j

(d) $\log_{10}\|w([y_j])\|$ and $\log_{10}\|\delta_j\|$ vs. t_j

Fig. 17. B has a dominant real eigenvalue

Example 16. B is normal, but neither symmetric nor skew-symmetric. We choose

$$B = \begin{pmatrix} 1 & -3 & -4 & -5 & -6 \\ 3 & 1 & -5 & -6 & -7 \\ 4 & 5 & 1 & -7 & -8 \\ 5 & 6 & 7 & 1 & -9 \\ 6 & 7 & 8 & 9 & 1 \end{pmatrix},$$

$k = 30$, and $h = 0.1$. The eigenvalues of $T = T_{29}(0.1B)$ are

$$\gamma(T) \approx \{ -0.3961 \pm 1.0317i, \ 1.0251 \pm 0.4129i, \ 1.1052 \},$$

all with a magnitude of ≈ 1.1052.

The plots are shown in Fig. 18. From Figs. 18(b) and 18(c), we expect the global errors in the interval methods to be a little bigger than in the PTS method. This is confirmed in Fig. 18(d), where we plot the base-10 logarithm of the global errors in these methods versus t_j.

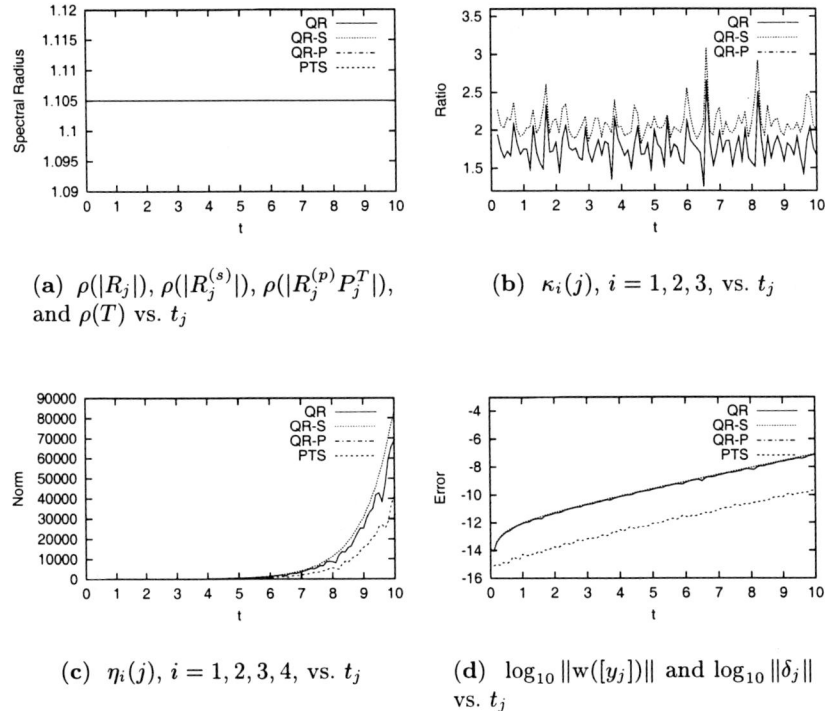

(a) $\rho(|R_j|)$, $\rho(|R_j^{(s)}|)$, $\rho(|R_j^{(p)}P_j^T|)$, and $\rho(T)$ vs. t_j

(b) $\kappa_i(j)$, $i = 1, 2, 3$, vs. t_j

(c) $\eta_i(j)$, $i = 1, 2, 3, 4$, vs. t_j

(d) $\log_{10}\|\mathrm{w}([y_j])\|$ and $\log_{10}\|\delta_j\|$ vs. t_j

Fig. 18. B is normal, but neither symmetric nor skew symmetric

Example 17. B is skew symmetric.

In this example,

$$B = \begin{pmatrix} 0 & -3 & -4 & -5 & -6 \\ 3 & 0 & -5 & -6 & -7 \\ 4 & 5 & 0 & -7 & -8 \\ 5 & 6 & 7 & 0 & -9 \\ 6 & 7 & 8 & 9 & 0 \end{pmatrix},$$

$k = 30$, and $h = 0.1$. The eigenvalues of $T = T_{29}(0.1B)$ are

$$\gamma(T) = \{ -0.3584 \pm 0.9336i, \ 0.9276 \pm 0.3736i, \ 1.0000 \},$$

all with a magnitude of ≈ 1; see Fig. 19.

For this problem, the global error of the PTS method increases slowly, as j increases, as do the global errors in the interval methods. Note that the errors in the interval methods do not become much bigger than the global error of the PTS method; see Fig. 19(d).

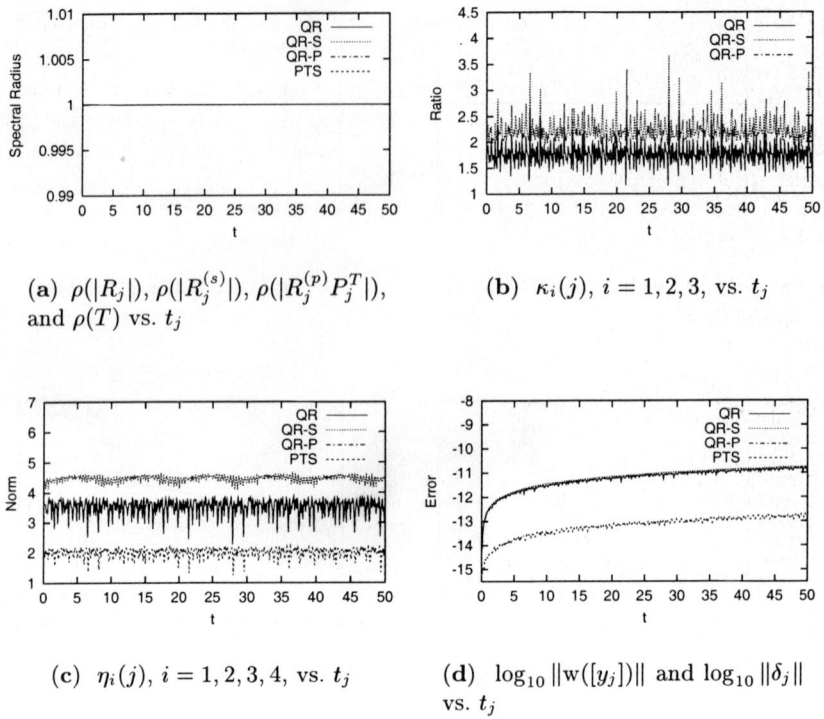

(a) $\rho(|R_j|)$, $\rho(|R_j^{(s)}|)$, $\rho(|R_j^{(p)} P_j^T|)$, and $\rho(T)$ vs. t_j

(b) $\kappa_i(j)$, $i = 1, 2, 3$, vs. t_j

(c) $\eta_i(j)$, $i = 1, 2, 3, 4$, vs. t_j

(d) $\log_{10} \|w([y_j])\|$ and $\log_{10} \|\delta_j\|$ vs. t_j

Fig. 19. B is skew symmetric

The numerical results in this subsection suggest that, if T is nonsingular and has some eigenvalues of the same magnitude, then the global errors in the QR, QR-S, and QR-P methods (at each t_j) are nearly the same and not much bigger than the corresponding global error of the PTS method. In addition, the global errors in the interval methods decrease or increase similarly to the global error of the PTS method.

If T has a dominant real eigenvalue, then the spectral radii in the interval methods usually settle to $\rho(T)$, as j becomes sufficiently large. Thus, this case is similar to the case that T is nonsingular with eigenvalues of distinct magnitudes.

9   Conclusions

We used eigenvalue techniques to develop a new perspective on the wrapping effect in interval methods for IVPs for ODEs. In our approach, we view the wrapping effect as a source of instability in interval methods and the problem of reducing the wrapping effect as one of finding a more stable scheme for advancing the solution. Eigenvalue techniques have proven useful in studying the stability of standard (point) methods; it seems natural to employ such techniques in the study of interval methods.

Our approach complements the geometric arguments used to date to study the wrapping effect in interval methods for IVPs for ODEs, providing a much more quantitative approach to the problem. If only geometric arguments are used in studying the wrapping effect, it is often difficult to draw sound conclusions about how an interval method will perform.

For example, the earlier geometric explanations of why Lohner's QR-factorization method works well [15,17] do give us an intuitive understanding of the scheme, but we believe that they do not provide as good an explanation as that developed in Sect. 8. To be more specific, we showed that, for linear constant-coefficient problems, the global error of an ITS method using Lohner's QR-factorization scheme is generally not much larger than the global error of the corresponding point method. It is hard to see how one could draw this conclusion from a geometric approach alone.

It does not appear that our stability analysis of linear constant-coefficient problems can be easily extended to more general IVPs. However, given an anti-wrapping scheme, we can study its stability for this restricted class of problems. If such a scheme does not perform well for these simple problems, it almost certainly will not work well in general. Conversely, if it performs well for linear constant-coefficient problems, we believe it is quite likely to work well in general.

A   Lemmas

Lemma 1. *Suppose that $A \in \mathbb{R}^{n \times n}$ is diagonalizable. That is, there exists a nonsingular matrix V and a diagonal matrix D such that $D = V A V^{-1}$. Let $E \in \mathbb{R}^{n \times n}$ be any matrix such that $\|E\| \leq \beta$. Then, for any integer $p \geq 0$,*

$$\|(A + E)^p\| \leq \operatorname{cond}(V)\big(\rho(A) + \beta \operatorname{cond}(V)\big)^p.$$

Proof.

$$
\begin{aligned}
\|(A + E)^p\| &= \|V^{-1}V(A + E)^p V^{-1}V\| \\
&\leq \|V^{-1}\|\|V(A + E)^p V^{-1}\|\|V\| \\
&= \operatorname{cond}(V)\|\big(V(A + E)V^{-1}\big)^p\| \\
&= \operatorname{cond}(V)\|(D + VEV^{-1})^p\| \\
&\leq \operatorname{cond}(V)\|D + VEV^{-1}\|^p \\
&\leq \operatorname{cond}(V)\big(\|D\| + \|VEV^{-1}\|\big)^p \\
&\leq \operatorname{cond}(V)\big(\rho(A) + \beta \operatorname{cond}(V)\big)^p,
\end{aligned}
$$

where we have used in the last line $\|D\| = \rho(D) = \rho(A)$.

Lemma 2. *Let*

$$C = \begin{pmatrix} c_{11} & c_{12} \\ c_{21} & c_{22} \end{pmatrix} \in \mathbb{R}^{2 \times 2}$$

have eigenvalues $\lambda_{1,2} = a \pm ib$ with $b \neq 0$ and let $C = QR$ be its QR decomposition. Then

$$R = \begin{pmatrix} s_1\alpha & s_1(c_{11}c_{12} + c_{21}c_{22})/\alpha \\ 0 & s_2|\lambda_1|^2/\alpha \end{pmatrix},$$

where

$$\alpha = \sqrt{c_{11}^2 + c_{21}^2} \neq 0,$$

s_1 *can be either $+1$ or -1, and s_2 can be either $+1$ or -1.*

Proof. First note that, since $C \in \mathbb{R}^{2 \times 2}$ and $\lambda_{1,2} \notin \mathbb{R}$, $c_{21} \neq 0$. Therefore, $\alpha = \sqrt{c_{11}^2 + c_{21}^2} \neq 0$.

It is straightforward to show that any orthogonal $Q \in \mathbb{R}^{2 \times 2}$ satisfying $C = QR$, or equivalently $Q^T C = R$, must be of the form

$$Q^T = \frac{1}{\sqrt{c_{11}^2 + c_{21}^2}} \begin{pmatrix} s_1 c_{11} & s_1 c_{21} \\ -s_2 c_{21} & s_2 c_{11} \end{pmatrix},$$

where s_1 can be either $+1$ or -1, and s_2 can be either $+1$ or -1. Hence,

$$R = Q^T C = \frac{1}{\sqrt{c_{11}^2 + c_{21}^2}} \begin{pmatrix} s_1(c_{11}^2 + c_{21}^2) & s_1(c_{11}c_{12} + c_{21}c_{22}) \\ 0 & s_2(c_{11}c_{22} - c_{12}c_{21}) \end{pmatrix}.$$

The eigenvalues of C are the roots of the quadratic equation

$$\lambda^2 - (c_{11} + c_{22})\lambda + c_{11}c_{22} - c_{12}c_{21} = 0.$$

Hence,

$$|\lambda_1|^2 = \lambda_1\lambda_2 = c_{11}c_{22} - c_{12}c_{21}.$$

Therefore,

$$R = \frac{1}{\sqrt{c_{11}^2 + c_{21}^2}} \begin{pmatrix} s_1(c_{11}^2 + c_{21}^2) & s_1(c_{11}c_{12} + c_{21}c_{22}) \\ 0 & s_2|\lambda_1|^2 \end{pmatrix}$$

$$= \begin{pmatrix} s_1\alpha & s_1(c_{11}c_{12} + c_{21}c_{22})/\alpha \\ 0 & s_2|\lambda_1|^2/\alpha \end{pmatrix}.$$

References

1. G. Alefeld and J. Herzberger. *Introduction to Interval Computations.* Academic Press, New York, 1983.
2. U. M. Ascher and L. R. Petzold. *Computer Methods for Ordinary Differential Equations and Differential-Algebraic Equations.* SIAM, Philadelphia, 1998.
3. C. Barbăroşie. Reducing the wrapping effect. *Computing*, 54(4):347–357, 1995.
4. M. Berz and K. Makino. Verified integration of ODEs and flows using differential algebraic methods on high-order Taylor models. *Reliable Computing*, 4:361–369, 1998.
5. P. Eijgenraam. *The Solution of Initial Value Problems Using Interval Arithmetic.* Mathematical Centre Tracts No. 144. Stichting Mathematisch Centrum, Amsterdam, 1981.
6. J. G. F. Francis. The QR transformation: A unitary analogue to the LR transformation—part 1. *The Computer Journal*, 4:265–271, 1961/1962.
7. T. Gambill and R. Skeel. Logarithmic reduction of the wrapping effect with application to ordinary differential equations. *SIAM J. Numer. Anal.*, 25(1):153–162, 1988.
8. E. Hairer, S. P. Nørsett, and G. Wanner. *Solving Ordinary Differential Equations I. Nonstiff Problems.* Springer-Verlag, 2nd revised edition, 1991.
9. L. W. Jackson. A comparison of ellipsoidal and interval arithmetic error bounds. In *Studies in Numerical Analysis, vol. 2, Proc. Fall Meeting of the Society for Industrial and Applied Mathematics*, Philadelphia, 1968. SIAM.
10. L. W. Jackson. Automatic error analysis for the solution of ordinary differential equations. Technical Report 28, Dept. of Computer Science, University of Toronto, 1971.
11. L. W. Jackson. Interval arithmetic error–bounding algorithms. *SIAM J. Numer. Anal.*, 12(2):223–238, 1975.
12. W. Kahan. A computable error bound for systems of ordinary differential equations. *Abstract in SIAM Review*, 8:568–569, 1966.
13. F. Krückeberg. Ordinary differential equations. In E. Hansen, editor, *Topics in Interval Analysis*, pages 91–97. Clarendon Press, Oxford, 1969.
14. W. Kühn. Rigorously computed orbits of dynamical systems without the wrapping effect. *Computing*, 61(1):47–67, 1998.
15. R. J. Lohner. *Einschließung der Lösung gewöhnlicher Anfangs- und Randwertaufgaben und Anwendungen.* PhD thesis, Universität Karlsruhe, 1988.
16. R. E. Moore. *Interval Analysis.* Prentice-Hall, Englewood Cliffs, N.J., 1966.
17. N. S. Nedialkov. *Computing Rigorous Bounds on the Solution of an Initial Value Problem for an Ordinary Differential Equation.* PhD thesis, Department of Computer Science, University of Toronto, Toronto, Canada, M5S 3G4, February 1999.
18. N. S. Nedialkov and K. R. Jackson. An interval Hermite-Obreschkoff method for computing rigorous bounds on the solution of an initial value problem for an ordinary differential equation. *Reliable Computing*, 5(3):289–310, 1999. Also in T. Csendes, editor, *Developments in Reliable Computing*, pp. 289–310, Kluwer, Dordrecht, Netherlands, 1999.
19. N. S. Nedialkov, K. R. Jackson, and G. F. Corliss. Validated solutions of initial value problems for ordinary differential equations. *Applied Mathematics and Computation*, 105(1):21–68, 1999.

20. N. S. Nedialkov, K. R. Jackson, and J. D. Pryce. An effective high-order interval method for validating existence and uniqueness of the solution of an IVP for an ODE, 2000. Accepted for publication in Reliable Computing, 17 pages.

21. A. Neumaier. Global, rigorous and realistic bounds for the solution of dissipative differential equations. Part I : Theory. *Computing*, 52(4):315–336, 1994.

22. K. Nickel. How to fight the wrapping effect. In K. Nickel, editor, *Interval Analysis 1985*, Lecture Notes in Computer Science No. 212, pages 121–132. Springer, Berlin, 1985.

23. R. Rihm. Interval methods for initial value problems in ODEs. In J. Herzberger, editor, *Topics in Validated Computations: Proceedings of the IMACS-GAMM International Workshop on Validated Computations, University of Oldenburg*, Elsevier Studies in Computational Mathematics, pages 173–207. Elsevier, Amsterdam, New York, 1994.

24. N. Stewart. A heuristic to reduce the wrapping effect in the numerical solution of $x' = f(t, x)$. *BIT*, 11:328–337, 1971.

25. D. S. Watkins. Understanding the QR algorithm. *SIAM Review*, 24(4):427–440, October 1982.

26. J. H. Wilkinson. *The Algebraic Eigenvalue Problem*. Oxford Science Publications, Oxford, England, 1965.

A Guaranteed Bound of the Optimal Constant in the Error Estimates for Linear Triangular Elements

Part II: Details

Mitsuhiro T. Nakao[1] and Nobito Yamamoto[2]

[1] Graduate School of Mathematics, Kyushu University 33
Fukuoka 812-8581, Japan.
mtnakao@math.kyushu-u.ac.jp
[2] Department of Computer Science, The University of Electro-Communications
1-5-1 Chofugaoka, Chofu, Tokyo, 182-8585 Japan.
yamamoto@im.uec.ac.jp

Abstract. In the previous paper([6]), we formulated a numerical method to get a guaranteed bound of the optimal constant in the error estimates with linear triangular elements in R^2. We describe, in this paper, detailed computational procedures for obtaining a rigorous upper bound of that constant with sufficient sharpness. The numerical verification method for solutions of nonlinear elliptic problems is successfully applied to the present purpose. A constructive error estimate for the triangular element with Neumann boundary condition plays an important role to implement the actual verified computations. Particularly, some special kind of techniques are utilized to improve the computational cost for the algorithm. As a result, we obtained a sufficiently sharp upper bound from the practical viewpoint.

1 Introduction

In [6], we described the formulation of the problem and presented our main results on a numerical method to get a guaranteed bound of the optimal constant in the error estimates of finite element method with linear triangular elements. The present paper is a continuation and a detailed version of [6], and we will mention about the following.

1. The method to get a rigorous lower bound of the smallest value of λ in the following problem.
   Find $\xi \in H^1(\Omega) \bigcap L_0^2(\Omega)$ and $\lambda \in R$ such that

   $$\begin{cases} -\Delta \xi &= \lambda \xi + \lambda \psi_0 \quad \text{in } \Omega, \\ \dfrac{\partial \xi}{\partial n} &= 0 \quad \text{on } \partial\Omega, \end{cases} \tag{1}$$

   with

   $$\int_{\Gamma_3} \xi \, ds = \frac{2}{\lambda} - \frac{1}{3}, \tag{2}$$

where Ω is the reference triangle $A_1 A_2 A_3$ in R^2, $A_1 = (0,0)$, $A_2 = (1,0)$, $A_3 = (0,1)$, Γ_3 is the edge $A_1 A_3$, and

$$\psi_0 = \frac{1}{2}\{(1-x)^2 + y^2\} - \frac{1}{3},$$

which belongs to $L_0^2(\Omega)$. We denote the smallest value of λ by λ^*.

2. Some special techniques to reduce the computing time in the calculation of lower bounds of λ^* using an iterative method.

Then, notice that the desired optimal constant C_0 is bounded as $C_0 \leq \dfrac{1}{\sqrt{\lambda^*}}$. Furthermore, it is expected that such an upper bound is particularly sharp by the consideration in [6].

2   Strategy

Note that, for a given λ, satisfying $0 < \lambda < \pi^2$, (1) has a unique solution ξ. Indeed, Proposition 1 in the below shows that π^2 coincides with the smallest positive eigenvalue of $-\Delta$ on Ω with Neumann boundary condition. Then, the well known Fredholm alternative theory yields the desired conclusion. We first calculate rigorously the solution ξ to (1) for a given λ, and then check whether the condition (2) holds. Lower bounds of the smallest value λ^* are obtained by the following iterative method.

1. Calculate an approximation $\hat{\lambda}$ to λ^*.
2. Set an interval $\Lambda \equiv [\underline{\Lambda}, \overline{\Lambda}]$ where $\underline{\Lambda}$ equals 0 and $\overline{\Lambda}$ is taken as some small positive number. Try to prove that there exists no solution ξ which satisfies (1) and (2) for any $\lambda \in \Lambda$. Then $\overline{\Lambda}$ gives a lower bound of λ^*.
3. If we can prove the above 2, then take a new interval Λ, extended to positive direction, so that it intersects with the old Λ. And try to prove again that there is no solution for the new Λ. If we can prove it, then $\overline{\Lambda}$ of the new Λ also gives a lower bound.
4. Repeat the extension described above until the Λ contains as large $\overline{\Lambda}$ as possible, which is expected to approach to $\hat{\lambda}$ step by step.
5. If we are unable to prove 3 for an interval Λ very close to $\hat{\lambda}$, then try to prove that there exists a solution pair (ξ, λ) satisfying (1) and (2) simultaneously.

In the actual computation, we can start from $\underline{\Lambda} = \lambda_N \equiv \dfrac{1}{0.81^2}$ because of Natterer's result ([7]).

Proposition 1. π^2 is the smallest positive eigenvalue for the problem

$$\begin{cases} -\Delta u = \lambda u & \text{in } \Omega, \\ \dfrac{\partial u}{\partial n} = 0 & \text{on } \partial\Omega. \end{cases} \tag{3}$$

Proof. By some simple calculations it follows that $\lambda = \pi^2$ and $u(x, y) = \cos \pi x - \cos \pi y$ satisfy (3). We denote the smallest positive eigenvalue for (3) by $\hat{\lambda}$ and the associated eigenfunction by $\hat{u}$, and suppose that $0 < \hat{\lambda} < \pi^2$. We now extend $\hat{u}$ to the function on the rectangle $\tilde{\Omega} = (0, 1) \times (0, 1)$ by setting $\tilde{u} := \hat{u}(1 - y, 1 - x)$ for $y > 1 - x$. Then, it is easily seen that $\tilde{u} \in H^1(\tilde{\Omega})$ and observe that the equalities which follow from the well known characterization of the smallest positive eigenvalue of (3)

$$
\hat{\lambda} = \min_{\int_\Omega v = 0} \frac{\int_\Omega |\nabla v|^2}{\int_\Omega |v|^2}
$$

$$
= \frac{\int_\Omega |\nabla \hat{u}|^2}{\int_\Omega |\hat{u}|^2} \tag{4}
$$

$$
= \frac{\int_{\tilde{\Omega}} |\nabla \tilde{u}|^2}{\int_{\tilde{\Omega}} |\tilde{u}|^2}.
$$

The last equality implies that the smallest positive eigenvalue for the operator $-\Delta$ with Neumann condition is less than π^2, which contradicts the well known result.

3   The Method to Calculate a Rigorous Solution

We define the functional subspace $H \equiv H^1(\Omega) \cap L_0^2(\Omega)$ with norm $\|\phi\|_H \equiv \|\nabla \phi\|$ where $\|\cdot\|$ is the usual L^2-norm on Ω. Let $\mathcal{T}_1$ be a uniform triangulation of Ω with mesh size h, and let $S_h^* \subset H^1(\Omega)$ be a linear finite element subspace on $\mathcal{T}_1$. And we define the projection P_h from H to $S_h^* \cap L_0^2(\Omega)$ by

$$
\begin{cases}
(\nabla P_h u, \nabla v_h) = (\nabla u, \nabla v_h) & \forall v_h \in S_h^*, \\
\\
(P_h u, 1) = 0,
\end{cases} \tag{5}
$$

for $u \in H^1(\Omega)$.

For any $f \in L_0^2(\Omega)$, let $u = Kf \in H$ denote a unique solution of the following problem:

$$
\begin{cases}
-\Delta u = f & in \quad \Omega, \\
\\
\dfrac{\partial u}{\partial n} = 0 & on \quad \partial \Omega, \\
\\
\displaystyle\int_\Omega u \, d\Omega = 0.
\end{cases} \tag{6}
$$

When we define $F\xi \equiv K\lambda(\xi + \psi_0)$, the affine operator F is clearly a compact operator on H and (1) is rewritten as the following fixed point equation

$$\xi = F\xi. \tag{7}$$

Applying Schauder's fixed point theorem, if we find a set Ξ which is a bounded convex and closed subset $\Xi \subset H$ such that

$$F\Xi \subset \Xi$$

holds, where $F\Xi \equiv \{f \in H \mid f = F\xi, \xi \in \Xi\}$, it implies that there exists an exact solution to the fixed point equation (7) within the set Ξ.

Actually we deal with an equivalent form to (7) as below.
For $\xi \in H$, define the mapping $T\xi$ by

$$T\xi = t_h + t_\perp, \tag{8}$$
$$(\nabla t_h \, \nabla \phi_h^*) - \lambda(t_h, \phi_h^*) = (\lambda((I - P_h)\xi + \psi_0), \phi_h^*), \qquad \forall \phi_h^* \in S_h^*, \tag{9}$$
$$t_\perp = (I - P_h)K(\lambda\xi + \lambda\phi), \tag{10}$$

where $t_h \in S_h^*$, $t_\perp \in (S_h^*)^\perp$ and I denotes the identity operator on H.
Note that we can solve the equation (9) with respect to t_h for $0 < \lambda < \pi^2$ if h is sufficiently small. It is easy to see that a fixed point equation

$$\xi = T\xi \tag{11}$$

is equivalent to (7) and the operator T is also compact on H. We try to apply Schauder's fixed point theorem to (11).

Now we describe the procedure to construct the set Ξ the so called candidate set.
First define the set $\Xi_\perp$ by

$$\Xi_\perp = \{\xi_\perp \in H \mid \|\nabla\xi_\perp\| \leq \alpha\}, \tag{12}$$

for a certain small positive number α. Then define

$$\Xi_h = \{\xi_h \in S_h^* \mid (\nabla\xi_h \, \nabla\phi_h^*) - \lambda(\xi_h, \phi_h^*) = (\lambda(\xi_\perp + \psi_0), \phi_h^*), \\ \forall \phi_h^* \in S_h^*, \quad \xi_\perp \in \Xi_\perp\}, \tag{13}$$

and set

$$\Xi = \Xi_h + \Xi_\perp.$$

In order to verify the inclusion $T\Xi \subset \Xi$, it is sufficient to show that

$$P_h T\Xi \subset \Xi_h,$$
$$(I - P_h)T\Xi \subset \Xi_\perp,$$

where $P_h T \Xi = \{ P_h T \xi \mid \xi \in \Xi \}$ and $(I - P_h) T \Xi = \{ (I - P_h) T \xi \mid \xi \in \Xi \}$. From the definition of the operator T and the set Ξ, we can see that $P_h T \Xi = \Xi_h$. Therefore we merely confirm the inclusion

$$(I - P_h) T \Xi \subset \Xi_\perp,$$

which means the radius of the set $(I - P_h) T \Xi$ is less than α.

Note that $K(\lambda \xi + \lambda \phi) \in H^2(\Omega)$. From the well known arguments($cf.$ [6]), there exists a numerically determined constant C_1 such that

$$\|\nabla (u - P_h u)\| \leq C_1 h \|\Delta u\| \tag{14}$$

holds for any solution u of (6). At first we may take $C_1 = 0.81$ from the result of Natterer, and once we obtain a $\overline{\Lambda}$ as a lower bound of λ^* then we can take $C_1 = \dfrac{1}{\sqrt{\overline{\Lambda}}}$.

Using the error estimate (14), we have

$$\|\nabla t_\perp\| \leq C_1 h \lambda \|\xi + \psi_0\|,$$

for an arbitrary $\xi \in \Xi$ and $t_\perp = T\xi - t_h$. Therefore the inclusion $(I - P_h) T \Xi \subset \Xi_\perp$ comes from

$$C_1 h \lambda \sup_{\xi \in \Xi} \|\xi + \psi_0\| \leq \alpha. \tag{15}$$

We will check (15) for a given α with the candidate set Ξ. More concrete explanations will be shown in Section 5.

4   Checking the Condition

For a certain interval Λ and a candidate set Ξ, after confirming the validity of condition (15) for all $\lambda \in \Lambda$, we verify that the condition (2) **does not** hold, namely,

$$\int_{\Gamma_3} \xi \, ds \neq \frac{2}{\lambda} - \frac{1}{3}, \quad \forall \xi \in \Xi, \forall \lambda \in \Lambda. \tag{16}$$

If all $\xi \in \Xi$ satisfy the above condition, then it proves that any $\lambda \in \Lambda$ is not equal to λ^*.

We write the left-hand side of (16) as

$$\int_{\Gamma_3} \xi \, ds = \int_{\Gamma_3} \xi_h \, ds + \int_{\Gamma_3} \xi_\perp \, ds, \tag{17}$$

where $\xi_h = P_h \xi$ and $\xi_\perp = (I - P_h) \xi$. Since $\xi_h \in \Xi_h$, the first term in the right-hand side of (17) can be estimated through the definition (13) of Ξ_h.

In order to estimate the second term, we introduce an equality which holds for any element $\tau_m \in \mathcal{T}_1$ having an edge on Γ_3.

Let $\tau_m = \triangle ABC$ with $A = (0, mh)$, $B = (h, mh)$ and $C = (0, (m+1)h)$ for $m = 0, \cdots, N-1$, where $N = 1/h$. Moreover, let $f(x,y) = 1 - \dfrac{x}{h} - \dfrac{y-mh}{h}$, $g(x,y) = \dfrac{y-mh}{h}$, $\mathbf{a} = \overrightarrow{BA}$, and $\mathbf{b} = \overrightarrow{CB}$. Then the following lemma holds.

Lemma 1. *On each element τ_m defined above, the equality*

$$\int_{AC} u \, ds \;=\; \frac{2}{h}\int_{\tau_m} u \, dxdy + \frac{1}{h}\int_{\tau_m}(f(x,y)\mathbf{a} - g(x,y)\mathbf{b}) \cdot \nabla u \, dxdy \quad (18)$$

holds for an arbitrary function $u \in H^1(\tau_m)$.

Proof.   Using Green's formula, observe that

$$\int_{\tau_m} \mathbf{a} \cdot \nabla(u\,f)\, dxdy \;=\; \int_{AC}(\mathbf{a}\cdot\mathbf{n})u\,f\,ds$$

$$=\; h\int_{AC} u\,f\,ds,$$

$$\int_{\tau_m} \mathbf{b} \cdot \nabla(u\,g)\, dxdy \;=\; \int_{AC}(\mathbf{b}\cdot\mathbf{n})u\,g\,ds$$

$$=\; -h\int_{AC} u\,g\,ds,$$

where $\mathbf{n}$ denotes the outer normal vector on the edge AC, namely, $\mathbf{n} = (-1,0)$. From these equalities, we have

$$\int_{AC} u\,ds \;=\; \int_{AC}(f+g)\,u\,ds$$

$$=\; \frac{1}{h}\int_{\tau_m}\{\mathbf{a}\cdot\nabla(u\,f) - \mathbf{b}\cdot\nabla(u\,g)\}\,dxdy.$$

Taking notice that $\mathbf{a}\cdot\nabla f = 1$ and $\mathbf{b}\cdot\nabla g = -1$, the conclusion of the lemma is proved.

Let E_3 be a set of elements that have their edges on Γ_3. We know that the area of E_3 equals $\dfrac{h}{2}$. Using $\|\xi_\perp\| \leq C_1 h\|\nabla\xi_\perp\|$ which comes from Aubin-Nitsche's trick (see [6]), we have

$$\left|\int_{E_3}\xi_\perp dxdy\right| \;\leq\; \sqrt{\int_{E_3} 1\,dxdy}\sqrt{\int_{E_3}\xi_\perp^2\,dxdy}$$

$$\leq\; \sqrt{\frac{h}{2}}C_1 h\|\nabla\xi_\perp\|.$$

From this estimate and (18), it follows that

$$|\int_{\Gamma_3} \xi_\perp ds| \le (\sqrt{\frac{2}{h}}C_1 h + \frac{1}{h\sqrt{h}}\|f\mathbf{a} - g\mathbf{b}\|_{\tau_m})\|\nabla\xi_\perp\|.$$

Note that

$$\|f\mathbf{a} - g\mathbf{b}\|_{\tau_m} = \sqrt{\int_{\tau_m} (f(x,y)\mathbf{a} - g(x,y)\mathbf{b})^2 dxdy}$$

$$= \frac{h^2}{\sqrt{3}}$$

holds for each $m = 0, \cdots, N - 1$, and we have

$$|\int_{\Gamma_3} \xi_\perp ds| \le (\sqrt{2}C_1 + \frac{1}{\sqrt{3}})\sqrt{h}\alpha. \tag{19}$$

From (19), a sufficient condition for (16) is written as

$$\frac{2}{\lambda} - \frac{1}{3} \notin \int_{\Gamma_3} \xi_h ds + [-1,1](\sqrt{2}C_1 + \frac{1}{\sqrt{3}}\sqrt{h})\alpha, \tag{20}$$

where $[-1, 1]$ denotes an interval in R.

5   Some Computational Techniques for Efficient Enclosure Methods

In this section, we show some actual computational techniques used in the verification procedures for (15) and (20), which enable us to reduce the CPU cost.

To explain our techniques for calculation, we introduce several notations on $\{\phi_i^*\}_{i=1,\cdots,M}$, a basis of S_h^*, as below.

$$\mathbf{\Phi} \equiv (\phi_1^*, \phi_2^*, \cdots, \phi_M^*)^T,$$

$$\|\mathbf{\Phi}\| \equiv \sqrt{\sum_{i=1}^{M} \|\phi_i^*\|^2},$$

$$D^* \in R^{M \times M} \quad : \quad D_{ij}^* = (\nabla\phi_i^*, \nabla\phi_j^*),$$

$$L^* \in R^{M \times M} \quad : \quad L_{ij}^* = (\phi_i^*, \phi_j^*),$$

$$G_\lambda \equiv D^* - \lambda L^*,$$

$$\|G_\lambda^{-1}\|_2 \equiv \sup_{\mathbf{x} \in R^M, \|\mathbf{x}\| \ne 0} \frac{\mathbf{x}^T G_\lambda^{-1}\mathbf{x}}{\mathbf{x}^T \mathbf{x}}.$$

Taking

$$\xi_h = \mathbf{z}^T\mathbf{\Phi},$$

where $\mathbf{z} = (z_1, z_2, \cdots, z_M)^T \in R^M$ which is called a coefficient vector, we have

$$G_\lambda \mathbf{z} = \lambda(\xi_\perp + \psi_0, \mathbf{\Phi}),$$

from the definition of Ξ_h. Here, $(\xi_\perp + \psi_0, \mathbf{\Phi})$ means a vector whose elements are $(\xi_\perp + \psi_0, \phi_i^*)$, $i = 1, \cdots, M$. Setting

$$\mathbf{r} \equiv G_\lambda^{-1}(\psi_0, \mathbf{\Phi}),$$

the coefficient vector $\mathbf{z}$ can be written by

$$\mathbf{z} = \lambda \mathbf{r} + \lambda G_\lambda^{-1}(\xi_\perp, \mathbf{\Phi}),$$

and we estimate the second term in the right-hand side as

$$\begin{aligned} \|G_\lambda^{-1}(\xi_\perp, \mathbf{\Phi})\|_2 &\leq \|G_\lambda^{-1}\|_2 \|\xi_\perp\| \|\mathbf{\Phi}\| \\ &\leq C_1 h\alpha \|G_\lambda^{-1}\|_2 \|\mathbf{\Phi}\|, \end{aligned} \qquad (21)$$

using $\|\xi_\perp\| \leq C_1 h \|\nabla \xi_\perp\|$. Here $\|\cdot\|_2$ in the left-hand side denotes the Euclidean norm in R^M. With the above notations and (21), we obtain a sufficient condition for (15) as follows.

$$\frac{C_1 h\lambda \|\lambda \mathbf{r}^T \mathbf{\Phi} + \psi_0\|}{1 - (C_1 h)^2 \lambda(1 + \lambda \|G_\lambda^{-1}\|_2 \|\mathbf{\Phi}\| \|L^*\|_2^{1/2})} \leq \alpha. \qquad (22)$$

Note that (22) gives a lower bound of α and we can use this value in checking the condition (20). Let us introduce some more notations.

$$\begin{aligned} \mathbf{p} &\equiv (\|\phi_1^*\|, \|\phi_2^*\|, \cdots, \|\phi_M^*\|)^T, \\ \mathbf{q} &\equiv \left(\int_{\Gamma_3} \phi_1^* \, ds, \int_{\Gamma_3} \phi_2^* \, ds, \cdots, \int_{\Gamma_3} \phi_M^* \, ds\right)^T, \\ \mathbf{s} &\equiv (|s_1|, |s_2|, \cdots, |s_M|)^T, s_i = (G_\lambda^{-1} \mathbf{q})_i. \end{aligned}$$

The following condition comes from (20).

$$\frac{2}{\lambda} - \frac{1}{3} - \lambda \mathbf{r}^T \mathbf{q} \notin [-1, 1] \times \left\{\lambda C_1 h \mathbf{p}^T \mathbf{s} + \left(\sqrt{2} C_1 + \frac{1}{\sqrt{3}}\right)\sqrt{h}\right\}\alpha.$$

Numerical experiments suggest that the left-hand side of the above relation is positive. Therefore we will check a sufficient condition as follows.

$$\frac{2}{\lambda} - \frac{1}{3} - \lambda \mathbf{r}^T \mathbf{q} > \left\{\lambda C_1 h \mathbf{p}^T \mathbf{s} + \left(\sqrt{2} C_1 + \frac{1}{\sqrt{3}}\right)\sqrt{h}\right\}\alpha. \qquad (23)$$

Combining (22) and (23), we define the following functions with respect to λ.

$$\begin{aligned} f_1(\lambda) &\equiv \frac{2}{\lambda} - \frac{1}{3} - \lambda \mathbf{r}^T \mathbf{q}, \\ f_2(\lambda) &\equiv 1 - (C_1 h)^2 \lambda(1 + \lambda \|G_\lambda^{-1}\|_2 \|\mathbf{\Phi}\| \|L^*\|_2^{1/2}), \\ f_3(\lambda) &\equiv \lambda C_1 h \mathbf{p}^T \mathbf{s} + \left(\sqrt{2} C_1 + \frac{1}{\sqrt{3}}\right)\sqrt{h}, \\ f_4(\lambda) &\equiv C_1 h\lambda \|\lambda \mathbf{r}^T \mathbf{\Phi} + \psi_0\|. \end{aligned}$$

What we have to do for proving $\lambda < \lambda^*$ is to show

$$f_2(\lambda) \; > \; 0, \tag{24}$$

and

$$f_1(\lambda)f_2(\lambda) - f_3(\lambda)f_4(\lambda) \; > \; 0. \tag{25}$$

Moreover, we may use the result of [8], that is, for $\lambda_N \leq \lambda \leq \hat{\lambda}$, the norm of G_λ^{-1} can be estimated by

$$\|G_\lambda^{-1}\|_2 \; \leq \; \frac{\theta}{\lambda}, \tag{26}$$

where

$$\theta \; \equiv \; \frac{1}{\mathbf{x_1}^T L^* \mathbf{x_1}},$$

$$\mathbf{x_1} \; \equiv \; (\frac{1}{\sqrt{M}}, \frac{1}{\sqrt{M}}, \cdots, \frac{1}{\sqrt{M}})^T.$$

We give a rough sketch of the idea for derivation of (26) in [8].

Let μ_1 and μ_2 be the first and the second smallest eigenvalue of G_λ, respectively. Since (1) is a Neumann problem, the matrix D^* which corresponds to the Laplacian is semi-positive definite and 0 is the smallest eigenvalue. Actually,

$$D^* \mathbf{x_1} \; = \; 0 \tag{27}$$

holds.

In what follows, we restrict λ within the range $\lambda_N \leq \lambda \leq \hat{\lambda}$ in which the matrix G_λ is nonsingular. From (27) and the positive definiteness of L, we have

$$\mu_1 \; < \; 0 \; < \; \mu_2.$$

Thus one of μ_1 and μ_2, having the smallest absolute value, gives the smallest singular value of G_λ, namely, $\dfrac{1}{\|G_\lambda^{-1}\|_2}$.

Since μ_1 is the smallest eigenvalue of G_λ, we have

$$\mu_1 \; \leq \; -\lambda \mathbf{x_1}^T L^* \mathbf{x_1} \tag{28}$$
$$=: \; -\chi_1(\lambda).$$

from (27) and the symmetry of G_λ. Moreover, letting ρ_2 be the second smallest eigenvalue of the matrix D^* and using Weyl's lemma which gives error estimation of eigenvalues of symmetric matrices, we have the following estimate for a lower bound of μ_2.

$$\mu_2 \; \geq \; \rho_2 - \lambda\|L^*\|_2 \tag{29}$$
$$=: \; \chi_2(\lambda).$$

Therefore we obtain an upper bound of $\|G_\lambda^{-1}\|_2$ by

$$\|G_\lambda^{-1}\|_2 \leq \max(\frac{1}{\chi_1(\lambda)}, \frac{1}{\chi_2(\lambda)}).$$

Actually, we can prove that $\dfrac{1}{\chi_1(\lambda)} \geq \dfrac{1}{\chi_2(\lambda)}$ for $\lambda_N \leq \lambda \leq \hat{\lambda}$.

Using (26), we can replace the definition of f_2 by

$$f_2(\lambda) \equiv 1 - (C_1 h)^2 \lambda (1 + \theta \|\Phi\| \|L^*\|_2^{1/2}).$$

For reduction of the CPU time, we adopt the following technique. Once we verify that (24) and (25) hold for some λ_1, then we can determine a λ_0, such that (24) and (25) hold for any $\lambda \in \Lambda \equiv [\lambda_0, \lambda_1]$.

Now, for a positive constant $\eta > 0$ such that $\lambda_1 > \eta$, set $\lambda = \lambda_1 - \eta$. Let $\mathbf{r}_1$ and $\mathbf{s}_1$ denote the vectors $\mathbf{r}$ and $\mathbf{s}$ for λ_1, respectively. It can be seen that

$$\mathbf{r} - \mathbf{r}_1 = \eta G_\lambda^{-1} L^* \mathbf{r}_1.$$

From this and (26), it follows that

$$\|\mathbf{r} - \mathbf{r}_1\|_2 \leq \frac{\eta}{\lambda}\theta \|L^*\|_2 \|\mathbf{r}_1\|_2. \tag{30}$$

In order to derive a similar estimate for $\mathbf{s}$, taking account of

$$|(G_\lambda^{-1}\mathbf{q})_i| \leq (\mathbf{s}_1)_i + |((G_\lambda^{-1} - G_{\lambda_1}^{-1})\mathbf{q})_i|,$$

for $i = 1, \cdots, M$ to obtain

$$\|\mathbf{s} - \mathbf{s}_1\|_2 \leq \frac{\eta}{\lambda}\theta \|L^*\|_2 \|\mathbf{s}_1\|_2. \tag{31}$$

Using (30) and (31), we choose η so that the following inequalities hold.

$$\begin{aligned}
f_1(\lambda) &\geq f_1(\lambda_1) + k_1\eta, \\
f_2(\lambda) &\geq f_2(\lambda_1) - k_2\eta, \\
f_3(\lambda) &\leq f_3(\lambda_1) + k_3\eta, \\
f_4(\lambda) &\leq f_4(\lambda_1) + k_4\eta,
\end{aligned}$$

where,

$$\begin{aligned}
k_1 &= \frac{2}{\lambda_1^2} - \|\mathbf{r}_1\|_2 \|\mathbf{q}\|_2 \|L^*\|_2 \theta + \mathbf{r}_1^T\mathbf{q}, \\
k_2 &= (C_1 h)^2 (1 + \|\Phi\| \|L^*\|_2^{1/2} \theta), \\
k_3 &= C_1 h\, (\|\mathbf{s}_1\|_2 \|\mathbf{p}\|_2 \|L^*\|_2 \theta - \mathbf{s}_1^T\mathbf{p}), \\
k_4 &= C_1 h\, (\lambda_1 \|\mathbf{r}_1\|_2 \|L^*\|_2^{3/2}\theta + \|(\psi_0, \Phi)\|_2 \|L^*\|_2^{1/2} \theta - \|\lambda_1 \mathbf{r}_1^T\Phi + \psi_0\|).
\end{aligned}$$

Note that, in actual calculation, it is not necessary to underestimate the term $\|\lambda_1 \mathbf{r_1}^T \Phi + \psi_0\|$ in k_4, because the same expression appears in the righthand side of the fourth inequality as $f_4(\lambda_1) = C_1 h \lambda_1 \|\lambda_1 \mathbf{r_1}^T \Phi + \psi_0\|$.

Thus, through these inequalities together with (24) and (25), we can estimate the value of λ_0, and it is proved that any $\lambda \in \Lambda = [\lambda_0, \lambda_1]$ gives no solution ξ to (1) satisfying (2). Applying this method to several points of $\lambda_1 \in [\lambda_N, \lambda^*]$, we can obtain a lower bound of λ^* with low computational cost.

6   Numerical Results

An approximate value of λ^* that we obtained was 4.111128. We tried to prove that there is no solution in $\lambda_N \leq \lambda \leq 4.1$.

I. For $\lambda_N \leq \lambda \leq 4.0$, we have proved that there is no solution by taking S_h^* with $h = \dfrac{1}{40}$.

II. For $4.0 \leq \lambda \leq 4.09$, we have proved that there is no solution with $h = \dfrac{1}{80}$.

III. For $4.09 \leq \lambda \leq 4.0995$, we have proved that there is no solution with $h = \dfrac{1}{120}$.

IV. For $4.0995 \leq \lambda \leq 4.1$, we have failed to prove that there is no solution.

Thus we proved that there is a solution ξ satisfying (1) and (2) within an interval $\Lambda^* = [4.09940, 4.12611]$ by our numerical verification method (cf. [3]–[5]).

Conclusion

From the numerical results mentioned above, we obtain a rigorous lower bound of λ^* as 4.0995, which gives a rigorous upper bound of C_1 as 0.4939. This value should be sufficiently sharp for our practical use, because a lower bound of C_0 is given as $0.467 \leq C_0$ (see, [6]).

In order to avoid the rounding error in the floating point computations, we used Fortran 90 with verified interval library INTLIB[2], coded by Kearfott, on the SUN Workstation Ultra Enterprise 450, Computing Center, Kyushu University.

References

1. Grisvard, P., Elliptic problems in nonsmooth domain, Pitman, Boston, 1985.
2. Kearfott, R.B. & Kreinovich, V. (eds.), *Applications of Interval Computations*, Kluwer Academic Publishers, Dordrecht, The Netherlands 1996.
3. Nakao, M.T., A numerical approach to the proof of existence of solutions for elliptic problems, Japan Journal of Applied Mathematics 5, (1988) 313 - 332.

4. Nakao, M.T. & Yamamoto, N., Numerical verification of solutions for nonlinear elliptic problems using L^∞ residual method, Journal of Mathematical Analysis and Applications **217**, (1998), 246-262.

5. Nakao, M.T., Yamamoto, N. & Nagatou, K., Numerical verifications of eigenvalues of second-order elliptic operators, Japan Journal of Industrial and Applied Mathematics 16 (1999), 307-320.

6. Nakao, M.T. & Yamamoto, N., A guaranteed bound of the optimal constant in the error estimates for linear triangular element, to appear in Computing Supplementum (2001).

7. Natterer, F., Berechenbare Fehlerschranken für die Methode der Finiten Elemente, International Series of Numerical Mathematics, vol. 28, Birkhäuser Verlag, Basel (1975), 109-121.

8. Yamamoto, N., Nakao, M.T. & Watanabe, Validated computation for a linear elliptic problem with a parameter, Advances in Numerical Mathematics ; Proc. Fourth Japan-China Joint Seminar on Numerical Mathematics, Aug. 24-28, 1998 (H. Kawarada, M. Nakamura, Z. Shi, eds.), GAKUTO International Series Mathematical Sciences and Applications, Volume 12, Gakkotosho, Tokyo, Japan (1999), 155-162.

Nonsmooth Global Optimization

Dietmar Ratz

Institut für Angewandte Informatik und Formale Beschreibungsverfahren,
Universität Karlsruhe (TH),
D-76128 Karlsruhe, Germany.
dra@aifb.uni-karlsruhe.de

Abstract. What can interval analysis do for Nonsmooth Global Optimization?
We will answer this question by presenting an overview on pruning techniques
based on interval slopes in the context of interval branch-and-bound methods for
global optimization. So, this paper is intended to guide interested researchers to
future research and improvements or to ways of using the techniques in different
contexts.

We show that it is possible to replace the frequently used monotonicity test
by a pruning step, and we demonstrate the theoretical and practical effect of this
pruning step within a first-order model algorithm for global optimization. It is
underlined how the technique provides considerable improvement in efficiency for
a model algorithm.

1 Introduction

1.1 Global Optimization

In almost all scientific disciplines using mathematical models, there are many
practical problems which can be formulated as multiextremal *global* optimization problems. By definition, these problems are concerned with the computation and characterization of global maxima or minima of a real-valued
objective function that possesses different local optima in the feasible set.
These tasks belong to the complexity class of NP-hard problems [28]. Such
problems are very difficult to solve.

Traditional nonlinear programming techniques based on local information
have not been successful in solving global optimization problems. Their de-
ficiency is due to the intrinsic multiextremality of the problem formulation.
Local tools, such as descent methods, cannot be expected to yield more than
local solutions. Moreover, even the determination of a local optimizer (i.e. a
local optimal point) is not always easy. Apart from this, classical methods
do not recognize conditions for global optimality. In many practical applications the number of local optima increases exponentially with the dimension
of the problem. Thus, most of the traditional approaches fail to escape from
a local optimum in order to continue the search for the global solution. For
these reasons, global solution methods must be significantly different from
standard nonlinear programming techniques. Of course, they can and must
be much more expensive computationally.

In practical problems, even if they are of global character, it is often sufficient to use algorithms from local optimization and to achieve approximative solutions. However, sometimes the outcome of results produced via local optimization is not a satisfactory answer. Especially when several local optimizers do appear, the additional question arises whether the solution produced by a local optimization procedure can be improved. Without being sure of having obtained a global optimum, it is hard to decide whether such an improvement is possible or not. Thus, finding a local solution might be adequate in some cases, but in others it might mean incurring a significant cost penalty or (even worse) getting an incorrect solution for a physical problem.

Therefore, global optimization can clearly be useful. It has received a lot of attention, due to the success of new algorithms for solving large classes of problems from diverse areas such as engineering design and control, computational chemistry and biology, structural optimization, computer science, operations research, and economics [7].

During the past 10 years the field of global optimization has been growing rapidly. The number of publications on all aspects of global optimization has been increasing steadily. Many new theoretical, algorithmic, and computational contributions have resulted, which can be found in the *Journal of Global Optimization*, for example. This is also underlined by the increase of research activities with several specialized conferences on global optimization and applications. With the development and implementation of practical global optimization algorithms, more and more scientists in diverse disciplines have been using global optimization techniques to solve their problems. An important fact is, that there is a noticeable gain in interest and importance of interval methods. This can be recognized in several conference proceedings and survey books (cf. [2], [7], [8], [17], [20], and [40]). Interval methods are almost predestined for addressing the *global* optimization problem, because interval analysis is an excellent tool for obtaining *global* information.

1.2   Interval Methods

In [40], Törn and Žilinskas gave a primary classification of global optimization methods. They divided the methods in two non-overlapping classes with respect to the accuracy, i.e. those with guaranteed accuracy and those without. The class of methods with guaranteed accuracy is called *covering methods*. The residual class of methods is divided into *direct methods* and *indirect methods*. Direct methods utilize only local information (such as random search methods, clustering methods, generalized descent methods). Indirect methods use local information to build a global model (methods approximating the level sets, methods approximating the objective function). Interval methods belong to the group of covering methods. In contrast to direct and indirect methods, covering methods can obtain solutions with guaranteed accuracy by exhaustive search. One aspect of those methods is the reliable elimination of subregions which do not contain a global optimizer. Interval arithmetic is ideally suited for this elimination technique.

An interval, even though representable by only two points, is an infinite set (a continuum). Thus, it is a carrier of an infinite amount of information, which means global information. Interval arithmetic can handle expressions for intervals. This means that the interval expression collects the information on the corresponding real expression for any real value within the interval. For example, one might address the question whether a function has zeros in a given interval. Then a single interval evaluation of the function may suffice to solve this question with complete mathematical certainty. If the resulting interval of the evaluation does not contain zero, then the range of the function over the interval also cannot contain zero. Thus, the function does not have any zero in the interval.

Very often, people misunderstand the use of interval arithmetic only as a means to control the rounding errors and to deliver bounds for them. In the context of global optimization methods, intervals are used very efficiently to keep global information. The error control is only a comfortable side effect which guarantees the accuracy of the results produced on a computer. Many people also believe that interval methods would lead to unrealistically wide intervals, but that is not true either.

Interval methods for global optimization combine interval arithmetic with the so-called *branch-and-bound technique*. This technique subdivides the search region in subregions (*branches*) and uses *bounds* for the objective function over the complete subregion to cut away some of these regions. Interval arithmetic provides the possibility to compute such bounds almost automatically. Additionally, interval techniques of higher order using gradient and Hessian information as well as local search methods can be included to improve the convergence speed and the overall efficiency of the methods. Such interval branch-and-bound methods are able to compute verified enclosures of highest accuracy for all global optimizers and for the global optimum value. The first basic interval approaches to global optimization go back to Moore [27] and Skelboe [39]. An excellent introduction, a reference for the techniques, and a guide to the history and the literature of interval methods for global optimization can be found in the recent book of Kearfott [20].

1.3   Nonsmooth Problems and Interval Slopes

Many practical optimization problems are nonsmooth optimization problems [25]. Typical examples are objective functions containing expressions such as $|\ldots|$ or $\max\{\ldots\}$. The derivatives of nonsmooth objective functions have jump discontinuities. For such problems, gradient type methods cannot be applied. One approach to overcome this difficulty is to replace the gradient by the generalized gradient [3] for which it is also possible to find an interval enclosure [31]. A similar approach is to specify formulas for the "elementary" nonsmooth functions to enable the computation of interval extensions for nonsmooth functions [20]. The main advantage of this approach is that nonsmooth problems can be treated with the same techniques as smooth problems.

In this paper another approach based on the so-called *interval slopes* introduced by Krawczyk and Neumaier [23] is investigated. Interval slopes together with centered forms offer the possibility to achieve better enclosures for the function ranges (independently from the objective function being continuously differentiable or not). Thus, they can improve the performance of interval branch-and-bound methods. Although, since slopes cannot replace the derivatives needed in the monotonicity test, the necessity of alternative box-discarding techniques arises. The most interesting fact concerning these new techniques (applicable to nonsmooth problems) is, that in many cases their application is superior to the application of derivatives (when applied to smooth problems).

The model algorithms presented in this paper can of course be extended to become more efficient and faster. For the numerical studies it was important to test all new tools in the absence of additional accelerating devices. Nevertheless, the techniques and algorithms are presented to guide interested readers to future research and improvements or to ways of using the techniques in different contexts.

Section 2 contains introductory material on interval extensions of real functions, centered forms, interval slopes, and interval branch-and-bound methods for global optimization, including basic algorithms used throughout the subsequent sections.

Section 3 introduces a global optimization method which incorporates a special pruning step generated by interval slopes in the context of onedimensional problems. The pruning step offers the possibility to cut away a large part of the currently investigated subregion. The theory for this slope pruning step is developed, hints for its implementation are described, and several examples are presented to underline its efficiency. For additional numerical examples with differentiable functions, the advantages of the pruning step over the monotonicity test are demonstrated.

The onedimensional case is treated separately for didactical reasons. Thus, Section 4 describes the extension of the pruning technique to the multidimensional case. First, a general multidimensional approach is discussed, which makes use of interval slope vectors. Then, a more successful componentwise approach is treated in detail. At the end of the section, several numerical examples are used to check the efficiency of the new technique. For the test problems in two variables, plots are given to clarify the structure of the objective functions.

Details on the subjects of this paper together with theory, algorithmic descriptions and implementations for automatic slope computation and for nonsmooth gloabl optimization in the onedimensional and multidimensional case can be found in [38]. The source code of all modules and programs described in this book are available by anonymous ftp from

$$\texttt{ftp://ftp.iam.uni-karlsruhe.de/}$$

in subdirectory `pub/slopego`.

2 Preliminaries

2.1 Notation

In the following, we denote *real numbers* by lower-case letters, e.g. $x \in \mathbb{R}$, and real bounded and closed *intervals* by capitals, e.g.

$$X = [\underline{x}, \overline{x}] = [\inf(X), \sup(X)] \in I\mathbb{R},$$

where $I\mathbb{R} = \{[a, b] \mid a \le b, \, a, b \in \mathbb{R}\}$ denotes the set of compact intervals. $d(X)$ and $m(X)$ denote the *diameter* (or width) and the *midpoint* of X, respectively. They are defined by

$$d(X) := \operatorname{diam}(X) := \overline{x} - \underline{x},$$

$$m(X) := \operatorname{mid}(X) \quad := \frac{\underline{x} + \overline{x}}{2}.$$

The *smallest* and the *greatest absolute value* of an interval X are denoted by

$$\langle X \rangle := \min\{|x| \mid x \in X\}, \text{ and}$$
$$|X| := \max\{|x| \mid x \in X\} = \max\{|\underline{x}|, |\overline{x}|\}, \tag{1}$$

i.e. both values are real values. However, the *absolute value* of an interval is an interval and is defined by

$$|X|_{[]} = \operatorname{abs}(X) := \{|x| \mid x \in X\} = [\langle X \rangle, |X|]. \tag{2}$$

We define the *relative diameter* of an interval X by

$$d_{\text{rel}}(X) := \begin{cases} \dfrac{d(X)}{\langle X \rangle} & \text{if } 0 \notin X \\ d(X) & \text{otherwise.} \end{cases} \tag{3}$$

Finally, the *hull* of two intervals is defined by

$$X \sqcup Y := [\min\{\underline{x}, \underline{y}\}, \max\{\overline{x}, \overline{y}\}].$$

We do not distinguish between scalar and vector quantities. Therefore, we also use lower-case letters for *real vectors* e.g.

$$y = (y_1, y_2, \dots, y_n)^\mathsf{T} \in \mathbb{R}^n$$

and capitals for *interval vectors* (also called *boxes*) e.g.

$$Y = [\underline{y}, \overline{y}] = [\inf(Y), \sup(Y)] \in I\!I\!R^n,$$

where

$$Y = (Y_1, Y_2, \ldots, Y_n)^{\mathsf{T}} \quad \text{and} \quad Y_i = [\underline{y_i}, \overline{y_i}] = [\inf(Y_i), \sup(Y_i)], \ i = 1, 2, \ldots, n.$$

For interval vectors the values $d(Y)$, $m(Y)$, $d_{\text{rel}}(Y)$, and the hull operation are defined componentwise. Finally, we introduce the notations

$$d_\infty(X) := \max_{1 \le i \le n} d(X_i), \text{ and}$$

$$d_{\text{rel},\infty}(X) := \max_{1 \le i \le n} d_{\text{rel}}(X_i)$$

for $X \in I\!I\!R^n$ and $e^{(k)} \in I\!R^n$ denoting the k-th unit vector.

2.2 Interval Extensions of Real Functions

We call a function $F : I\!R \to I\!R$ an *inclusion function* or an *interval extension* of $f : I\!R \to I\!R$ in X, if $x \in X$ implies $f(x) \in F(X)$. In other words,

$$f(X) = f_{\text{rg}}(X) = \{f(x) \mid x \in X\} \subseteq F(X),$$

where $f(X)$ or $f_{\text{rg}}(X)$ both denote the *range* of the function f on X. Inclusion functions (or interval extensions) for $f : I\!R^n \to I\!R^m$ are defined similarly. The inclusion function of the derivative f' of a function $f : I\!R \to I\!R$ is denoted by F', and the gradient ∇f of a function $f : I\!R^n \to I\!R$ is denoted by ∇F.

Interval extensions can be computed via interval arithmetic ([1],[9],[24],[27],[29]) for almost all functions specified by a finite algorithm (i.e. not only for given expressions). For example, we obtain the so-called *natural interval extension* by replacing all real operations ($+$, $-$, $\cdot$, and $/$) and elementary function calls within a real function by the corresponding interval operations and elementary interval function calls. A usual set of elementary functions within interval environments such as [21] or [9] is

$$\mathcal{F}_\mathcal{E} = \{ \text{sqr, sqrt, exp, ln,}$$
$$\text{sin, cos, tan, cot, arcsin, arccos, arctan, arccot,}$$
$$\text{sinh, cosh, tanh, coth, arsinh, arcosh, artanh, arcoth} \}.$$

If there is a nonnegative constant $\alpha \in I\!R$, independent of the box X, such that the interval extension F of $f : D \subseteq I\!R^n \to I\!R$ satisfies

$$d(F(X)) - d(f_{\text{rg}}(X)) \le \alpha d_\infty(X) \qquad \forall X \subseteq D,$$

then F is called a *first-order interval extension* of f. If there is a nonnegative constant $\beta \in I\!R$, independent of the box X, such that the interval extension F of $f : D \subseteq I\!R^n \to I\!R$ satisfies

$$d(F(X)) - d(f_{\text{rg}}(X)) \le \beta(d_\infty(X))^2 \qquad \forall X \subseteq D,$$

then F is called a *second-order interval extension* of f.

Natural interval extensions are first-order extensions and they have the *isotonicity* property, i.e. $X \subseteq Y$ implies $F(X) \subseteq F(Y)$ (cf. [1]). Second-order extensions may be obtained via so-called centered forms using derivatives and slopes.

2.3   Centered Forms and Interval Slopes

Centered forms (see [1], [20], or [29]) are special interval extensions and serve to reduce the overestimation in computing interval enclosures of the range of a function f over some interval X. Usually, a centered form is derived from the mean-value theorem. Suppose $f : D \subseteq I\!R^n$ is differentiable on its domain D. Then $f(x) = f(c) + \nabla f(\xi)(x - c)$ with some fixed $c \in D$ and ξ between x and c. Let $c, x \in X$, so $\xi \in X$. Therefore

$$
\begin{aligned}
f(x) = f(c) + \nabla f(\xi)(x - c) \ &\in \ f(c) + \nabla f(X) \cdot (x - c) \\
&\subseteq \ \underbrace{f(c) + \nabla F(X) \cdot (X - c)}_{=:F_c(X)}.
\end{aligned}
\tag{4}
$$

$F_c(X)$ is called *centered form* or *generalized mean-value form* of f over X with *center* c, and we have $f_{\mathrm{rg}}(X) \subseteq F_c(X)$ (note that the center is not necessarily the midpoint of X). Here, f is extended with respect to every $x \in X$, since $G = \nabla F(X)$ is an interval evaluation of the derivative (gradient) of f over the entire interval X. If the inclusion function ∇F for the gradient ∇f is of at least first order, then $F_c(X)$ is a second-order extension of f (cf. [22]).

Krawczyk and Neumaier ([23],[29]) showed that if we have an interval vector $S \in I\!R^n$ such that, for all $x \in X$ we have

$$
f(x) = f(c) + s \cdot (x - c) \quad \text{for some } s \in S,
\tag{5}
$$

then the interval $F_s(X) := f(c) + S \cdot (X - c)$ encloses the range of f over X, that is $f_{\mathrm{rg}}(X) \subseteq F_s(X)$. Such an interval vector S can be calculated by means of an interval slope and not only with an interval derivative. If we use a slope, then f is extended with respect to an arbitrary but fixed $c \in X$.

Definition 1. The function $s_f : D \times D \to I\!R^n$ with

$$
f(x) = f(c) + s_f(c, x) \cdot (x - c)
$$

is called a **slope** (or slope vector) of $f : D \subseteq I\!R^n \to I\!R$ (between c and x). In the one-dimensional case ($D \subseteq I\!R$), we have

$$
s_f(c, x) = \begin{cases} \dfrac{f(x) - f(c)}{x - c} & \text{if } x \neq c \\[2mm] \widetilde{s} & \text{if } x = c, \end{cases}
$$

where $\tilde{s} \in \mathbb{R}$ may be arbitrarily chosen. Assuming f to be differentiable and the slope to be continuous, we can define $\tilde{s} := f'(c)$.

Moreover, we define the **interval slope** of f over the interval X by

$$s_f(c, X) := \{s_f(c, x) \mid x \in X, \quad x \neq c\},$$

where it is not necessary that f is differentiable.

Remarks: (i) It is easy to see that $S = s_f(c, X)$ satisfies (5) and

$$f(x) \in f(c) + S \cdot (x - c) \subseteq f(c) + S \cdot (X - c). \tag{6}$$

(ii) Often $c = m(X)$ is used to compute the interval slope.
(iii) If we assume f to be continuously differentiable, then we have (cf. [29])

$$s_f(c, X) \subseteq s_f(X, X) = \nabla F(X). \tag{7}$$

In general, slopes lead to narrower intervals than derivatives, if the latter are applicable. For $f : D \subseteq \mathbb{R} \to \mathbb{R}$ and $c \in X \subseteq D$ we have (cf. [20])

$$\lim_{d(X) \to 0} \frac{d(f'_{rg}(X))}{d(s_f(c, X))} = 2.$$

Example 1. Let $f(x) = \frac{1}{4}x^2 - x + \frac{1}{2}$. Then we have $f'(x) = \frac{1}{2}x - 1$. Therefore, the range of f' over $X = [1, 7]$ is $f'(X) = [-0.5, 2.5]$ (cf. Figure 1) and its diameter is 3.

On the other hand,

$$s_f(c, x) = \frac{\frac{1}{4}(x^2 - c^2) - x + c}{x - c} = \frac{1}{4}(x + c) - 1,$$

and therefore, the range of s_f over $X = [1, 7]$ for $c = 4$ is $s_f(c, X) = [0.25, 1.75]$ (cf. Figure 2) and its diameter is 1.5.

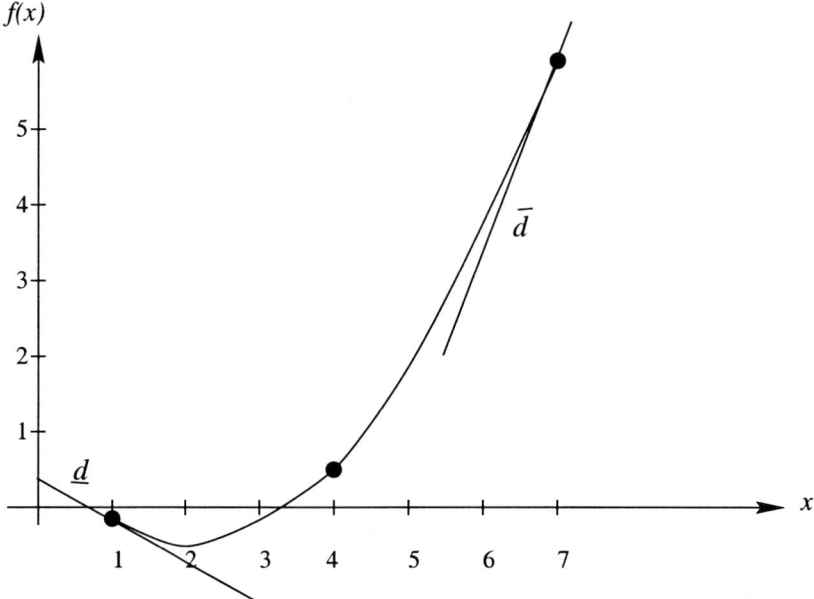

Fig. 1. $D = f'(X)$ for $f(x) = \frac{1}{4}x^2 - x + \frac{1}{2}$ and $X = [1, 7]$

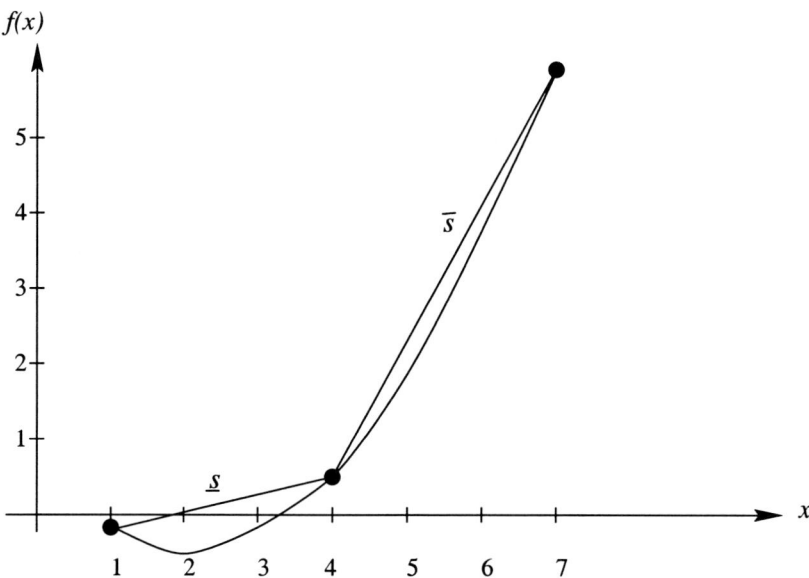

Fig. 2. $S = s_f(c, X)$ for $f(x) = \frac{1}{4}x^2 - x + \frac{1}{2}$, $c = 4$, and $X = [1, 7]$

Usually, it is not possible to compute the exact ranges of the slopes (or the derivatives), but it is possible to compute enclosures for them. Slopes as well as interval slopes can be calculated by means of an automatic process similar to the process of computing derivatives by so-called automatic differentiation (cf. [9], [20], [29], [30]). The main advantage of this process is that only the algorithm or formula for the function must be available. No explicit formulas for the derivatives or slopes are required. In [38], these slope computing technique is treated in detail.

2.4   Branch-and-Bound Methods for Global Optimization

Branch-and-bound methods for global optimization address the task of finding guaranteed and reliable solutions of the *global optimization problem*

$$\min_{x \in X} f(x). \tag{8}$$

The function $f : D \to \mathbb{R}$ with $D \subseteq \mathbb{R}^n$ is called the *objective function*, and $X \in I\mathbb{R}^n$ with $X \subseteq D$ is called the *search box*. The latter represents bound constraints for the unknowns x_i, $i = 1, \ldots, n$, i.e.

$$x \in X \iff \underline{x} \leq x \leq \overline{x} \iff \underline{x}_i \leq x_i \leq \overline{x}_i, \ i = 1, \ldots, n.$$

The interval branch-and-bound methods treated in this paper aim at computing tight enclosures of the set X^* of all *global minimizers* x^* and of the *global minimum value* $f^* = f(x^*)$ (i.e. $X^* = \{x \in X \mid f(x) = f^*\}$). We assume f to be continuous throughout this paper to assure the existence of a global minimizer. The global minimizers need not be unique. In fact, the minimizers may include continua of points.

Many authors have contributed with a significant amount of work in developing and investigating interval branch-and-bound methods for global optimization. An excellent detailed overview on background and history of these methods can be found in [20, Section 5.1]. Very early and simple algorithms are from Moore and Skelboe [39], Ichida and Fujii [18], and from Hansen ([11],[12]). Comprehensive overviews on these methods can be found in the books of Ratschek and Rokne [31] and and of Hansen [13]. More recent algorithms and practical implementations are the developments of Csendes [4], Jansson [19], Kearfott [20], and Ratz [32]. Most of these involve interval Newton methods for computational existence and uniqueness proofs of critical points and for quadratic convergence properties.

All of these methods usually apply several interval techniques to reject regions in which the optimum can be guaranteed not to lie. For this reason, the original box X gets subdivided, and subregions which cannot contain a global minimizer of f are discarded. The other subregions get subdivided again until the desired accuracy (width) of the boxes is achieved. We shortly review the overall structure of such branch-and-bound algorithms for global optimization.

Basic Branch-and-Bound Pattern Starting from the initial search box X, the algorithm subdivides X and stores the subintervals (branches) $Y \subset X$ in a pending list L. For each box Y in L, additional information on the lower bound of F on Y is stored. We use the notation $\underline{f_Y}$ as abbreviation for the lower interval bound of the interval function evaluation $F_Y := F(Y)$. Thus the elements of L are pairs $(Y, \underline{f_Y})$.

Additionally, the algorithm uses a value $\widetilde{f}$ representing a *guaranteed upper bound* of the global minimum value, i.e. $\widetilde{f} \geq f^*$. The algorithm proceeds as follows (where $L^{(i)}$ denotes the list L within the i-th iteration):

Initial phase:
- Initialize L with search box X:
$$L = L_0 := \boxed{\genfrac{}{}{1pt}{}{X}{\underline{f_X}}}, \text{ where } [\underline{f_X}, \overline{f_X}] = F_X = F(X)$$
- Initialize guaranteed upper bound $\widetilde{f} \geq f^*$ by interval evaluation with point argument:
  Choose $c \in X$ (e.g. the midpoint), compute $F_c := F(c)$ and set $\widetilde{f} := \overline{f_c}$.

Iteration:
- In iteration step $(i+1)$, remove the first box Y from $L^{(i)}$ (the final list of iteration step i) and subdivide it. Apply several tests (*range check*, *monotonicity test*, etc.) to assure nonexistence of global minimizers in the subboxes of Y. If none of these tests is successful, then append the subboxes at the end of the list forming $L^{(i+1)}$.

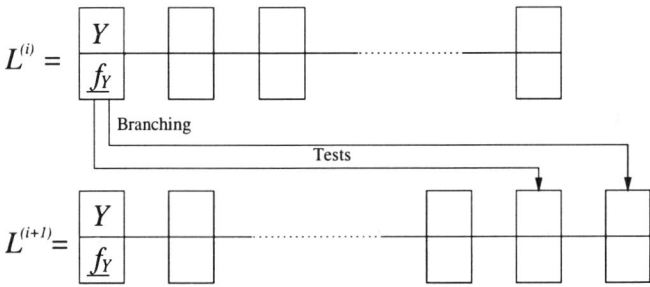

- Choose new $c := m(Y)$ in the new first element Y of $L^{(i+1)}$, compute $F_c := F(c)$, and update $\widetilde{f}$ by $\widetilde{f} := \min\{\widetilde{f}, \overline{f_c}\}$.
- Optionally, use $\widetilde{f}$ to execute the so-called *cut-off test*.

Termination:
- Terminate if the diameters of the intervals Y or $F(Y)$ in the list fulfill some criterion.
  Result: For the global minimum value and the global minimizers we have
$$f^* \in [\min\{\underline{f_Y} \in L\}, \widetilde{f}] \quad \text{and} \quad X^* \subseteq \bigcup_{Y \in L} Y.$$

In addition to the cut-off test, the range check, and the monotonicity test mentioned above, second-order methods also use the concavity test and an interval Newton-like step. For a detailed description of these steps requiring an enclosure of the Hessian matrix of f see [32] or [9], for example. Subsequently, we give a short overview of the tests relevant for this paper.

Cut-Off Test By comparing the guaranteed upper bound $\widetilde{f}$ for the global minimum value f^* with the lower bound $\underline{f_Y}$, we can discard all subintervals Y in L for which

$$\underline{f_Y} > \widetilde{f} \geq f^*. \tag{9}$$

The cut-off test is relatively inexpensive, and it often allows to discard from consideration large portions of the original interval X. Figure 3 illustrates this procedure, which deletes the intervals Y^2, Y^3, Y^7, and Y^8 in this special case.

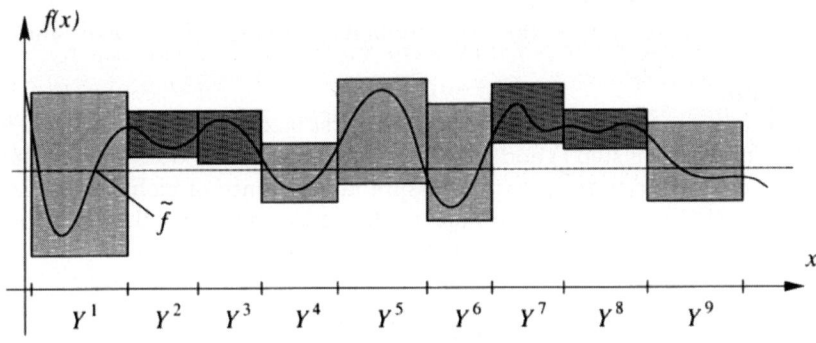

Fig. 3. Cut-off test

The value $\widetilde{f}$ can be improved by using a local approximate search to find a $c \in Y$ that is likely to give a smaller upper bound for f than $\widetilde{f}$ gives. In Figure 3, we evaluate f at $c \in Y^1$ to get $\widetilde{f}$. A value c near the left end of Y^1 would have yielded a more effective $\widetilde{f}$.

In an algorithmic context we use the notation CutOffTest for the cut-off test defined by

Algorithm 2.1: CutOffTest $(L, \widetilde{f})$

1.   **for all** $(Y, \underline{f_Y}) \in L$ **do**

2.   **if** $\widetilde{f} < \underline{f_Y}$ **then** $L := L \uplus (Y, \underline{f_Y})$;

3.   **return** L;

where $L \uplus (Y, \underline{f_Y})$ removes the element $(Y, \underline{f_Y})$ from L.

Remark: We use the algorithmic notation described in [9, Section 1.4] for all algorithms presented in this paper.

Range Check The value $\tilde{f}$ is also used when newly subdivided intervals W are tested in a so-called *range check*. If we know that $F_W = F(W)$ satisfies $\underline{f}_W > \tilde{f}$, then W cannot contain a global minimizer. Thus, we must only enter intervals W that satisfy $\underline{f}_W \leq \tilde{f}$ in the list L. The range check can be improved by incorporating centered forms for computing F_W.

Monotonicity Test For a continuously differentiable function f the monotonicity test determines whether the function is *strictly monotone* in an entire subinterval $Y \subset X$. If f is strictly monotone in Y, then Y cannot contain a global minimizer in its interior. Furthermore, a global minimizer can only lie on a boundary point of Y if this point is also a boundary point of X. Therefore, if the interval enclosure $G = \nabla F(Y)$ of ∇f evaluated over Y satisfies

$$0 \notin G_i \quad \text{for some } i = 1, \ldots, n,$$

then the subinterval Y can be deleted (with the exception of boundary points of X).

Figure 4 demonstrates the monotonicity test in the onedimensional case for four subintervals of X. In this special case, Y^1 can be reduced to the boundary point $\underline{x}$, Y^2 remains unchanged, and Y^3 can be deleted. Since f is monotonically increasing in Y^4, this entire interval can also be deleted because we are looking for a minimum.

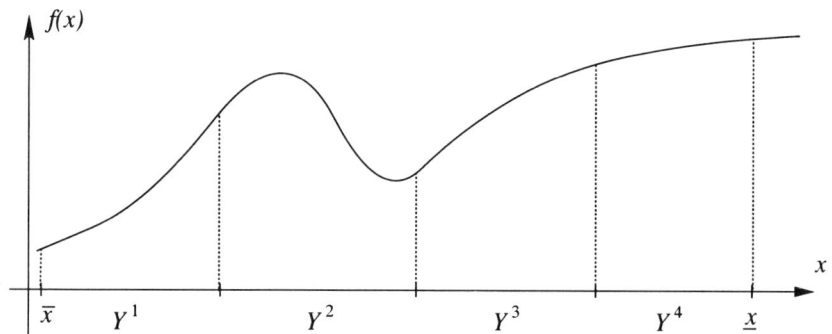

Fig. 4. Monotonicity test

In an algorithmic context we use the notation CheckMonotonicity for the onedimensional monotonicity test defined by

Algorithm 2.2: CheckMonotonicity (Y, G, X, U)

1. **if** $0 \in G$ **then**
2. $\quad U := Y;$ { leave Y unchanged }
3. **else if** $\underline{g} > 0$ **and** $\underline{x} = \underline{y}$ **then**

4. $U := [\underline{y}, \underline{y}];$ { reduce Y to left boundary point }

5. **else if** $\overline{g} < 0$ **and** $\overline{x} = \overline{y}$ **then**

6. $U := [\overline{y}, \overline{y}];$ { reduce Y to right boundary point }

7. **else**

8. $U := \emptyset;$ { delete Y }

9. **return** $U;$

where $Y \subseteq X$ is the current box to be checked, $X \in I\!R$ is the global search box, $G \supseteq F'(Y)$, and U is the updated value for Y. Note, that Algorithm 2.2 takes care of the boundary points of the original search region X which may be global minimizers without being stationary points.

For the multidimensional monotonicity test, which is defined componentwise, we use the notation MultiCheckMonotonicity, defined by

Algorithm 2.3: MultiCheckMonotonicity $(\boldsymbol{Y, G, X, U, m})$

1. $m := 1;$

2. **for** $i := 1$ **to** n **do**

3.   CheckMonotonicity (Y_i, G_i, X_i, U_i)

4.   **if** $U_i = \emptyset$ **then return** $m := 0;$

5. **endfor**

6. **return** $U,\; m;$

where $Y \subseteq X$ is the current box to be checked, $X \in I\!R^n$ is the global search box, $G \supseteq \nabla F(Y)$, and U is the updated value for Y with $m = 0$ signaling that $U = \emptyset$.

Sometimes it makes sense to apply interval slopes instead of interval derivatives within an interval branch-and-bound method. On one hand, interval slopes (together with centered forms) offer the possibility to achieve better enclosures for the function range. On the other hand, interval slopes can be applied even if the objective function is not everywhere differentiable and derivatives cannot be applied. However, since interval slopes cannot be used for the monotonicity test (see Section 3 for details), the need of an alternative test arises. Such new techniques which can replace the monotonicity test are the subject of Sections 3 and 4.

List Ordering Interval branch-and-bound methods benefit from an "optimal" ordering of the boxes in the pending list L. For the variants of the techniques treated in this paper we use the list ordering proposed in [32] and [33]. The boxes Y are stored in L sorted in nondecreasing order with respect to the $\underline{f_Y}$ values (the primary ordering criterion) and in decreasing order with respect to the age of the boxes (the secondary ordering criterion).

Therefore, a newly computed pair $(Y, \underline{f_Y})$ is stored in the list L according to the following ordering rule:

$$
\left.
\begin{array}{ll}
\bullet \text{ either } & \underline{f_W} \leq \underline{f_Y} < \underline{f_Z} \text{ holds,} \\
\bullet \text{ or } & \underline{f_Y} < \underline{f_Z} \text{ holds, and } (Y, \underline{f_Y}) \text{ is the first element of the list,} \\
\bullet \text{ or } & \underline{f_W} \leq \underline{f_Y} \text{ holds, and } (Y, \underline{f_Y}) \text{ is the last element of the list,} \\
\bullet \text{ or } & (Y, \underline{f_Y}) \text{ is the only element of the list,}
\end{array}
\right\} \quad (10)
$$

where $(w, \underline{f_W})$ is the predecessor and $(Z, \underline{f_Z})$ is the successor of $(Y, \underline{f_Y})$ in L.

That is, the second components of the list elements may not decrease, and a new pair is inserted behind all other pairs with the same second component. Since the first element of the list has the smallest second component, we can directly use the corresponding box to compute $F(m(Y))$ for the improvement of $\tilde{f}$ before performing the cut-off test. Due to this special ordering we can also save some work when deleting elements in the cut-off test. The order allows to delete the whole rest of the list when we have reached the first element to be deleted.

3 A Pruning Technique for Global Optimization

Many interval branch-and-bound methods for global optimization incorporate the monotonicity test (cf. Section 2.4) to discard subboxes of the global search box, provided that the objective function is continuously differentiable. This test uses first-order information of the objective function by means of an interval evaluation of the derivative over the current box. Depending on this enclosure containing zero or not, the current box must be treated further or can be deleted, respectively. Moreover, the interval derivative evaluation together with a centered form is often used to improve the enclosure of the function range.

On the other hand, interval slopes (together with centered forms) offer the possibility to achieve better enclosures for the function range, as described in [38], for example. Thus, they might improve the performance of interval branch-and-bound methods. Although, since slopes cannot be used within the monotonicity test (see Section 3.1 for details), the need of a global optimization method with an alternative box-discarding technique arises. This applies to the smooth case if we do not have access to derivative values as well as to the nonsmooth case where the slope values are the only available "first order informations".

In this section, we introduce a method which incorporates a special pruning step generated by interval slopes in the context of onedimensional global optimization problems. The pruning step offers the possibility to cut away a large part of the current box, independently of the slope interval containing zero or not. We develop the theory for this slope pruning step and we give several examples underlining its efficiency. For several numerical examples with differentiable functions we demonstrate the advantages of the pruning step over the monotonicity test. We treat the onedimensional case separately for didactical reasons.

3.1   A Pruning Technique Using Interval Slopes

In first-order interval methods for global optimization, the monotonicity test determines whether the function f is *strictly monotone* in an entire subinterval $Y \subset X$. If this is the case, then Y cannot contain a global minimizer in its interior. Furthermore, a global minimizer can only lie on a boundary point of Y if this point is also a boundary point of X. Therefore, the subinterval Y can be deleted (with the exception of boundary points of X), if f satisfies

$$0 \notin F'(Y). \tag{11}$$

If we want to apply slopes instead of derivatives, we cannot use this monotonicity test, since we have $s_f(c, X) \subseteq F'(X)$, but in general it is *not* true that $f'(x) \in s_f(c, X)$ $\forall c, x \in X$. Therefore, it might happen that $0 \notin s_f(c, Y)$, although $x^* \in Y$ is a local (or even global) minimizer with $f'(x^*) = 0$. So, $0 \notin s_f(c, Y)$ cannot be used as a criterion to discard the box Y.

Example 2. We consider the sample function

$$f(x) = x^2 - 4x + 2 = (x - 2)^2 - 2,$$

and we easily see that $x^* = 2$ is a local and global minimizer of f. With $Y = [1, 7]$ we have $s_f(c, Y) = [1, 7]$, and consequently $0 \notin s_f(c, Y)$ cannot be used as a criterion to discard Y, since $x^* \in Y$.

For the monotonicity test being an essential accelerating tool for an efficient interval global optimization method [5], the need of a corresponding tool in connection with slopes arises which is applicable to nonsmooth problems. In the following, we develop such a tool, which we call a *pruning step using slopes*. We assume the (possibly nonsmooth) objective function f to be continuous.

The main idea for this pruning step (cf. [37]) is based on the slope extension (6) and our subsequent Theorem 1. Improvements which take into account a known upper bound for the global minimum value are based on Theorems 2, 3, and 4.

Remark: A similar treatise in the context of at least three times differentiable functions and so-called cord-slopes generated by derivatives of higher order can be found in [14]. In contrast to the proof of [14, Test 2] which only proves that no global *minimum value* is lost by the test, we additionally prove that the pruning step does not delete any global *minimizer* (see [38]).

Theorem 1. *Let $f : D \to \mathbb{R}$, $Y = [\underline{y}, \overline{y}] \in I\mathbb{R}$, $c \in Y \subseteq D \subseteq \mathbb{R}$. Moreover, let $S = [\underline{s}, \overline{s}] = s_f(c, Y)$ with $\underline{s} > 0$. Then*

$$p := c + (\underline{y} - c) \cdot \underline{s}/\overline{s}$$

satisfies

$$\underline{y} \le p \le c \tag{12}$$

and

$$\min_{x \in Y} f(x) = \min_{x \in [\underline{y}, p]} f(x) < \min_{p < x \le \overline{y}} f(x). \tag{13}$$

Proof. See [38].

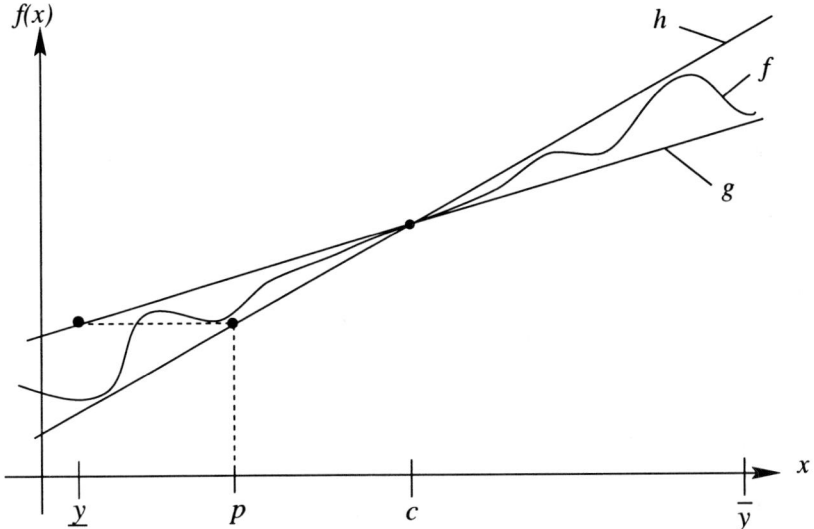

Fig. 5. Generation of the pruning point p for a positive interval slope

Figure 5 illustrates the geometrical interpretation for finding the point p. First of all, we define the two lines

$$g : \mathbb{R} \to \mathbb{R} \qquad g(x) := f(c) + \underline{s} \cdot (x - c) \tag{14}$$

and

$$h : \mathbb{R} \to \mathbb{R} \qquad h(x) := f(c) + \overline{s} \cdot (x - c). \tag{15}$$

Then we know that $g(\underline{y})$ is an upper bound for $f(\underline{y})$ and thus for $\min_{x \in Y} f(x)$. Now we can locate p as the leftmost point in Y, for which f can not fall below $g(\underline{y})$. Since h is a lower bound for f in $[\underline{y}, c]$, we can do this very simply by computing the intersection point of h and the horizontal line z with $z(x) = g(\underline{y})$.

Using the value p of Theorem 1 within a global optimization method, we can prune a subinterval $Y \subseteq X$, if $0 < \underline{s} \leq \overline{s}$ for $S = s_f(c, Y)$ to

$$Y_p := [\underline{y}, c + (\underline{y} - c) \cdot \underline{s}/\overline{s}].$$

Example 3. We consider $f(x) = \frac{1}{2}x^2$, and we assume the current interval to be $Y = [-1, 4]$. First of all, we try to apply the monotonicity test. We evaluate the derivative $f'(x) = x$ over Y, and we get $F'(Y) = Y = [-1, 4]$. Since $0 \in F'(Y)$, we cannot discard Y from further consideration. We must subdivide Y and treat parts of it in the same manner.

Now, we apply the pruning step. We first evaluate the interval slope $S = s_f(c, Y) = \frac{1}{2}(c + Y)$. With $c = 1.5$ we get $S = [0.25, 2.75]$. Since $0 \notin S$ we can prune Y to

$$Y_p = [\underline{y}, c + (\underline{y} - c) \cdot \underline{s}/\overline{s}] = [-1, 1.5 + (-1 - 1.5) \cdot 0.25/2.75] = [-1, 1.273]$$

using four significant digits and rounding outwards.

If we recall the situation in Figure 5, we see that we are able to improve the pruning of an interval Y. We can improve the point p (by moving it to the left), if we know a better (smaller) upper bound $\widetilde{f}$ for $f(x)$ on Y than $g(\underline{y})$ was. Moreover, if $\widetilde{f}$ is an upper bound for the global minimum value f^* on the whole search box X, then we can locate p as the leftmost point in Y, for which f can not fall below $\widetilde{f}$. Since h is a lower bound for f near $\underline{y}$, we can do this by computing the intersection point of h and the horizontal line z with $z(x) = \widetilde{f}$. In the context of a global optimization method using branch-and-bound techniques such as the cut-off test, an improved upper bound $\widetilde{f}$ for the global minimum value f^* is usually known. Therefore, we can state

Theorem 2. *Let* $f : D \to \mathbb{R}$, $Y = [\underline{y}, \overline{y}] \in I\mathbb{R}$, $c \in Y \subseteq X \subseteq D \subseteq \mathbb{R}$. *Moreover, let* $S = [\underline{s}, \overline{s}] = s_f(c, Y)$ *with* $\underline{s} > 0$ *and*

$$\widetilde{f} \geq f^* = \min_{x \in X} f(x). \tag{16}$$

Then $p := c + m/\overline{s}$ *with* $m = \min\{\widetilde{f} - f(c), (\underline{y} - c) \cdot \underline{s}\}$ *satisfies*

$$p \leq c, \tag{17}$$

and

$$\min_{p < x \leq \overline{y}} f(x) > f^* \quad \textit{for} \ \ \underline{y} \leq p \tag{18}$$

or

$$\min_{x \in Y} f(x) > f^* \quad \textit{for} \ \ p < \underline{y}, \tag{19}$$

respectively.

Proof. See [38].

Remark: It is easy to see, that in the case $\widetilde{f} < f(c) + (\underline{y} - c) \cdot \underline{s}$ the value p computed in Theorem 1 is smaller than that computed in Theorem 2.

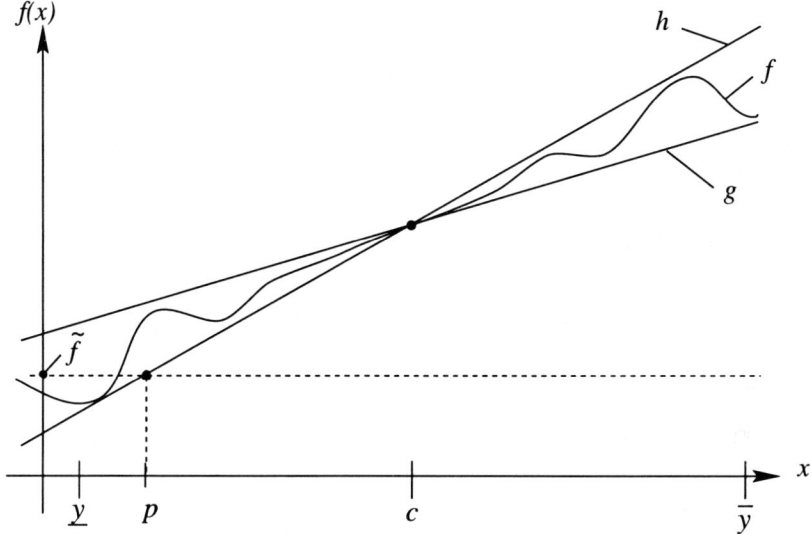

Fig. 6. Generation of the pruning point p with known $\widetilde{f}$

Figure 6 illustrates the geometrical interpretation for finding the point p when using the known upper bound $\widetilde{f}$ for the global minimum value. Again we use the two lines

$$g : I\!R \to I\!R \qquad g(x) := f(c) + \underline{s} \cdot (x - c)$$

and

$$h : I\!R \to I\!R \qquad h(x) := f(c) + \overline{s} \cdot (x - c).$$

Then we know that $\widetilde{f}$ is an upper bound for $\min_{x \in Y} f(x)$. Now we can locate p as the leftmost point in Y, for which f cannot fall below $\widetilde{f}$. Since h is a lower bound for f in $[\underline{y}, c]$, we can do this very simply by computing the intersection point of h and the horizontal line z with $z(x) = \widetilde{f}$.

So, we can use Theorem 2 within a global optimization method to prune or delete a subinterval $Y \subseteq X$, if $0 < \underline{s} \leq \overline{s}$ for $S = [\underline{s}, \overline{s}] = s_f(c, Y)$. That is, we first compute

$$m = \min\{\widetilde{f} - f(c), (\underline{y} - c) \cdot \underline{s}\}$$

and then

$$p = c + m/\overline{s} \leq c.$$

Then, if $p \geq \underline{y}$, we replace Y by

$$Y := [\underline{y}, p],$$

otherwise we delete the whole subbox Y.

It is easy to see, that we can apply a similar procedure for pruning in the case $\overline{s} < 0$, since we immediately have

Theorem 3. *Let* $f : D \to \mathbb{R}$, $Y = [\underline{y}, \overline{y}] \in I\mathbb{R}$, $c \in Y \subseteq X \subseteq D \subseteq \mathbb{R}$. *Moreover, let* $S = [\underline{s}, \overline{s}] = s_f(c, Y)$ *with* $\overline{s} < 0$ *and*

$$\widetilde{f} \geq f^* = \min_{x \in X} f(x). \tag{20}$$

Then $q := c + m/\underline{s}$ *with* $m = \min\{\widetilde{f} - f(c), (\overline{y} - c) \cdot \overline{s}\}$ *satisfies*

$$c \leq q \tag{21}$$

and

$$\min_{\underline{y} \leq x < q} f(x) > f^* \quad \textit{for} \quad q \leq \overline{y} \tag{22}$$

or

$$\min_{x \in Y} f(x) > f^* \quad \textit{for} \quad \overline{y} < q, \tag{23}$$

respectively.

Proof. See [38].

Up to now, we have treated the case $0 \notin s_f(c, Y)$, which corresponds to the (successful) case $0 \notin F'(Y)$ in the usual monotonicity test. A further advantage of the pruning technique with slopes is, that we also can apply this technique successfully in the case $0 \in s_f(c, Y)$, which corresponds in a sense to the (unsuccessful) case $0 \in F'(Y)$ for the usual monotonicity test. For this purpose we can use the following

Theorem 4. *Let* $f : D \to \mathbb{R}$, $Y = [\underline{y}, \overline{y}] \in I\mathbb{R}$, $c \in Y \subseteq X \subseteq D \subseteq \mathbb{R}$. *Moreover, let* $S = [\underline{s}, \overline{s}] = s_f(c, Y)$ *with* $0 \in S$ *and*

$$f(c) > \widetilde{f} \geq f^* = \min_{x \in X} f(x). \tag{24}$$

Then

$$p := \begin{cases} c + (\widetilde{f} - f(c))/\overline{s} & \text{if } \overline{s} \neq 0, \\ -\infty & \text{otherwise,} \end{cases}$$

$$q := \begin{cases} c + (\widetilde{f} - f(c))/\underline{s} & \text{if } \underline{s} \neq 0, \\ +\infty & \text{otherwise,} \end{cases}$$

and

$$\mathcal{Z} := (p, q) \cap Y$$

satisfy

$$p < c < q \qquad (25)$$

and

$$\min_{x \in Z} f(x) > f^*. \qquad (26)$$

Proof. See [38].

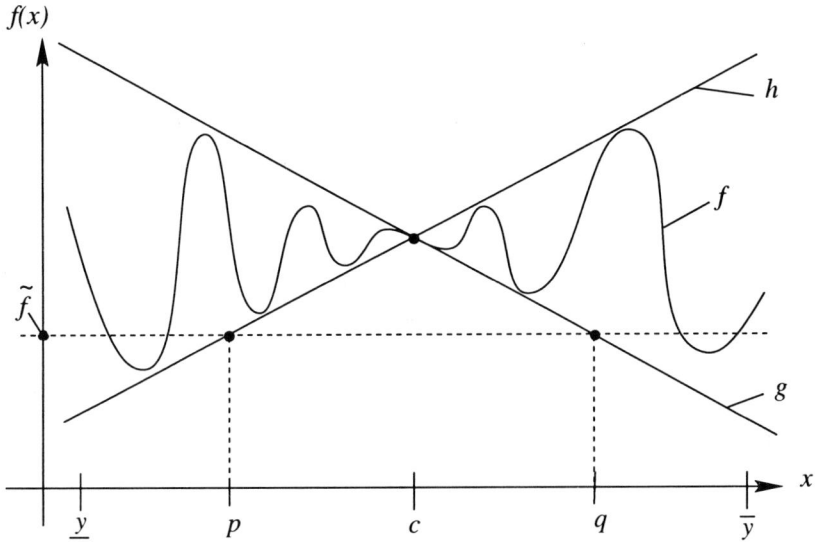

Fig. 7. Generation of pruning points p and q with known $\widetilde{f} < f(c)$

We illustrate the geometrical interpretation for finding the points p and q in Figure 7. Again we use the two lines

$$g : \mathbb{R} \to \mathbb{R} \qquad g(x) := f(c) + \underline{s} \cdot (x - c)$$

and

$$h : \mathbb{R} \to \mathbb{R} \qquad h(x) := f(c) + \overline{s} \cdot (x - c),$$

assuming $\underline{s} < 0 < \overline{s}$. Now we can locate p as the leftmost point and q as the rightmost point in Y, for which f can not fall below $\widetilde{f}$ according to the bounding by g and h. Since h is a lower bound for f in $[\underline{y}, c]$ and since g is a lower bound for f in $[c, \overline{y}]$, we can do this very simply by computing the intersection points of h and g with the horizontal line z with $z(x) = \widetilde{f}$.

So, we can use Theorem 4 within a global optimization method to prune or delete a subinterval $Y \subseteq X$, if $\underline{s} \leq 0 \leq \overline{s}$ for $S = s_f(c, Y)$ and if $\widetilde{f} < f(c)$. That is, we first compute

$$p = c + (\widetilde{f} - f(c))/\overline{s} \quad \text{and} \quad q = c + (\widetilde{f} - f(c))/\underline{s}.$$

Then, we replace Y by

$$
\begin{array}{ll}
[\underline{y},p] \cup [q,\overline{y}] & \text{if } \underline{y} \le p \wedge q \le \overline{y}, \\
[\underline{y},p] & \text{if } \underline{y} \le p \wedge q > \overline{y}, \\
[q,\overline{y}] & \text{if } \underline{y} > p \wedge q \le \overline{y},
\end{array}
$$

and otherwise we delete the whole subbox Y.

3.2   An Algorithmic Pruning Step

We are now able to give an algorithmic formulation of a pruning step, which can be applied to a subinterval $Y \subseteq X$ when globally minimizing $f : D \to \mathbb{R}$ on $X \subseteq D$. The algorithm uses

$$
\begin{aligned}
Y &= [\underline{y},\overline{y}], \\
c &\in Y, \\
f_c &= f(c), \\
S &= [\underline{s},\overline{s}] = s_f(c,Y), \quad \text{and} \\
\widetilde{f} &\ge \min_{x \in X} f(x)
\end{aligned}
$$

as input, and it delivers the pruned (and possibly empty) subset $U_1 \cup U_2$ of Y with $U_1, U_2 \in I\!\!R \cup \{\emptyset\}$ and a possibly improved $\widetilde{f}$ as output. Note that the algorithm also performs a bisection if no pruning is possible.

Algorithm 3.1: SlopePruning $(Y, c, f_c, S, \widetilde{f}, U_1, U_2)$

1. $U_1 := \emptyset; \quad U_2 := \emptyset;$
2.   **if** $0 \in S$ **then**   { pruning from the center }
3.   **if** $\widetilde{f} < f_c$ **then**   { a pruning is possible }
4.   **if** $\overline{s} > 0$ **then**   { pruning from the center to the left }
5. $p := c + (\widetilde{f} - f_c)/\overline{s};$
6.   **if** $p \ge \underline{y}$ **then** $U_1 := [\underline{y},p];$ { compute remaining left part }
7.   **if** $\underline{s} < 0$ **then**   { pruning from the center to the right }
8. $q := c + (\widetilde{f} - f_c)/\underline{s};$
9.   **if** $q \le \overline{y}$ **then** $U_2 := [q,\overline{y}];$ { compute remaining right part }
10.   **else**   { a pruning is *not* possible }
11. $U_1 := [\underline{y},m(Y)]; \quad U_2 := [m(Y),\overline{y}];$ { bisection }
12. **else if** $\underline{s} > 0$ **then**   { pruning from the right }
13. $\widetilde{f} := \min\{\widetilde{f}, \ (\underline{y} - c) \cdot \underline{s} + f_c\};$ { update $\widetilde{f}$ }
14. $p := c + (\widetilde{f} - f_c)/\overline{s};$
15.   **if** $p \ge \underline{y}$ **then** $U_1 := [\underline{y},p];$ { compute remaining left part }

16.   **else** { $\overline{s} < 0$ }   { pruning from the left }
17. $\widetilde{f} := \min\{\widetilde{f},\ (\overline{y} - c) \cdot \overline{s} + f_c\};$ { update $\widetilde{f}$}
18. $q := c + (\widetilde{f} - f_c)/\underline{s};$
19.   **if** $q \leq \overline{y}$ **then** $U_2 := [q, \overline{y}];$ { compute remaining right part }
20.   **return** $U_1,\quad U_2,\quad \widetilde{f};$

Summarizing the properties of this pruning step, we can state

Theorem 5. *Let* $f : D \to I\!\!R,\ Y \in I\!I\!R,\ c \in Y \subseteq X \subseteq D \subseteq I\!\!R.$ *Moreover, let* $f_c = f(c),\ S = s_f(c, Y),$ *and* $\widetilde{f} \geq \min_{x \in X} f(x).$ *Then Algorithm 3.1 applied as* SlopePruning $(Y, c, f_c, S, \widetilde{f}, U_1, U_2)$ *has the following properties:*

1. *$U_1 \cup U_2 \subseteq Y.$*
2. *Every global minimizer x^* of f in X with $x^* \in Y$ satisfies $x^* \in U_1 \cup U_2.$*
3. *If $U_1 \cup U_2 = \emptyset$, then there is no global (w.r.t. X) minimizer of f in Y.*

Proof. See [38].

It is obvious that the success of Algorithm 3.1 in pruning Y depends on the quality of $\widetilde{f}$. Therefore, the pruning step within a global optimization method can benefit greatly from a fast local search method delivering a good (small) value $\widetilde{f}$ on a very early stage of the method. For our further studies in this paper, we do not use an additional local method.

Now we integrate Algorithm 3.1 in a global optimization algorithm to study the effect of the application of the slope pruning step (replacing the monotonicity test).

3.3   A Global Optimization Algorithm Using Pruning Steps

Subsequently, we give a simple first-order model algorithm to demonstrate the advantages of the pruning step. Our model algorithm uses the cut-off test, but it includes no local search procedure, no concavity test, and no Newton-like steps, since the latter would require smoothness.

Algorithm 3.2: GlobalOptimize $(f, X, \varepsilon, F^*, L_{\mathbf{res}})$

1. $c := m(X);\quad \widetilde{f} := f(c);$ { initialize upper bound }
2. $F_X := (f(c) + s_f(c, X) \cdot (X - c)) \cap F(X);$ { centered form }
3. $L := \{(X, \underline{f_X})\};\quad L_{\mathrm{res}} := \{\ \};$ { initialize working list and result list }
4.   **while** $L \neq \{\ \}$ **do**
5. $(Y, \underline{f_Y}) := \mathsf{PopHead}\,(L);\quad c := m(Y);$ { get first element of the list }
6.   SlopePruning $(Y,\ c,\ f(c),\ s_f(c, Y),\ \widetilde{f},\ U_1,\ U_2);$
7.   **for** $i := 1$ **to** 2 **do**

8.   **if** $U_i = \emptyset$ **then** next$_i$;

9. $c := m(U_i)$;

10.   **if** $f(c) < \tilde{f}$ **then** $\tilde{f} := f(c)$;   { update $\tilde{f}$ }

11. $F_U := (f(c) + s_f(c, U_i) \cdot (U_i - c)) \cap F(U_i)$;   { centered form }

12.   **if** $\underline{f_U} \leq \tilde{f}$ **then**

13.   **if** $d_{\mathrm{rel}}(F_U) \leq \varepsilon$ **or** $d_{\mathrm{rel}}(U_i) \leq \varepsilon$ **then**

14. $L_{\mathrm{res}} := L_{\mathrm{res}} \uplus (U_i, \underline{f_U})$ { accept U_i for the result list }

15.   **else**

16. $L := L \uplus (U_i, \underline{f_U})$;   { store U_i in the working list }

17.   **endfor**

18.   CutOffTest $(L, \tilde{f})$;

19. **endwhile**

20. $(Y, \underline{f_Y}) := $ Head (L_{res}); $F^* := [\underline{f_Y}, \tilde{f}]$;   CutOffTest $(L_{\mathrm{res}}, \tilde{f})$;

21. **return** F^*, L_{res}.

Algorithm 3.2 first computes an upper bound $\tilde{f}$ for the global minimum value and initializes the working list L and the result list L_{res}. The main iteration (from Step 4 to Step 19) starts with the pruning step applied to the leading interval of the working list. Then we apply a range check using a centered form to the resulting boxes U_1 and U_2 if they are non-empty. If the current box is still a candidate for containing a global minimizer, we store it in L_{res} (if it can be accepted with respect to the tolerance ε) or in L (if it must be treated further).

Note that by the operation $\uplus$ the boxes are stored in list L as pairs $(Y, \underline{f_Y})$ sorted in *nondecreasing* order with respect to the lower bounds $\underline{f_Y} \leq \underline{f_{\mathrm{rg}}(Y)}$ and in *decreasing* order with respect to the ages of the boxes in L as specified in Section 2.4. Thus, the leading box of L is the oldest element with the smallest $\underline{f_Y}$ value. When the iteration stops because the working list L is empty, we compute a final enclosure F^* for the global minimum value and return L_{res} and F^*.

The method can be improved by incorporating an approximate local search procedure to try to decrease the value $\tilde{f}$. For our studies in this paper, we do not apply any local method. The definition and algorithmic description of the cut-off test is given in Section 2.4.

For our global optimization model algorithm (Algorithm 3.2) we can state

Theorem 6. *Let* $f : D \to \mathbb{R}$, $X \subseteq D \subseteq \mathbb{R}$, *and* $\varepsilon > 0$. *Then Algorithm 3.2 has the following properties:*

1. $f^* \in F^*$.

2. $X^* \subseteq \displaystyle\bigcup_{(Y, \underline{f_Y}) \in L_{\mathrm{res}}} Y$.

Proof. See [38].

Example 4. To demonstrate the performance of our global optimization algorithm using pruning steps, we give an extract (about the first 9 steps) of the protocol of the pruning steps when applying Algorithm 3.2 on function

$$f(x) = \frac{(x-a)^2}{20} - \cos(x-a) + 2$$

with $a = 1.125$ and starting interval $X = [-5, 5]$.

For each current box Y (chosen in Step 5 of Algorithm 3.2), we list its value, the value of the slope $S = s_f(c, Y)$, the chosen pruning step, and the resulting boxes U_1 and U_2. We use four significant digits and rounding outwards.

Y	S	pruning	U_1	U_2
$[-5, 5]$	$[-1.363, 1.138]$	$\Rightarrow$ bisection	$\Rightarrow [-5, 0]$	$[0, 5]$
$[0, 5]$	$[-0.8898, 1.263]$	$\Rightarrow$ from center	$\Rightarrow [0, 2.288]$	$[2.801, 5]$
$[0, 2.288]$	$[-0.9576, 0.9771]$	$\Rightarrow$ bisection	$\Rightarrow [0, 1.144]$	$[1.143, 2.288]$
$[0, 1.144]$	$[-0.9862, -0.007798]$	$\Rightarrow$ from left	$\Rightarrow \emptyset$	$[0.7384, 1.144]$
$[0.7384, 1.144]$	$[-0.4056, 0.01067]$	$\Rightarrow$ from center	$\Rightarrow \emptyset$	$[0.9863, 1.144]$
$[0.9863, 1.144]$	$[-0.1481, 0.01687]$	$\Rightarrow$ from center	$\Rightarrow \emptyset$	$[1.077, 1.144]$
$[1.077, 1.144]$	$[-0.05098, 0.01913]$	$\Rightarrow$ bisection	$\Rightarrow [1.077, 1.111]$	$[1.110, 1.144]$
$[1.110, 1.144]$	$[-0.01.510, 0.01997]$	$\Rightarrow$ bisection	$\Rightarrow [1.110, 1.128]$	$[1.127, 1.144]$
$[1.110, 1.128]$	$[-0.01552, 0.002018]$	$\Rightarrow$ from center	$\Rightarrow \emptyset$	$[1.120, 1.128]$

3.4   Modifications for an Implementation

In an implementation of Algorithms 3.1 and 3.2, we have to be aware of the fact, that the evaluation of $f(c)$ must be carried out in interval arithmetic to bound all rounding errors. Moreover, in practical computations we only have $S \supset s_f(c, Y)$, i.e. S is possibly an overestimation of the true interval slope. Therefore, the formulas for computing the values p and q in the pruning step differ from the theoretical ones, and machine-oriented versions of Theorems 1 to 4 need a somewhat different approach. Also, we have to take special care of correct rounding when computing the values p and q in the pruning step and of the correct use of interval evaluations instead of real evaluations where necessary. Subsequently, we give the necessary details for a practical realization of the algorithm.

Figure 8 illustrates the situation arising in practical computations. Here we have $S \supset s_f(c, Y)$ and $f(c) \in Z$. Therefore, in contrast to Figure 5, the two lines

$$g : \mathbb{R} \to \mathbb{R} \qquad g(x) := \overline{z} + \underline{s} \cdot (x - c) \tag{27}$$

and

$$h : \mathbb{R} \to \mathbb{R} \qquad h(x) := \underline{z} + \overline{s} \cdot (x - c). \tag{28}$$

are used to generate the pruning point p. The situation for S containing only negative values or containing zero is similar.

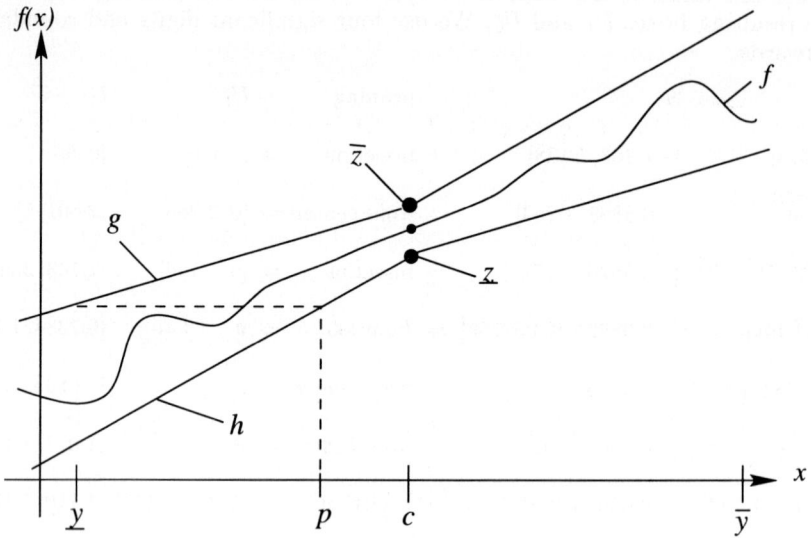

Fig. 8. Generation of the pruning point p for $f(c) \in Z$

Thus, machine-oriented versions of Theorem 1 to 4 are the following.

Theorem 7. *Let* $f : D \to \mathbb{R}$, $Y = [\underline{y}, \overline{y}] \in I\mathbb{R}$, *and* $c \in Y \subseteq D \subseteq \mathbb{R}$. *Moreover, let* $f(c) \in Z = [\underline{z}, \overline{z}] \in I\mathbb{R}$ *and* $S = [\underline{s}, \overline{s}] \supseteq s_f(c, Y)$ *with* $\underline{s} > 0$. *Then*

$$p := \min\{c, \, c + ((\underline{y} - c) \cdot \underline{s} + d(Z))/\overline{s}\}$$

satisfies

$$\underline{y} \le p \le c \tag{29}$$

and

$$\min_{x \in Y} f(x) = \min_{x \in [\underline{y}, p]} f(x) < \min_{p < x \le \overline{y}} f(x). \tag{30}$$

Proof. See [38].

Theorem 8. *Let $f : D \to \mathbb{R}$, $Y = [\underline{y}, \overline{y}] \in I\mathbb{R}$, and $c \in Y \subseteq X \subseteq D \subseteq \mathbb{R}$. Moreover, let $f(c) \in Z = [\underline{z}, \overline{z}] \in I\mathbb{R}$, $S = [\underline{s}, \overline{s}] \supseteq s_f(c, Y)$ with $\underline{s} > 0$, and*

$$\tilde{f} \geq f^* = \min_{x \in X} f(x). \tag{31}$$

Then $p := c + (m + d(Z))/\overline{s}$ with $m = \min\{-d(Z), \tilde{f} - \overline{z}, (\underline{y} - c) \cdot \underline{s}\}$ satisfies

$$p \leq c \tag{32}$$

and

$$\min_{p < x \leq \overline{y}} f(x) > f^* \quad \text{for} \quad \underline{y} \leq p \tag{33}$$

or

$$\min_{x \in Y} f(x) > f^* \quad \text{for} \quad p < \underline{y}, \tag{34}$$

respectively.

Proof. See [38].

Remark: The case $\overline{s} < 0$ can be treated completely analogous.

Theorem 9. *Let $f : D \to \mathbb{R}$, $Y = [\underline{y}, \overline{y}] \in I\mathbb{R}$, $c \in Y \subseteq X \subseteq D \subseteq \mathbb{R}$, $f(c) \in Z = [\underline{z}, \overline{z}] \in I\mathbb{R}$. Moreover, let $S = [\underline{s}, \overline{s}] \supseteq s_f(c, Y)$ with $0 \in S$ and*

$$\underline{z} > \tilde{f} \geq f^* = \min_{x \in X} f(x). \tag{35}$$

Then

$$p := \begin{cases} c + (\tilde{f} - \underline{z})/\overline{s} & \text{if } \overline{s} \neq 0, \\ -\infty & \text{otherwise,} \end{cases}$$

$$q := \begin{cases} c + (\tilde{f} - \underline{z})/\underline{s} & \text{if } \underline{s} \neq 0, \\ +\infty & \text{otherwise,} \end{cases}$$

and

$$Z := \begin{cases} (p, q) \cap Y & \text{if } p < q, \\ \emptyset & \text{otherwise,} \end{cases}$$

satisfy

$$\min_{x \in Z} f(x) > f^*. \tag{36}$$

Proof. See [38].

Remark: When implementing the pruning step on a machine, we must guarantee that all rounding errors are taken into account. Thus, in all three pruning cases the values p and q must be computed with upwardly-directed and downwardly-directed roundings, respectively.

3.5   Practical Algorithms

We specify a machine-oriented version of Algorithm 3.1. It is applicable to a subinterval $Y \subseteq X$ when globally minimizing $f : D \to I\!R$ on $X \subseteq D$. The algorithm uses

$$
\begin{aligned}
Y &= [\underline{y}, \overline{y}], \\
c &\in Y, \\
Z &= [\underline{z}, \overline{z}] \ni f(c), \\
S &= [\underline{s}, \overline{s}] \supseteq s_f(c, Y), \quad \text{and} \\
\widetilde{f} &\geq \min_{x \in X} f(x)
\end{aligned}
$$

as input, and it delivers the pruned (and possibly empty) subset $U_1 \cup U_2$ of Y with $U_1, U_2 \in I\!R \cup \{\emptyset\}$ and a possibly improved $\widetilde{f}$ as output. We use $\triangle(expr)$ and $\triangledown(expr)$ to indicate that a guaranteed upper and lower bound for the expression $expr$ is computed, respectively.

Algorithm 3.3: SlopePruning $(Y, c, Z, S, \widetilde{f}, U_1, U_2)$

1. $U_1 := \emptyset; \quad U_2 := \emptyset;$
2.   **if** $0 \in S$ **then**   { pruning from the center }
3.   **if** $\widetilde{f} < \underline{z}$ **then**   { a pruning is possible }
4.   **if** $\overline{s} > 0$ **then**   { pruning from the center to the left }
5. $p := \triangle(c + (\widetilde{f} - \underline{z})/\overline{s});$
6.   **if** $p \geq \underline{y}$ **then** $U_1 := [\underline{y}, p];$ { compute remaining left part }
7.   **if** $\underline{s} < 0$ **then**   { pruning from the center to the right }
8. $q := \triangledown(c + (\widetilde{f} - \underline{z})/\underline{s});$
9.   **if** $q \leq \overline{y}$ **then** $U_2 := [q, \overline{y}];$ { compute remaining right part }
10.   **else**   { a pruning is *not* possible }
11. $U_1 := [\underline{y}, m(Y)]; \quad U_2 := [m(Y), \overline{y}];$ { bisection }
12. **else if** $\underline{s} > 0$ **then**   { pruning from the right }
13. $\widetilde{f} := \min\{\widetilde{f}, \ \triangle((\underline{y} - c) \cdot \underline{s} + \overline{z})\};$ { update $\widetilde{f}$}
14.   **if** $\widetilde{f} < \underline{z}$ **then**   { a pruning with $p < c$ is possible }
15. $p := \min\{c, \ \triangle(c + (\widetilde{f} - \underline{z})/\overline{s})\};$
16.   **if** $p \geq \underline{y}$ **then** $U_1 := [\underline{y}, p];$ { compute remaining left part }
17.   **else**   { a pruning with $p = c$ is possible }

18. $U_1 := [\underline{y}, c]$;

19. **else** $\{\bar{s} < 0\}$ { pruning from the left }

20. $\tilde{f} := \min\{\tilde{f}, \ \triangle((\bar{y} - c) \cdot \bar{s} + \bar{z})\}$;   { update $\tilde{f}$}

21.   **if** $\tilde{f} < \underline{z}$ **then**   { a pruning with $q > c$ is possible }

22. $q := \max\{c, \ \triangledown(c + (\tilde{f} - \underline{z})/\underline{s})\}$;

23.   **if** $q \leq \bar{y}$ **then** $U_2 := [q, \bar{y}]$;   { compute remaining right part }

24.   **else**   { a pruning with $q = c$ is possible }

25. $U_2 := [c, \bar{y}]$;

26. **return** $U_1, \ U_2, \ \tilde{f}$;

The following corollary corresponds to Theorem 5. It summarizes the properties of this pruning step algorithm.

Corollary 1. *Let* $f : D \to I\!\!R$, $Y \in I I\!\!R$, $c \in Y \subseteq X \subseteq D \subseteq I\!\!R$. *Moreover, let* $f(c) \in Z$, $s_f(c, Y) \subseteq S$, *and* $\tilde{f} \geq \min_{x \in X} f(x)$. *Then Algorithm 3.3 applied as* SlopePruning $(Y, c, Z, S, \tilde{f}, U_1, U_2)$ *has the following properties:*

1. $U_1 \cup U_2 \subseteq Y$.
2. *Every global minimizer* x^* *of* f *in* X *with* $x^* \in Y$ *satisfies* $x^* \in U_1 \cup U_2$.
3. *If* $U_1 \cup U_2 = \emptyset$, *then there is no global (w.r.t.* X*) minimizer of* f *in* Y.

Proof. See [38].

A machine-oriented version of our first-order model algorithm using the pruning step can be specified as follows. Note that all interval operations and evaluations must be executed in machine interval arithmetic (cf. [9, Section 3.6]), even though this is not explicitly marked in our algorithmic notation.

Algorithm 3.4: GlobalOptimize $(f, X, \varepsilon, F^*, L_{res})$

1. $c := m(X)$; $\tilde{f} := \sup(F(c))$;   { initialize upper bound }

2. $F_X := (F(c) + S_f(c, X) \cdot (X - c)) \cap F(X)$;   { centered form }

3. $L := \{(X, \underline{f_X})\}$; $L_{res} := \{\}$;   { initialize working list and result list }

4.   **while** $L \neq \{\}$ **do**

5. $(Y, \underline{f_Y}) := \text{PopHead}(L)$; $c := m(Y)$;   { get first element of the list }

6.   SlopePruning $(Y, \ c, \ F(c), \ S_f(c, Y), \ \tilde{f}, \ U_1, \ U_2)$;

7.   **for** $i := 1$ **to** 2 **do**

8.   **if** $U_i = \emptyset$ **then** next_i;

9. $c := m(U_i)$;

10.   **if** $\sup(F(c)) < \tilde{f}$ **then** $\tilde{f} := \sup(F(c))$;   { update $\tilde{f}$ }

11. $F_U := (F(c) + S_f(c, U_i) \cdot (U_i - c)) \cap F(U_i)$;   { centered form }

12.   **if** $\underline{f_U} \leq \tilde{f}$ **then**

13.   **if** $d_{\mathrm{rel}}(F_U) \le \varepsilon$ **or** $d_{\mathrm{rel}}(U_i) \le \varepsilon$ **then**

14. $L_{\mathrm{res}} := L_{\mathrm{res}} \uplus (U_i, \underline{f_U})$ { accept U_i for the result list }

15.   **else**

16. $L := L \uplus (U_i, \underline{f_U})$;   { store U_i in the working list }

17.   **endfor**

18.   CutOffTest $(L, \widetilde{f})$;

19. **endwhile**

20. $(Y, \underline{f_Y}) :=$ Head (L_{res}); $F^* := [\underline{f_Y}, \widetilde{f}]$;   CutOffTest $(L_{\mathrm{res}}, \widetilde{f})$;

21. **return** F^*, L_{res}.

For our practical global optimization algorithm (Algorithm 3.4) we can state a corollary corresponding to Theorem 2.

Corollary 2. *Let $f : D \to \mathbb{R}$, $X \subseteq D \subseteq \mathbb{R}$, and $\varepsilon > 0$. Then Algorithm 3.4 has the following properties:*

1. $f^* \in F^*$.

2. $X^* \subseteq \displaystyle\bigcup_{(Y, \underline{f_Y}) \in L_{\mathrm{res}}} Y.$

Proof. See [38].

Finally, we list Algorithm 3.5 combining Algorithms 3.3 and 3.4 to compute enclosures for all global minimizers x^* of the function f and for the global minimum value f^* within the input interval X. The desired accuracy (relative diameter) of the interval enclosures is specified by the input parameter ε. To guarantee termination, 1 ulp accuracy (cf. [9, Section 3.6]) is chosen if the specified value of ε is too small (for example 0). The enclosures for the global minimizers of f are returned in the interval vector Opt. The number of enclosures computed is returned in the integer variable N. F^* encloses the global minimum value.

We use a function called CheckParameters as an abbreviation for the error checks for the parameters of AllGOp1 which are necessary in an implementation. If no error occurs, AllGOp1 delivers the N enclosures Opt_i, $i = 1, 2, \ldots, N$, and it holds

$$X^* \subseteq \bigcup_{i=1}^{N} Opt_i \quad \text{and} \quad f^* \in F^*.$$

Algorithm 3.5: AllGOp1 $(f, X, \varepsilon, Opt, N, F^*, Err)$

1. $Err :=$ CheckParameters;
2. **if** $Err \neq$ "No Error" **then return** Err;
3. **if** ε is too small **then** set ε to "1 ulp accuracy";
4. GlobalOptimize $(f, X, \varepsilon, F^*, L_{\mathrm{res}})$;
5. $N :=$ Length (L_{res});
6. **for** $i := 1$ **to** N **do**
7. $Opt_i :=$ Head (L_{res});
8. **return** Opt, N, F^*, Err;

A detailed description of an implementation of this algorithm can be found in [38].

3.6   Examples and Tests

Smooth Problems In the following examples we demonstrate that (for smooth problems) the method using the pruning technique is superior to a similar method using first order information in form of derivative evaluations. We compare the pruning method with a corresponding method using the monotonicity test. For this alternative method we use

Algorithm 3.4′:

Algorithm 3.4 with the modification that Steps 2, 6, and 10 were replaced by

2. $F_X := (f(c) + F'(X) \cdot (X - c)) \cap F(X)$;
6a.   CheckMonotonicity $(Y, F'(Y), X, U)$;
6b.   **if** $U = Y$ **then** Bisect (Y, U_1, U_2);
10. $F_U := (F(c) + F'(U_i) \cdot (U_i - c)) \cap F(U_i)$;

where CheckMonotonicity is Algorithm 2.2 given in Section 2.4.

We denote this alternative version by Algorithm 3.4′ to symbolize the use of derivatives.

Now, we compare the two methods for some examples. We list the numerical results (the computed enclosures) and the evaluation efforts, the number of bisections, the necessary storage space, and the run-time.

Example 5. We minimize the Shekel function

$$f(x) = -\sum_{i=1}^{10} \frac{1}{(k_i(x - a_i))^2 + c_i},$$

with a, c, and k as given below in the full description of the test set (cf. [40]).

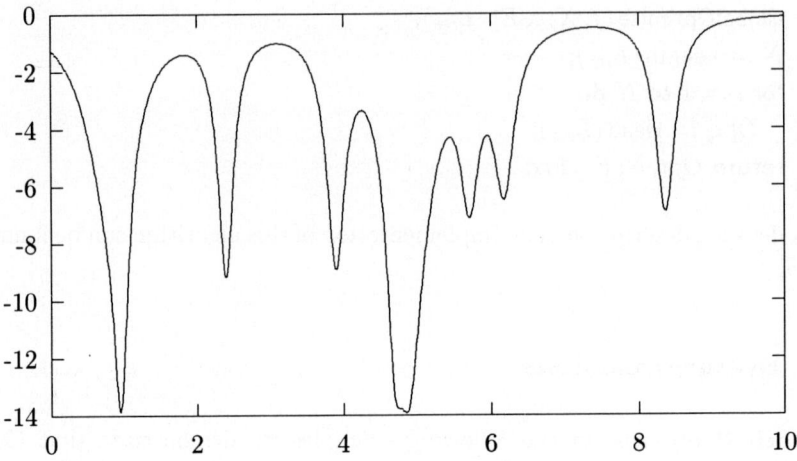

Applying Algorithm 3.4′ *using derivatives and monotonicity tests*, we get

```
Search interval          : [0,10]
Tolerance (relative)     :    1E-8
No. of function eval.    :     310
No. of derivative eval. :     155
No. of bisections        :      77
Necessary list length    :      10
Run-time (in sec.)       :   0.360
Global minimizer in      : [  4.855546951294E+000,   4.855585098262E+000 ]
Global minimum value in : [     -1.39223450E+001,      -1.39223448E+001 ]
```

Applying Algorithm 3.4 *using slopes and the pruning steps*, we get

```
Search interval          : [0,10]
Tolerance (relative)     :    1E-8
No. of function eval.    :     162
No. of slope eval.       :      81
No. of bisections        :       7
Necessary list length    :       7
Run-time (in sec.)       :   0.140
Global minimizer in      : [  4.855551999519E+000,   4.855585573167E+000 ]
Global minimum value in : [ -1.392234492356E+001, -1.392234487477E+001 ]
```

Example 6. We minimize $f(x) = (x - 1)^2(1 + 10\sin^2(x + 1)) + 1$.

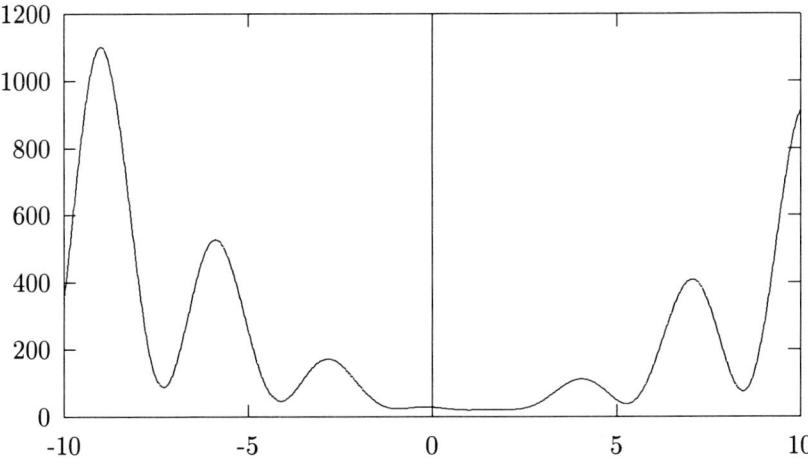

Applying Algorithm 3.4$'$ *using derivatives and monotonicity tests*, we get

```
Search interval        : [-10,10]
Tolerance (relative)   :     1E-8
No. of function eval.  :       78
No. of derivative eval. :      39
No. of bisections      :       19
Necessary list length  :        2
Run-time (in sec.)     :    0.030
Global minimizer in    : [  9.984741209375E-001,  1.002288183594E+000 ]
Global minimum value in : [  1.000000000000E+000,  1.000000001000E+000 ]
```

Applying Algorithm 3.4 *using slopes and the pruning steps*, we get

```
Search interval        : [-10,10]
Tolerance (relative)   :     1E-8
No. of function eval.  :       62
No. of slope eval.     :       31
No. of bisections      :        9
Necessary list length  :        3
Run-time (in sec.)     :    0.020
Global minimizer in    : [  9.999710616245E-001,  1.000019004318E+000 ]
Global minimum value in : [  1.000000000000E+000,  1.000000000200E+000 ]
```

Example 7. We minimize another function of Hansen (cf. [13])

$$f(x) = x^6 - 15x^4 + 27x^2 + 250.$$

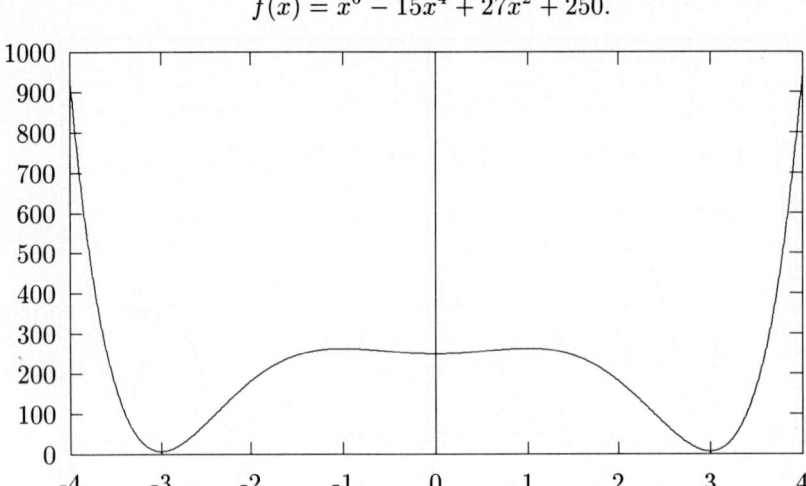

Applying Algorithm 3.4′ *using derivatives and monotonicity tests*, we get

```
Search interval          : [-4,4]
Tolerance (relative)     :   1E-8
No. of function eval.    :    582
No. of derivative eval. :    291
No. of bisections        :    145
Necessary list length    :     12
Run-time (in sec.)       :  0.240
Global minimizer in      : [ -3.000007629394E+000,  -2.999992370605E+000 ]
Global minimizer in      : [  2.999992370605E+000,   3.000007629394E+000 ]
Global minimum value in : [       6.99999998E+000,       7.00000000E+000 ]
```

Applying Algorithm 3.4 *using slopes and the pruning steps*, we get

```
Search interval          : [-4,4]
Tolerance (relative)     :   1E-8
No. of function eval.    :    252
No. of slope eval.       :    126
No. of bisections        :      3
Necessary list length    :      6
Run-time (in sec.)       :  0.090
Global minimizer in      : [ -3.000006841743E+000,  -2.999993933729E+000 ]
Global minimizer in      : [  2.999995403717E+000,   3.000001210429E+000 ]
Global minimum value in : [  6.999999970929E+000,   7.000000000000E+000 ]
```

The results for the run-time listed here are in some cases (if elementary functions are involved) much better than those reported in [37]. This is due to the fact that we carried out the numerical tests on a HP 9000/730 using a PASCAL–XSC version equipped with a run-time library including fast interval standard functions. This library is based on the thorough work of Hofschuster and Krämer described in [15] and [16].

For this reason, we present the new results (for the complete test set used in [37]) in the following. The test set consists of 20 onedimensional problems collected in [40] (which "well represent practical problems"), problems presented in [9] and [13], and some new problems. The test functions, the corresponding search boxes, and the solutions are as follows.

1. $f(x) = (x + \sin x) \cdot e^{-x^2}$, $X = [-10, 10]$,   (taken from [40])

   $X^* = \{-0.6795...\}$, $f^* = -0.8242....$

2. $f(x) = -\sum_{k=1}^{5} k\sin((k+1)x + k)$, $X = [-10, 10]$,   (taken from [40])

   $X^* = \{-6.774..., -0.4913..., 5.791...\}$, $f^* = -12.03....$

3. $f(x) = \sin x$, $X = [0, 20]$,

   $X^* = \{\frac{3\pi}{2}, \frac{7\pi}{2}, \frac{11\pi}{2}\}$, $f^* = -1$.

4. $f(x) = e^{-3x} - \sin^3 x$, $X = [0, 20]$,   (taken from [9])

   $X^* = \{\frac{9\pi}{2}\}$, $f^* = e^{\frac{27\pi}{2}} - 1$.

5. $f(x) = \sin\frac{1}{x}$, $X = [0.02, 1]$,

   $X^* = \{\frac{2}{(4k-1)\pi} \mid k = 1, \ldots, 6\}$, $f^* = -1$.

6. $f(x) = x^4 - 10x^3 + 35x^2 - 50x + 24$, $X = [-10, 20]$,

   $X^* = \{\frac{5-\sqrt{5}}{2}, \frac{5+\sqrt{5}}{2}\}$, $f^* = -1$.

7. $f(x) = 24x^4 - 142x^3 + 303x^2 - 276x + 93$, $X = [0, 3]$, (taken from [13])

   $X^* = \{2\}$, $f^* = 1$.

8. $f(x) = \sin x + \sin\dfrac{10x}{3} + \ln x - 0.84x$, $X = [2.7, 7.5]$, (taken from [40])

   $X^* = \{5.1997...\}$, $f^* = -4.601....$

9. $f(x) = 2x^2 - \dfrac{3}{100}e^{-(200(x-0.0675))^2}$, $X = [-10, 10]$,   (taken from [9])

   $X^* = \{0.06738...\}$, $f^* = -0.02090$.

10. $f(x) = -\sum_{i=1}^{10}\dfrac{1}{(k_i(x - a_i))^2 + c_i}$, $X = [0, 10]$,   (taken from [40])

$$
\begin{aligned}
a &= (3.040, 1.098, 0.674, 3.537, 6.173, 8.679, 4.503, 3.328, 6.937, 0.700), \\
k &= (2.983, 2.378, 2.439, 1.168, 2.406, 1.236, 2.868, 1.378, 2.348, 2.268), \\
c &= (0.192, 0.140, 0.127, 0.132, 0.125, 0.189, 0.187, 0.171, 0.188, 0.176),
\end{aligned}
$$

$X^* = \{0.6858...\}$, $f^* = -14.59....$

11. $f(x) = -\sum_{i=1}^{10} \dfrac{1}{(k_i(x - a_i))^2 + c_i}$, $X = [0, 10]$,   (taken from [40])

$\begin{aligned}
a &= (4.696, 4.885, 0.800, 4.986, 3.901, 2.395, 0.945, 8.371, 6.181, 5.713), \\
k &= (2.871, 2.328, 1.111, 1.263, 2.399, 2.629, 2.853, 2.344, 2.592, 2.929), \\
c &= (0.149, 0.166, 0.175, 0.183, 0.128, 0.117, 0.115, 0.148, 0.188, 0.198),
\end{aligned}$

$X^* = \{4.855...\}$, $f^* = -13.92....$

12. $f(x) = \dfrac{x^2}{20} - \cos x + 2$, $X = [-20, 20]$,

$X^* = \{0\}$, $f^* = 1$.

13. $f(x) = -\dfrac{1}{(x - 2)^2 + 3}$, $X = [0, 10]$,

$X^* = \{2\}$, $f^* = \frac{1}{3}$.

14. $f(x) = x^2 - \cos(18x)$, $X = [-5, 5]$,

$X^* = \{0\}$, $f^* = -1$.

15. $f(x) = (x - 1)^2(1 + 10\sin^2(x + 1)) + 1$, $X = [-10, 10]$,

$X^* = \{1\}$, $f^* = 1$.

16. $f(x) = e^{x^2}$, $X = [-10, 10]$,

$X^* = \{0\}$, $f^* = 1$.

17. $f(x) = x^4 - 12x^3 + 47x^2 - 60x - 20e^{-x}$, $X = [-1, 7]$, (taken from [9])

$X^* = \{0.7136...\}$, $f^* = -32.78....$

18. $f(x) = x^6 - 15x^4 + 27x^2 + 250$, $X = [-4, 4]$,   (taken from [13])

$X^* = \{-3, 3\}$, $f^* = 7$.

19. $f(x) = \sin^2\left(1 + \dfrac{x - 1}{4}\right) + \left(\dfrac{x - 1}{4}\right)^2$, $X = [-10, 10]$,

$X^* = \{-0.7878...\}$, $f^* = 0.4756....$

20. $f(x) = (x - x^2)^2 + (x - 1)^2$, $X = [-10, 10]$,

$X^* = \{1\}$, $f^* = 0$.

In Table 1, we give an overview of the results obtained with $\varepsilon = 10^{-8}$. For each test function we list the number of function evaluations, the number of derivative or slope evaluations, the number of bisections, the necessary list length, and the required CPU time. The columns with the header M contain the values for the variant with <u>M</u>onotonicity test, those with the header P contain the values for the variant with slope <u>P</u>runing step, and those with the header (P/M) contain the percentage of the slope method with respect to the derivative method. The last row of the table gives average values for the complete test set.

Table 1. Results for the complete test set when using the monotonicity test (M) and the slope pruning step (P), respectively

no.	F eval.			D or S eval.			bisections			list length			execution time		
	M	P	(P/M)	M	P	(P/M)	M	P	(P/M)	M	P	(P/M)	M	P	(P/M)
1	142	98	(69%)	71	49	(69%)	35	13	(37%)	5	4	(80%)	060	030	(50%)
2	346	314	(91%)	173	157	(91%)	86	5	(6%)	17	18	(106%)	330	240	(73%)
3	194	154	(79%)	97	77	(79%)	48	7	(15%)	4	4	(100%)	060	040	(67%)
4	142	116	(82%)	71	58	(82%)	35	8	(23%)	3	4	(133%)	070	050	(71%)
5	526	400	(76%)	263	200	(76%)	131	9	(7%)	9	9	(100%)	180	130	(72%)
6	1234	618	(50%)	617	309	(50%)	308	6	(2%)	29	16	(55%)	580	220	(38%)
7	956	488	(51%)	478	244	(51%)	238	12	(5%)	25	15	(60%)	450	180	(40%)
8	82	82	(100%)	41	41	(100%)	20	5	(25%)	4	4	(100%)	040	040	(100%)
9	106	90	(85%)	53	45	(85%)	26	10	(38%)	4	3	(75%)	060	040	(67%)
10	106	98	(92%)	53	49	(92%)	26	9	(35%)	5	5	(100%)	120	090	(75%)
11	310	162	(52%)	155	81	(52%)	77	7	(9%)	10	7	(70%)	360	140	(39%)
12	150	58	(39%)	75	29	(39%)	37	1	(3%)	2	2	(100%)	050	020	(40%)
13	66	56	(85%)	33	28	(85%)	16	11	(69%)	2	2	(100%)	020	010	(50%)
14	166	78	(47%)	83	39	(47%)	41	1	(2%)	2	2	(100%)	060	030	(50%)
15	78	62	(79%)	39	31	(79%)	19	9	(47%)	2	3	(150%)	030	020	(67%)
16	142	54	(38%)	71	27	(38%)	35	1	(3%)	2	2	(100%)	040	010	(25%)
17	308	144	(47%)	154	72	(47%)	76	10	(13%)	11	7	(64%)	210	090	(43%)
18	582	252	(43%)	291	126	(43%)	145	3	(2%)	12	6	(50%)	240	090	(38%)
19	98	82	(84%)	49	41	(84%)	24	7	(29%)	5	4	(80%)	040	030	(75%)
20	78	70	(90%)	39	35	(90%)	19	9	(47%)	2	2	(100%)	030	020	(67%)
$\sum$	5812	3476	(60%)	2906	1738	(60%)	1442	143	(10%)	155	119	(77%)	3.03	1.52	(50%)
$\emptyset$			(69%)			(69%)			(21%)			(91%)			(57%)

According to our numerical tests, the method with pruning step is always better than or at least as good as the traditional method with monotonicity test (with the exception of only three cases where the list length was a bit worse). On avarage, we have more than 30% improvement in the computation time and the number of function or derivative/slope evaluations, if we use the pruning step. Moreover, there are many examples for which the required CPU time is reduced to around 2/5 of the time required by the derivative variant.

Therefore, the pruning technique based on interval slopes can be very successfully applied in interval branch-and-bound methods for global optimization. This technique can replace the often used monotonicity test with a considerable improvement in efficiency for our global optimization model algorithm.

Nonsmooth Problems The most important fact is that the pruning technique is also applicable to nonsmooth problems, because interval slopes are computable for nondifferentiable functions, too. Therefore, we are able to use first-order information of the function within a global optimization method when applying it to nondifferentiable functions. We conclude this section by some nonsmooth examples.

Example 8. We minimize

$$f(x) = \min\{|\cos(\pi x/2)| - 3\sin(\pi x/10), 50|x - 1| - 3\}.$$

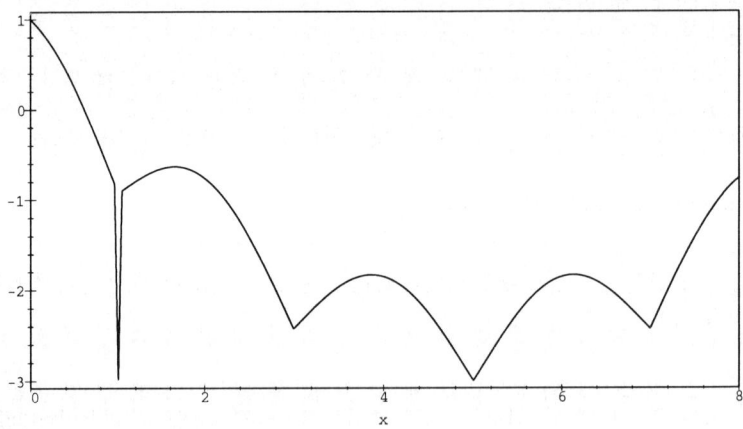

With Algorithm 3.4 we obtain:

```
Search interval        :  [0,8]
Tolerance (relative)   :  1E-8
No. of function eval.  :    58
No. of slope eval.     :    29
No. of bisections      :     5
Necessary list length  :     4
```

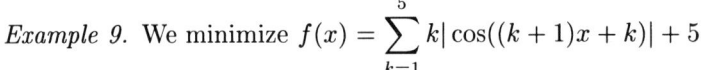

```
Run-time (in sec.)      : 0.040
Global minimizer in     : [  1.000000000000E+000,  1.000000000000E+000 ]
Global minimizer in     : [  4.999999986829E+000,  5.000000000008E+000 ]
Global minimum value in : [ -3.000000000000E+000, -3.000000000000E+000 ]
```

Example 9. We minimize $f(x) = \sum_{k=1}^{5} k|\cos((k+1)x + k)| + 5$.

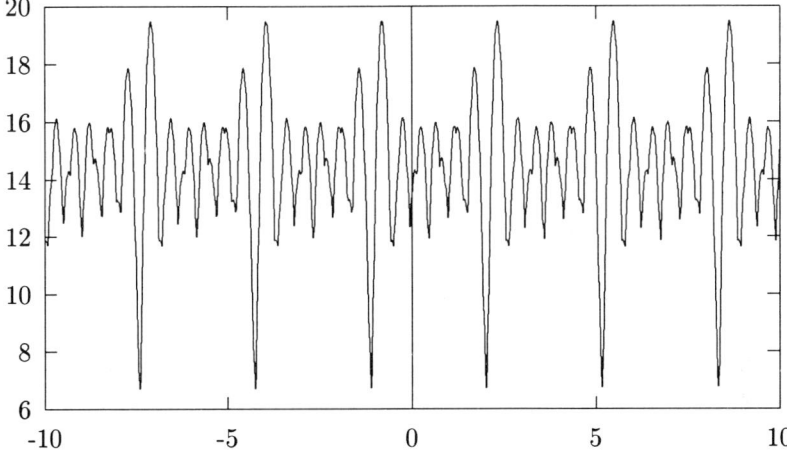

With Algorithm 3.4 we obtain:

```
Search interval         : [-10,10]
Tolerance (relative)    :      1E-8
No. of function eval.   :       412
No. of slope eval.      :       206
No. of bisections       :         6
Necessary list length   :        31
Run-time (in sec.)      :     0.430
Global minimizer in     : [ -7.397344586683E+000, -7.397344529213E+000 ]
Global minimizer in     : [ -4.255751923713E+000, -4.255751914035E+000 ]
Global minimizer in     : [ -1.114159270123E+000, -1.114159265284E+000 ]
Global minimizer in     : [  2.027433383466E+000,  2.027433388306E+000 ]
Global minimizer in     : [  5.169026037056E+000,  5.169026085146E+000 ]
Global minimizer in     : [  8.310618653412E+000,  8.310618700324E+000 ]
Global minimum value in : [  6.699793627703E+000,  6.699793776608E+000 ]
```

More examples can be found in [38].

4   Multidimensional Pruning Techniques for Global Optimization

In [38] we showed that interval slopes (together with centered forms) offer the possibility to achieve better enclosures for the range of multidimensional functions. Similar to the onedimensional case, this can improve the performance of interval branch-and-bound methods in the multidimensional case. Since slopes cannot be used within the monotonicity test, we introduced a pruning technique in Section 3, and we treated the onedimensional case separately for didactical reasons.

In this section, we now describe the extension of the pruning technique to the multidimensional case. We first treat a general multidimensional approach, which makes use of interval slope vectors. Then we describe in detail a more successful componentwise approach.

4.1   A First Approach Using Interval Slope Vectors

In order to generalize the pruning technique to the multidimensional case, we first recall the slope extension

$$F_s(Y) = f(c) + S \cdot (Y - c)$$

for a function $f : D \subseteq \mathbb{R}^n \to \mathbb{R}$, an interval vector $Y \subseteq D$, and a vector $c \in Y$, which we introduced in Section 2.3. $F_s(Y)$ encloses the range of f over Y, that is $f_{\mathrm{rg}}(Y) \subseteq F_s(Y)$, provided that the interval slope vector S satisfies

$$f(x) = f(c) + s \cdot (x - c) \quad \text{for some } s \in S,$$

for all $x \in Y$. An interval slope vector S can be calculated by means of the MASC process described in [38].

Given such an interval slope vector $S = (S_1, \ldots, S_n)$, we know that the graph of f over an interval Y is enclosed by 2^n hyperplanes $\widehat{h}$ defined by

$$\widehat{h}(x) = f(c) + \widehat{s}(x - c) = f(c) + \sum_{i=1}^{n} \widehat{s}_i(x_i - c_i),$$

where $x \in Y$ and $\widehat{s}_i \in \{\underline{s}_i, \overline{s}_i\}$, $i = 1, \ldots, n$. Figure 9 illustrates this situation for the twodimensional case. Here the graph of f is enclosed by four planes.

To clarify this, it is better to use an edited version of this figure given in Figure 10. In this figure, we deleted all unnecessary parts of the hyperplanes. With the help of this simplification, we can see how the graph of f is enclosed by two pyramidal boundaries from above and from below. This corresponds to the onedimensional case demonstrated by Figure 7.

The idea of the pruning technique (cf. Section 3.1) has been to prune away those parts of Y which cannot contain a global minimizer of f (with respect to the global search box X). As we see in Figure 10, we know that f is bounded from below by the lower pyramidal boundary. Moreover, if we have a guaranteed upper bound $\widetilde{f}$ for the global minimum value of f, then we can intersect the lower boundary with the plane corresponding to the $\widetilde{f}$ level. Projecting the lines of intersection onto the $x_1 x_2$-plane they enclose an area which is free of global minimizers of f.

Thus, the problem of pruning Y turns into the problem of computing these lines of intersection. In general, we must determine 2^n lines of intersection, that is we must compute the solution set of

$$\widetilde{f} = f(c) + h(x) \tag{37}$$

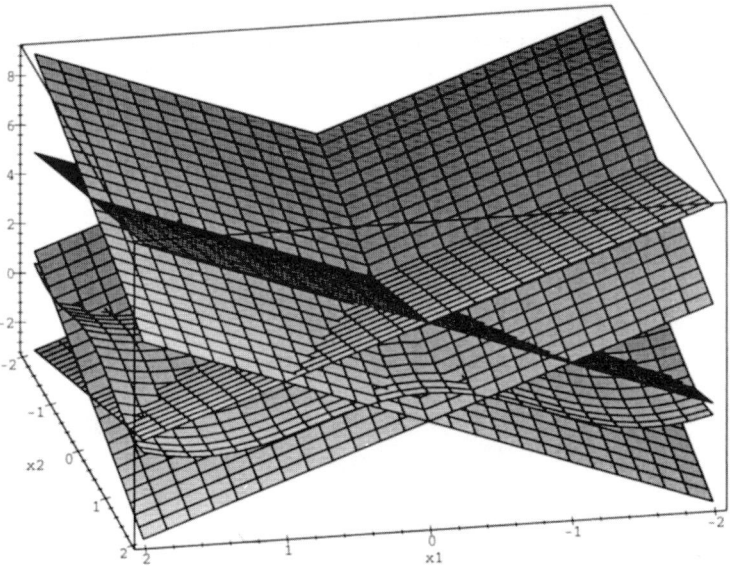

Fig. 9. Enclosure of f by the four hyperplanes defined by $\underline{S}$ and $\overline{S}$

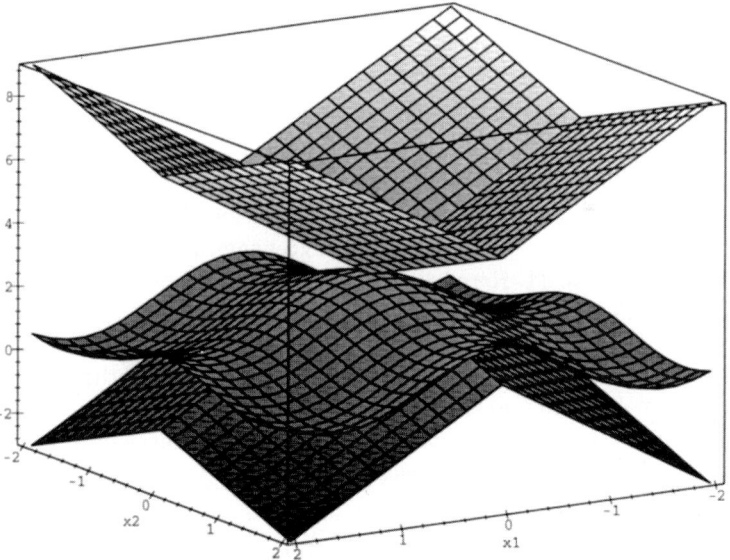

Fig. 10. Enclosure of f by the four hyperplanes defined by $\underline{S}$ and $\overline{S}$

where

$$h(x) = \min_{\widehat{s} \in \widehat{S}} \left(\widehat{s}(x - c) \right)$$

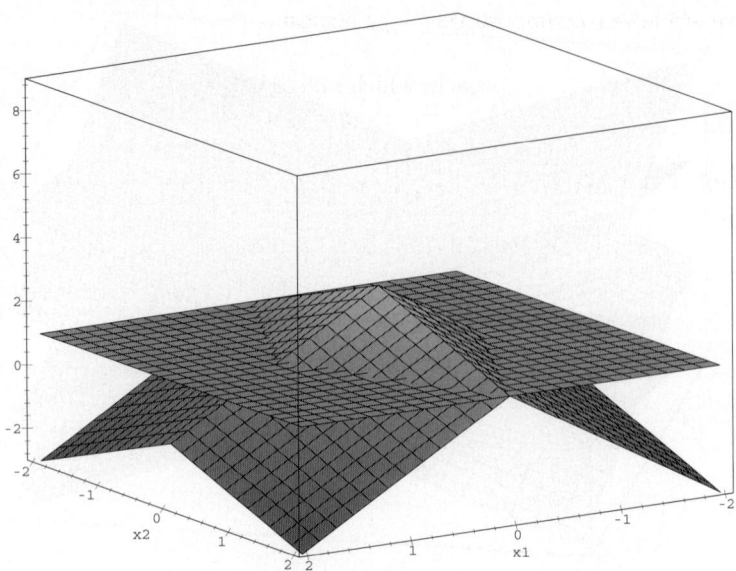

Fig. 11. Intersection of the lower boundary planes with the $\widetilde{f}$ level

and

$$\widehat{S} = \{\widehat{s} = (\widehat{s}_1, \ldots, \widehat{s}_n) \mid \widehat{s}_i \in \{\underline{s}_i, \overline{s}_i\}, \; i = 1, \ldots, n\}$$

for $x \in Y$. Such a pruning technique would be very costly.

An additional problem of this approach is the fact that the solution set of (37) representing the inner boundaries of the pruned Y are not parallel to the coordinate axes, in general. So, we must subdivide the pruned Y to enclose the subsets in boxes as small as possible. In general, this is not possible, as we can see in Figure 11. Here we would have to subdivide the pruned box Y in such a manner that the union of the resulting boxes would be equal to the original box Y.

It seams that this approach for a multidimensional pruning technique would be to expensive and the efficiency of a corresponding algorithm would be low. For this reason, we recall the possibility to compute interval slopes in a componentwise manner as discussed in [38].

4.2 An Alternative Componentwise Approach

Let $f : D \subseteq \mathbb{R}^n \to \mathbb{R}$, $c \in Y \subseteq X \subseteq D = D_1 \times D_2 \times \ldots \times D_n$, and $i \in \{1, \ldots, n\}$. Then we define $g_i : D_i \subseteq \mathbb{R} \to I\!\mathbb{R}$ on the currently treated interval vector Y by

$$g_i(w) := f(Y_1, \ldots, Y_{i-1}, w, Y_{i+1}, \ldots, Y_n), \quad w \in D_i. \tag{38}$$

Now we are able to compute the slope enclosure

$$S \supseteq s_{g_i}(c_i, Y_i) = \bigcup_{w \in Y_i} \frac{g_i(w) - g_i(c_i)}{w - c_i}$$

(for example by means of the MASC process described in [38]), and we have

$$f(Y) = g_i(Y_i) \subseteq g_i(c_i) + S(Y_i - c_i).$$

If we have $Z \supseteq g_i(c_i)$, then we know that the graph of f over an interval Y is enclosed by 4 hyperplanes h_j, $j = 1, 2, 3, 4$, defined by

$$\begin{aligned} h_1(x) &= \underline{z} + \underline{s}(x_i - c_i), \\ h_2(x) &= \underline{z} + \overline{s}(x_i - c_i), \\ h_3(x) &= \overline{z} + \underline{s}(x_i - c_i), \quad \text{and} \\ h_4(x) &= \overline{z} + \overline{s}(x_i - c_i), \end{aligned}$$

where $x \in Y$. These hyperplanes only depend on the i-th component of x. Figure 12 (for $\underline{s} > 0$) and Figure 13 (for $0 \in S$) illustrate this for two sample functions in the twodimensional case.

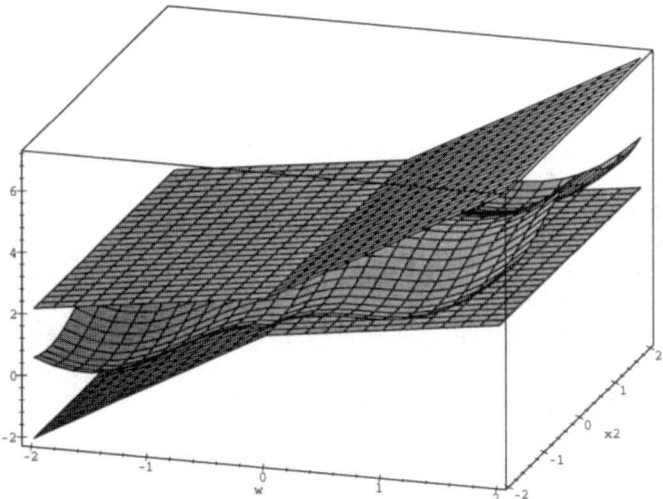

Fig. 12. Enclosure of f by the four hyperplanes defined by $\underline{z}$, $\overline{z}$, $\underline{s}$, and $\overline{s}$

To prune Y we must intersect the lower boundary planes with the plane corresponding to the $\tilde{f}$ level. Projecting the lines of intersection onto the x-plane they enclose an area which is free of global minimizers of f. Thus, the

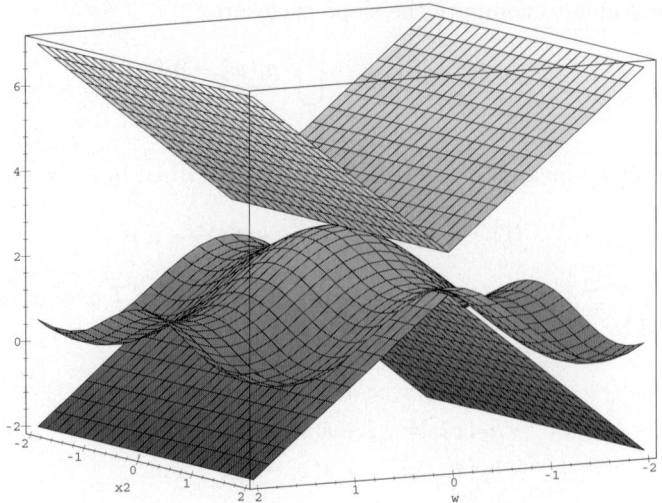

Fig. 13. Enclosure of f by the four hyperplanes defined by $\underline{z}$, $\overline{z}$, $\underline{s}$, and $\overline{s}$

problem of pruning Y turns into the problem of computing these two lines of intersection. That is we must compute the solution set of

$$\tilde{f} = \underline{z} + h(x)$$

where

$$h(x) = \min\{h_1(x), h_2(x)\}$$

for $x \in Y$. The idea of our componentwise pruning technique is to apply the onedimensional pruning technique to g_i for some or all $i \in \{1, \dots, n\}$. It is based on the following theorems.

Theorem 10. *Let* $f : D \to \mathbb{R}$, $Y = [\underline{y}, \overline{y}] \in I\mathbb{R}$, $c \in Y \subseteq X \subseteq D \subseteq \mathbb{R}^n$. *Moreover, let* $i \in \{1, \dots, n\}$, $g_i(w) = f(Y_1, \dots, Y_{i-1}, w, Y_{i+1}, \dots, Y_n)$, $g_i(c_i) \subseteq Z = [\underline{z}, \overline{z}] \in I\mathbb{R}$, $S = [\underline{s}, \overline{s}] \supseteq s_{g_i}(c_i, Y_i)$ *with* $\underline{s} > 0$, *and*

$$\tilde{f} \geq f^* = \min_{x \in X} f(x).$$

Then $p := c_i + (m + d(Z))/\overline{s}$ *with* $m = \min\{-d(Z), \tilde{f} - \overline{z}, (\underline{y} - c) \cdot \underline{s}\}$ *and* $\mathcal{Z} = (Y_1, \dots, Y_{i-1}, \mathcal{Z}_i, Y_{i+1}, \dots, Y_n)^\mathsf{T}$ *with* $\mathcal{Z}_i = Y_i \cap (p, \overline{y}_i]$ *satisfy*

$$p \leq c, \tag{39}$$

and

$$\min_{x \in \mathcal{Z}} f(x) > f^*. \tag{40}$$

Proof. See [38].

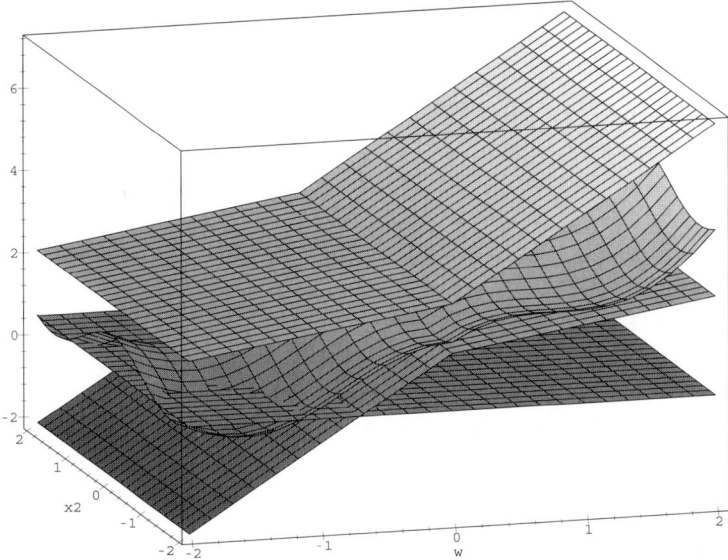

Fig. 14. Generation of the pruning point p

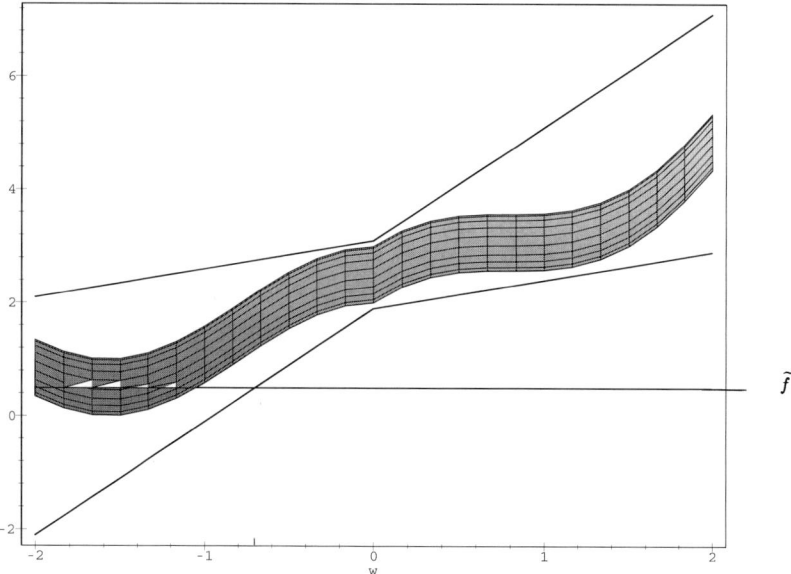

Fig. 15. Generation of the pruning point p (projection)

Figure 14 and its projection Figure 15 illustrate the geometrical interpretation for finding the point p for $i = 1$ in the twodimensional case.

It is easy to see, that we can apply a similar procedure for pruning in the case $\bar{s} < 0$, since we immediately have

Theorem 11. *Let $f : D \to \mathbb{R}$, $Y = [\underline{y}, \overline{y}] \in I\mathbb{R}$, $c \in Y \subseteq X \subseteq D \subseteq \mathbb{R}^n$. Moreover, let $i \in \{1, \ldots, n\}$, $g_i(w) = f(Y_1, \ldots, Y_{i-1}, w, Y_{i+1}, \ldots, Y_n)$, $g_i(c_i) \subseteq Z = [\underline{z}, \overline{z}] \in I\mathbb{R}$, $S = [\underline{s}, \overline{s}] \supseteq s_{g_i}(c_i, Y_i)$ with $\overline{s} < 0$, and*

$$\tilde{f} \geq f^* = \min_{x \in X} f(x).$$

Then $q := c_i + (m + d(Z))/\underline{s}$ with $m = \min\{-d(Z), \tilde{f} - \overline{z}, (\overline{y} - c) \cdot \overline{s}\}$ and $\mathcal{Z} = (Y_1, \ldots, Y_{i-1}, \mathcal{Z}_i, Y_{i+1}, \ldots, Y_n)^{\mathsf{T}}$ with $\mathcal{Z}_i = Y_i \cap [\underline{y}_i, q)$ satisfy

$$q \geq c, \tag{41}$$

and

$$\min_{x \in \mathcal{Z}} f(x) > f^*. \tag{42}$$

Proof. See [38].

For the case $0 \in S$ we have

Theorem 12. *Let $f : D \to \mathbb{R}$, $Y = [\underline{y}, \overline{y}] \in I\mathbb{R}$, $c \in Y \subseteq X \subseteq D \subseteq \mathbb{R}^n$. Moreover, let $i \in \{1, \ldots, n\}$, $g_i(w) = f(Y_1, \ldots, Y_{i-1}, w, Y_{i+1}, \ldots, Y_n)$, $g_i(c_i) \subseteq Z = [\underline{z}, \overline{z}] \in I\mathbb{R}$, $S = [\underline{s}, \overline{s}] \supseteq s_{g_i}(c_i, Y_i)$ with $0 \in S$, and*

$$\underline{z} > \tilde{f} \geq f^* = \min_{x \in X} f(x).$$

Then

$$p := \begin{cases} c_i + (\tilde{f} - \underline{z})/\overline{s} & \text{if } \overline{s} \neq 0, \\ -\infty & \text{otherwise}, \end{cases}$$

$$q := \begin{cases} c_i + (\tilde{f} - \underline{z})/\underline{s} & \text{if } \underline{s} \neq 0, \\ +\infty & \text{otherwise}, \end{cases}$$

and $\mathcal{Z} = (Y_1, \ldots, Y_{i-1}, \mathcal{Z}_i, Y_{i+1}, \ldots, Y_n)^{\mathsf{T}}$ with $\mathcal{Z}_i = Y_i \cap (p, q)$ satisfy

$$\min_{x \in \mathcal{Z}} f(x) > f^*. \tag{43}$$

Proof. See [38].

Figure 16 and its projection Figure 17 illustrate the geometrical interpretation for finding the points p and q for $i = 1$ in the twodimensional case.

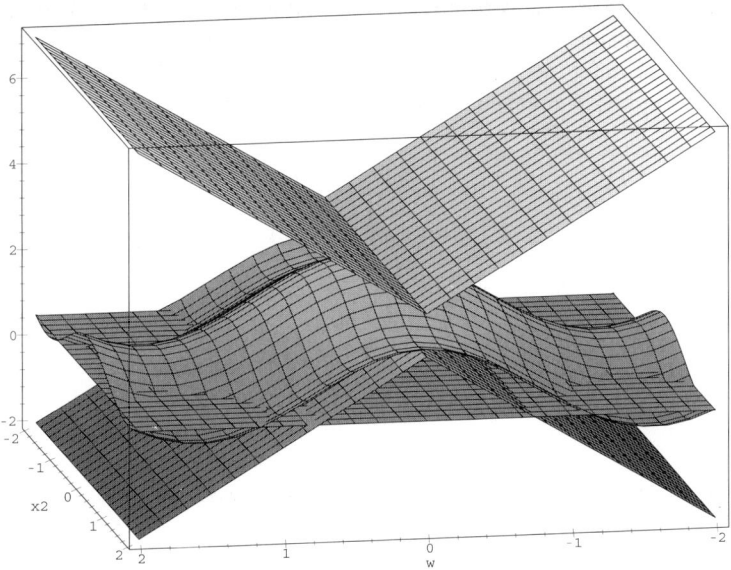

Fig. 16. Generation of the pruning points p and q

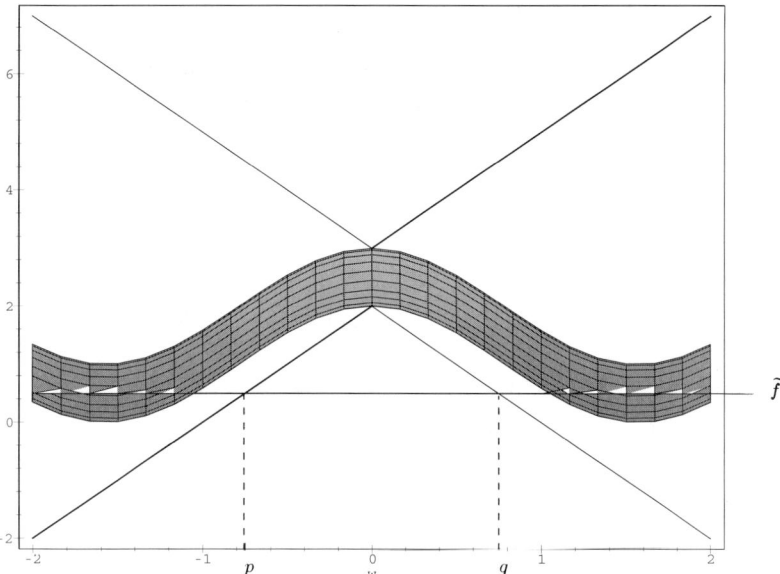

Fig. 17. Generation of the pruning points p and q (projection)

4.3   An Algorithmic Multidimensional Pruning Step

We are now able to give an algorithmic formulation of a pruning step, which can be applied to a subinterval $Y \subseteq X$ when globally minimizing $f : D \to \mathbb{R}$

on $X \subseteq D \subseteq \mathbb{R}^n$. In contrast to Section 3, we already use a machine-oriented description. Note that all interval operations and evaluations must be executed in machine interval arithmetic (cf. [9, Section 3.6]), even though this is not explicitly marked in our algorithmic notation.

The algorithm uses

$$
\begin{aligned}
g_i(w) &= f(Y_1, \ldots, Y_{i-1}, w, Y_{i+1}, \ldots, Y_n), \\
Y &= [\underline{y}, \overline{y}], \\
c &\in Y, \\
\widetilde{f} &\geq \min_{x \in X} f(x), \\
t &= (t_1, t_2, \ldots, t_n) \text{ with } t_k \in \{1, \ldots, n\} \text{ and } t_k \neq t_j \text{ for } k \neq j
\end{aligned}
$$

as input. It delivers the pruned (and possibly empty) subset $U_1 \cup \ldots \cup U_m$ of Y with $U_i \in I\mathbb{R}$ for $i = 1, \ldots, m$ and $0 \leq m \leq n+1$, and a possibly improved $\widetilde{f}$ as output. Note that the algorithm also performs a bisection if no pruning is possible.

If $0 \in S_{g_i}(c_i, Y_i)$ for *several* components i, then the method possibly produces *several* gaps in the current box Y. So we must split the result in several boxes U_i. In this case, different splitting techniques may be applied resulting in different values for U and m (cf. [32], [33], and [36]). We use the special splitting technique introduced in [32] resulting in at most $m = n + 1$ boxes. This technique uses each gap to store one part of the current box Y by using one part of the component Y_i and to update Y with the other part of Y_i, before continuing with the next component step of the pruning step.

That is, we perform one component step according to the following scheme:

1. Compute $Y_i = W \cup V$.

2. If $W = V = \emptyset$, then stop { no solution in Y }.

3. If $V \neq \emptyset$, then set $Y_i := V$ and store Y.

4. Set $Y_i := W$ and continue with next i.

Since it is not necessary to compute the Y_i in fixed order $i = 1, \ldots, n$, we use a sorted index vector t generated by the branching rule A (cf. [6], [34], or [36]) based on $D(i) = d(Y_i)$ for $i = 1, \ldots, n$. The index vector $t = (t_1, t_2, \ldots, t_n)$ with $t_k \in \{1, \ldots, n\}$ and $t_k \neq t_j$ for $k \neq j$, satisfies $D(t_k) \geq D(t_{k+1})$, $k = 1, \ldots, n - 1$.

Algorithm 4.1: MultiSlopePruning $(Y, c, \widetilde{f}, t, U, m)$

1. $m := 0$;
2.   **for** $k := 1$ **to** n **do**
3. $i := t_k$;   { use sorted indices }
4.   **if** $\sup(G_i(c_i)) < \widetilde{f}$ **then** $\widetilde{f} := \sup(G_i(c_i))$;   { update $\widetilde{f}$ }
5. $F_Y := (G_i(c_i) + S_{g_i}(c_i, Y_i) \cdot (Y_i - c_i)) \cap G_i(Y_i)$;   { centered form }
6.   **if** $\inf(F_Y) > \widetilde{f}$ **then return** ;   { no global optimizer in current Y }
7.   SlopePruning $(Y_i, c_i, G_i(c_i), S_{g_i}(c_i, Y_i), \widetilde{f}, W, V)$;   { apply Alg. 3.3 }
8.   **if** $W = V = \emptyset$ **then return** ;
9.   **if** $V \neq \emptyset$ **then**   { store part of Y in U_m }
10. $Y_i := V$; $m := m + 1$; $U_m := Y$;
11. $Y_i := W$;   { update current Y }
12.   **endfor**
13. $m := m + 1$; $[U]_m := Y$;   { store final Y in U_m }
14.   **return** $[U]$, m;

Summarizing the properties of this pruning step, we can state

Theorem 13. *Let* $f : D \to \mathbb{R}$, $Y \in I\,\mathbb{R}^n$, $c \in Y \subseteq X \subseteq D \subseteq \mathbb{R}^n$, $m \in \{0, \ldots, n + 1\}$, $U = (U_1, \ldots, U_m)$ *with* $U_i \in I\,\mathbb{R}^n$ *for* $i = 1, \ldots, m$, *and* $\widetilde{f} \geq \min_{x \in X} f(x)$, *then Algorithm 4.1 applied as* MultiSlopePruning $(Y, c, \widetilde{f}, t, U, m)$ *has the following properties:*

1. *$U_1 \cup \ldots \cup U_m \subseteq Y$.*
2. *Every global minimizer* x^* *of* f *in* X *with* $x^* \in Y$ *satisfies* $x^* \in U_1 \cup \ldots \cup U_m$.
3. *If* $m = 0$, *then there is no global (w.r.t. X) minimizer of* f *in* Y.

Proof. See [38].

As in the onedimensional case, it is obvious that the success of Algorithm 4.1 in pruning Y depends on the quality of $\widetilde{f}$. Therefore, the pruning step within a global optimization method can benefit greatly from a fast local search method delivering a good (small) value $\widetilde{f}$ on a very early stage of the method. For our further studies in this paper, we do not use an additional local method.

Now we integrate Algorithm 4.1 in a global optimization algorithm to study the effect of the application of the slope pruning step (replacing the monotonicity test).

4.4   A Multidimensional Global Optimization Algorithm Using Pruning Steps

Subsequently, we give a simple model algorithm for global minimization to demonstrate the application of the multidimensional pruning step. Our model algorithm uses the cut-off test, but it does not include neither a local search procedure, nor the concavity test, nor Newton-like steps, since the latter require smoothness.

Algorithm 4.2: GlobalOptimize $(f, X, \varepsilon, F^*, L_{\mathrm{res}})$

1. $c := m(X);$ $\widetilde{f} := \sup(F(c));$ { initialize upper bound }
2. $F_X := F(X);$ $L := \{(X, \underline{f_X})\};$ $L_{\mathrm{res}} := \{\};$ { initialize lists }
3. **while** $L \neq \{\}$ **do**
4. $(Y, \underline{f_Y}) := \mathsf{PopHead}\,(L);$ $c := m(Y);$ { get first element of the list }
5. $t := \mathsf{IndexSortVector}\,(Y);$ { compute sorted index vector }
6.   **if** $\sup(F(c)) < \widetilde{f}$ **then**   { update $\widetilde{f}$ and skip the pruning step }
7. $\widetilde{f} := \sup(F(c));$ $m := 1;$ $U_1 := Y;$
8.   **else**   { apply pruning step }
9. $\mathsf{MultiSlopePruning}\,(Y, c, \widetilde{f}, t, U, m);$
10.   **if** $m = 1 \wedge Y = U_1$ **then**   { no progress }
11. $\mathsf{Bisect}\,(Y, t_1, U_1, U_2);$ $m := 2;$
12.   **for** $i := 1$ **to** m **do**
13. $c := m(U_i);$
14.   **if** $\sup(F(c)) < \widetilde{f}$ **then** $\widetilde{f} := \sup(F(c));$ { update $\widetilde{f}$ }
15. $F_U := (F(c) + S_f(c, U_i) \cdot (U_i - c)) \cap F(U_i);$ { centered form }
16.   **if** $\underline{f_U} \leq \widetilde{f}$ **then**
17.   **if** $d_{\mathrm{rel}}(F_U) \leq \varepsilon$ **or** $d_{\mathrm{rel}, \infty}(U_i) \leq \varepsilon$ **then**
18. $L_{\mathrm{res}} := L_{\mathrm{res}} \uplus (U_i, \underline{f_U})$ { accept U_i for the result list }
19.   **else**
20. $L := L \uplus (U_i, \underline{f_U});$ { store U_i in the working list }
21.   **endfor**
22. $\mathsf{CutOffTest}\,(L, \widetilde{f});$
23.   **endwhile**
24. $(Y, \underline{f_Y}) := \mathsf{Head}\,(L_{\mathrm{res}});$ $F^* := [\underline{f_Y}, \widetilde{f}];$ $\mathsf{CutOffTest}\,(L_{\mathrm{res}}, \widetilde{f});$
25.   **return** F^*, L_{res}.

Algorithm 4.2 first computes an upper bound $\widetilde{f}$ for the global minimum value and initializes the working list L and the result list L_{res}. The main iteration (from Step 3 to Step 23) starts with the pruning step applied to the leading box of the working list. Then a centered form is used to apply a range check to the resulting boxes U_i, $i = 1, \ldots, m$. If the current box is still a candidate for containing a global minimizer, we store it in L_{res} (if it can be accepted with respect to the tolerance ε) or in L (if it must be treated further).

Note that by the operation $\uplus$ the boxes are stored in list L as pairs $(Y, \underline{f_Y})$ sorted in *nondecreasing* order with respect to the lower bounds $\underline{f_Y} \leq f_{\mathrm{rg}}(Y)$ and in *decreasing* order with respect to the ages of the boxes in L as specified in Section 2.4. Thus, the leading box of L is the oldest element with the smallest $\underline{f_Y}$ value. When the iteration stops because of the working list L being empty, we compute a final enclosure F^* for the global minimum value and return L_{res} and F^*.

The method can be improved by incorporating an approximate local search procedure to try to decrease the value $\tilde{f}$. For our studies in this paper, we do not apply any local method. The definition and algorithmic description of the cut-off test is given in Section 2.4.

For our global optimization model algorithm (Algorithm 4.2) we can state

Theorem 14. *Let* $f : D \to \mathbb{R}$, $X \subseteq D \subseteq \mathbb{R}^n$, *and* $\varepsilon > 0$. *Then Algorithm 4.2 has the following properties:*

1. $f^* \in F^*$.
2. $X^* \subseteq \displaystyle\bigcup_{(Y, \underline{f_Y}) \in L_{\mathrm{res}}} Y$.

Proof. See [38].

Finally, we list Algorithm 4.3 combining Algorithms 4.1 and 4.2 to compute enclosures for all global minimizers x^* of the function f and for the global minimum value f^* within the input interval X. The desired accuracy (relative diameter) of the interval enclosures is specified by the input parameter ε. To guarantee termination, 1 ulp accuracy (cf. [9, Section 3.6]) is chosen if the specified value of ε is too small (for example 0). The enclosures for the global minimizers of f are returned row by row in the interval matrix Opt. The number of enclosures computed is returned in the integer variable N. F^* encloses the global minimum value.

We use a function called CheckParameters as an abbreviation for the error checks for the parameters of AllGOp which are necessary in an implementation. If no error occurs, AllGOp delivers the N enclosures Opt_i, $i = 1, 2, \ldots, N$, and it holds

$$X^* \subseteq \bigcup_{i=1}^{N} Opt_i \quad \text{and} \quad f^* \in F^*.$$

Algorithm 4.3: AllGOp $(f, X, \varepsilon, Opt, N, F^*, \mathrm{Err})$

1. $\mathrm{Err} := \mathsf{CheckParameters}\,;$
2. **if** $\mathrm{Err} \neq$ "No Error" **then return** $\mathrm{Err};$
3. **if** ε is too small **then** set ε to "1 ulp accuracy";
4. $\mathsf{GlobalOptimize}\,(f, X, \varepsilon, F^*, L_{\mathrm{res}});$
5. $N := \mathsf{Length}\,(L_{\mathrm{res}});$
6. **for** $i := 1$ **to** N **do**
7. $Opt_i := \mathsf{Head}\,(L_{\mathrm{res}});$
8. **return** $Opt, N, F^*, \mathrm{Err};$

A detailed description of an implementation of this algorithm can be found in [38].

4.5   Examples and Tests

Smooth Problems In the following examples we solve some smooth standard test problems. These problems enable us to compare the method using slopes and the pruning step with a similar method using gradients and the monotonicity test. This alternative method is

Algorithm 4.2′:

Algorithm 4.2 with the modification that Steps 9 and 15 are replaced by

    9. $\mathsf{MultiCheckMonotonicity}\,(Y,\ \boldsymbol{\nabla}F(Y),\ X,\ U,\ m);$

   15. $F_U := (F(c) + \boldsymbol{\nabla}F(U_i) \cdot (U_i - c)) \cap F(U_i);$

where $\mathsf{MultiCheckMonotonicity}$ is Algorithm 2.3 given in Section 2.4.

We denote this alternative version by Algorithm 4.2′ to symbolize the use of derivatives.

Now, we compare the two methods for several examples, for which we list the numerical results (the computed enclosures) and the evaluation efforts, the number of bisections, the necessary storage space, and the run-time.

Comparing the efficiency of our two methods, there is no general trend according to our numerical tests. We cannot say that the method with pruning step is always better than the traditional method with monotonicity test. There are many test problems which can be solved much faster when slope techniques are applied, On the other hand, there are also test problems where the slope version of our model algorithm is slower.

Example 10. We minimize function G7 of Griewank [40] with $x \in I\!R^7$:

$$f(x) = \sum_{i=1}^{7} \frac{x_i^2}{4000} - \prod_{i=1}^{7} \cos\left(\frac{x_i}{\sqrt{i}}\right) + 1.$$

Applying Algorithm 4.2′ *using derivatives and monotonicity tests*, we get

```
Search interval        : [ -600,600 ]
                         [ -600,600 ]
                         [ -600,600 ]
                         [ -600,600 ]
                         [ -600,600 ]
                         [ -600,600 ]
                         [ -600,600 ]
Tolerance (relative)   : 1E-2
No. of function eval.  : 276456
No. of derivative eval.: 241899
No. of bisections      :   11519
Necessary list length  :     128
Run-time (h:m:sec)     : 0:32:34,630
Global minimizer in    : [              -7.4E-002,          7.4E-002 ]
                         [              -7.4E-002,          7.4E-002 ]
                         [              -7.4E-002,          7.4E-002 ]
                         [              -7.4E-002,          7.4E-002 ]
                         [              -7.4E-002,          7.4E-002 ]
                         [              -1.5E-001,          1.5E-001 ]
                         [              -1.5E-001,          1.5E-001 ]
Global minimum value in : [             0.0E+000,          1.1E-014 ]
```

Applying Algorithm 4.2 *using slopes and the pruning steps*, we get

```
Search interval        : [ -600,600 ]
                         [ -600,600 ]
                         [ -600,600 ]
                         [ -600,600 ]
                         [ -600,600 ]
                         [ -600,600 ]
                         [ -600,600 ]
Tolerance (relative)   : 1E-2
No. of function eval.  : 127131
No. of slope eval.     :  61733
No. of bisections      :     127
Necessary list length  :     128
Run-time (h:m:sec)     : 0:08:18,730
Global minimizer in    : [              -3.8E-002,          3.8E-002 ]
                         [              -3.9E-002,          1.1E-001 ]
                         [              -3.6E-002,          9.5E-002 ]
                         [              -3.0E-002,          8.1E-002 ]
                         [              -2.7E-002,          7.1E-002 ]
                         [              -2.8E-002,          7.3E-002 ]
                         [              -2.5E-002,          6.7E-002 ]
Global minimum value in : [             0.0E+000,          1.1E-014 ]
```

Example 11. We minimize function L3 of Levy [26] with $x \in \mathbb{R}^2$:

$$f(x) = \sum_{i=1}^{5} i \cos((i-1)x_1 + i) \sum_{j=1}^{5} j \cos((j+1)x_2 + j)$$

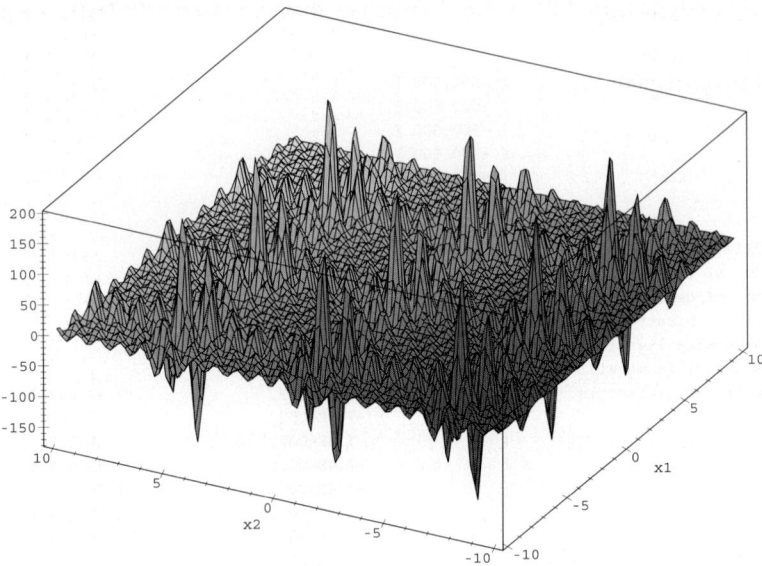

Applying Algorithm 4.2$'$ *using derivatives and monotonicity tests*, we get

```
Search interval          : [ -10, 10 ]
                           [ -10, 10 ]
Tolerance (relative)     : 1E-3
No. of function eval.    : 5769
No. of derivative eval.  : 3843
No. of bisections        :  607
Necessary list length    :  127
Run-time (h:m:sec)       : 0:00:13,690
Global minimizer in      : [        4.975E+000,           4.981E+000 ]
                           [         4.85E+000,            4.87E+000 ]
Global minimizer in      : [        4.975E+000,           4.981E+000 ]
                           [        -1.43E+000,           -1.42E+000 ]
Global minimizer in      : [        -1.31E+000,           -1.30E+000 ]
                           [         4.85E+000,            4.87E+000 ]
Global minimizer in      : [       -7.593E+000,          -7.587E+000 ]
                           [         4.85E+000,            4.87E+000 ]
Global minimizer in      : [        -1.31E+000,           -1.30E+000 ]
                           [        -1.43E+000,           -1.42E+000 ]
Global minimizer in      : [       -7.593E+000,          -7.587E+000 ]
                           [        -1.43E+000,           -1.42E+000 ]
Global minimizer in      : [        4.975E+000,           4.981E+000 ]
                           [        -7.72E+000,           -7.70E+000 ]
Global minimizer in      : [        -1.31E+000,           -1.30E+000 ]
                           [        -7.72E+000,           -7.70E+000 ]
Global minimizer in      : [       -7.593E+000,          -7.587E+000 ]
                           [        -7.72E+000,           -7.70E+000 ]
Global minimum value in  : [       -1.766E+002,          -1.765E+002 ]
```

Applying Algorithm 4.2 *using slopes and the pruning steps*, we get

```
Search interval        : [ -10, 10 ]
Tolerance (relative)   : 1E-3
No. of function eval.  : 8531
No. of slope eval.     : 3918
No. of bisections      :  310
Necessary list length  :  129
Run-time (h:m:sec)     : 0:00:15,440
Global minimizer in    : [        4.96E+000,          4.99E+000 ]
                         [       -7.72E+000,         -7.69E+000 ]
Global minimizer in    : [       -7.60E+000,         -7.57E+000 ]
                         [       -7.73E+000,         -7.69E+000 ]
Global minimizer in    : [       -1.32E+000,         -1.29E+000 ]
                         [        4.85E+000,          4.87E+000 ]
Global minimizer in    : [       -1.32E+000,         -1.29E+000 ]
                         [       -7.72E+000,         -7.70E+000 ]
Global minimizer in    : [        4.96E+000,          4.99E+000 ]
                         [       -1.44E+000,         -1.42E+000 ]
Global minimizer in    : [        4.96E+000,          4.99E+000 ]
                         [        4.85E+000,          4.87E+000 ]
Global minimizer in    : [       -1.32E+000,         -1.30E+000 ]
                         [       -1.44E+000,         -1.42E+000 ]
Global minimizer in    : [       -7.60E+000,         -7.58E+000 ]
                         [       -1.44E+000,         -1.42E+000 ]
Global minimizer in    : [       -7.60E+000,         -7.58E+000 ]
                         [        4.85E+000,          4.87E+000 ]
Global minimum value in : [     -1.767E+002,         -1.765E+002 ]
```

Example 12. We minimize function L5 of Levy [26] with $x \in \mathbb{R}^2$:

$$f(x) = \sum_{i=1}^{5} i \cos((i-1)x_1 + i) \sum_{j=1}^{5} j \cos((j+1)x_2 + j)$$
$$+(x_1 + 1.42513)^2 + (x_2 + 0.80032)^2.$$

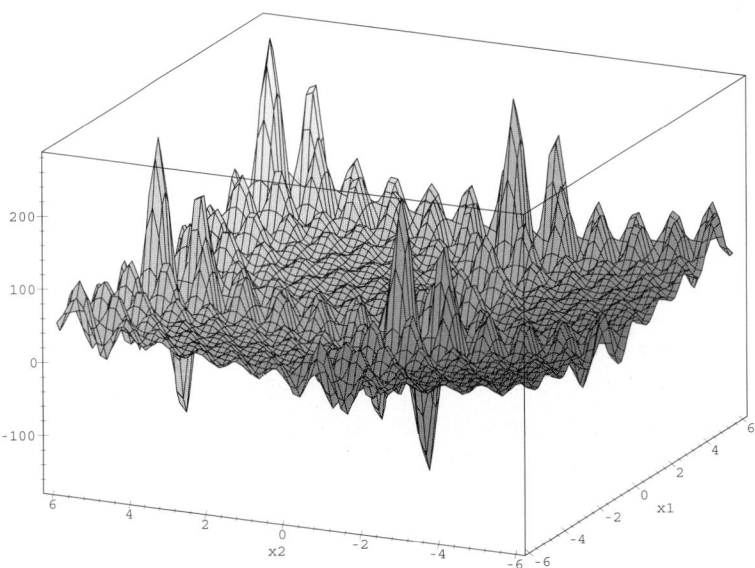

Applying Algorithm 4.2′ *using derivatives and monotonicity tests*, we get

```
Search interval         : [ -10, 10 ]
                          [ -10, 10 ]
Tolerance (relative)    : 1E-4
No. of function eval.   : 1586
No. of derivative eval. : 1057
No. of bisections       :  171
Necessary list length   :   33
Run-time (h:m:sec)      : 0:00:03,110
Global minimizer in     : [        -1.308E+000,         -1.306E+000 ]
                          [        -1.43E+000,          -1.42E+000 ]
Global minimum value in : [     -1.7615E+002,       -1.7613E+002 ]
```

Applying Algorithm 4.2 *using slopes and the pruning steps*, we get

```
Search interval         : [ -10, 10 ]
Tolerance (relative)    : 1E-4
No. of function eval.   : 2004
No. of slope eval.      :  920
No. of bisections       :   67
Necessary list length   :   36
Run-time (h:m:sec)      : 0:00:03,140
Global minimizer in     : [        -1.31E+000,          -1.30E+000 ]
                          [        -1.43E+000,          -1.42E+000 ]
Global minimum value in : [     -1.7615E+002,       -1.7613E+002 ]
```

Nonsmooth Problems Our main goal in developing the multidimensional pruning technique was to obtain a tool which can be applied to nonsmooth problems. Thus, we conclude the section with several nonsmooth examples.

Example 13. We minimize the function

$$f(x) = 10 \cdot |x_1 - 1| \cdot |\sin(\tfrac{1}{x_2})| + (x_2 + 2) \cdot |x_1 - 1 + 2x_2|.$$

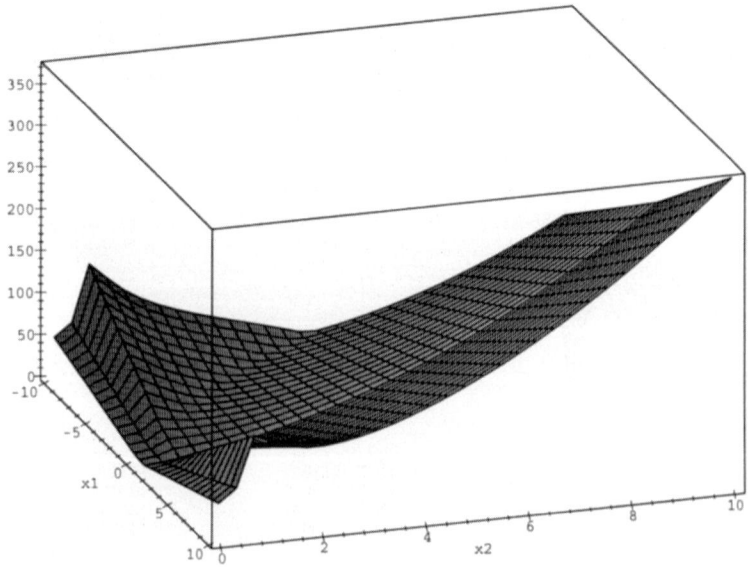

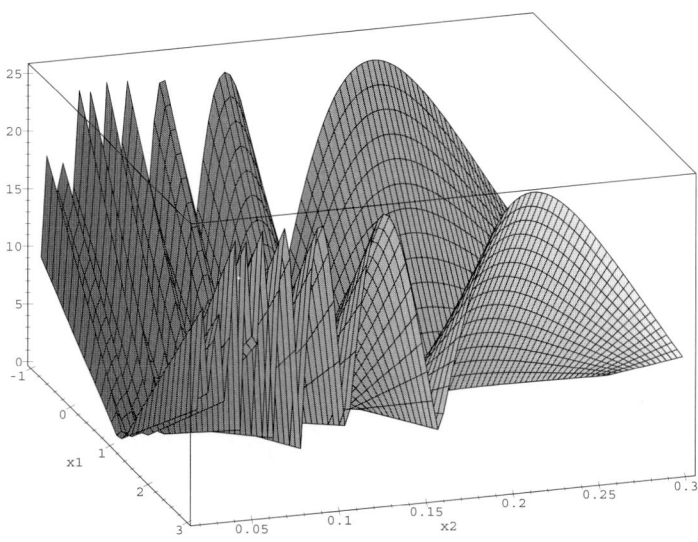

With Algorithm 4.2 we obtain for $x \in I\!R^2$

```
Search interval        : [ -100, 100 ]
                         [ 0.02, 100 ]
Tolerance (relative)   : 1E-8
No. of function eval.  : 5430
No. of slope eval.     : 2494
No. of bisections      :  242
Necessary list length  :   81
Run-time (h:m:sec)     : 0:00:05,130
Global minimizer in    : [        3.63380225E-001,          3.63380229E-001 ]
                         [        3.183098860E-001,         3.183098863E-001 ]
Global minimizer in    : [        6.81690112E-001,          6.81690116E-001 ]
                         [        1.591549430E-001,         1.591549432E-001 ]
Global minimizer in    : [        7.87793408E-001,          7.87793412E-001 ]
                         [        1.061032953E-001,         1.061032955E-001 ]
Global minimizer in    : [        8.40845052E-001,          8.40845058E-001 ]
                         [        7.957747153E-002,         7.957747157E-002 ]
Global minimizer in    : [        8.72676041E-001,          8.72676047E-001 ]
                         [        6.366197723E-002,         6.366197726E-002 ]
Global minimizer in    : [        8.93896702E-001,          8.93896706E-001 ]
                         [        5.305164768E-002,         5.305164771E-002 ]
Global minimizer in    : [        9.09054311E-001,          9.09054319E-001 ]
                         [        4.54728408E-002,          4.54728410E-002 ]
Global minimizer in    : [        9.20422527E-001,          9.20422531E-001 ]
                         [        3.978873576E-002,         3.978873579E-002 ]
Global minimizer in    : [        9.29264468E-001,          9.29264472E-001 ]
                         [        3.536776512E-002,         3.536776515E-002 ]
Global minimizer in    : [        9.36338020E-001,          9.36338024E-001 ]
                         [        3.183098861E-002,         3.183098863E-002 ]
Global minimizer in    : [        9.42125473E-001,          9.42125477E-001 ]
                         [        2.893726236E-002,         2.893726239E-002 ]
Global minimizer in    : [        9.4694834E-001,           9.4694836E-001 ]
                         [        2.652582382E-002,         2.652582386E-002 ]
Global minimizer in    : [        9.5102924E-001,           9.5102926E-001 ]
                         [        2.44853758E-002,          2.44853759E-002 ]
Global minimizer in    : [        9.5452715E-001,           9.5452717E-001 ]
                         [        2.273642042E-002,         2.273642045E-002 ]
Global minimizer in    : [        9.57558679E-001,          9.57558683E-001 ]
                         [        2.122065907E-002,         2.122065909E-002 ]
Global minimum value in : [       0.0E+000,                 4.4E-010 ]
```

Example 14. We minimize the function

$$f(x) = |x_1 - 1| + |x_2 - 1| + |\cos(18x_1 - 18)| + |\cos(18x_2 - 18)| + 1.$$

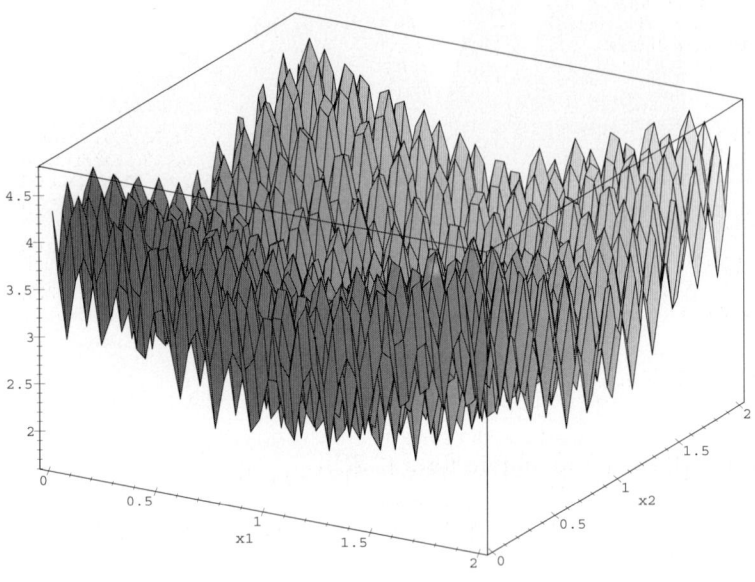

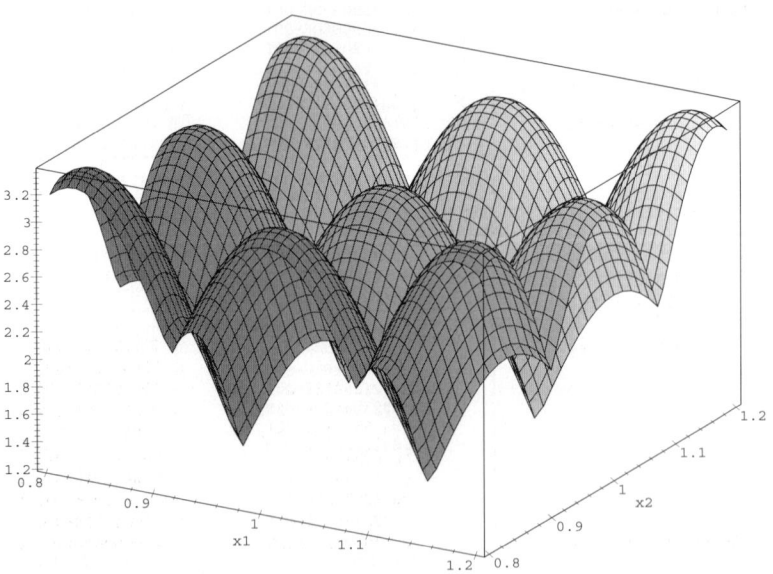

With Algorithm 4.2 we obtain for $x \in I\!\!R^2$

```
Search interval         : [ -10, 10 ]
                          [ -10, 10 ]
Tolerance (relative)    : 1E-8
No. of function eval.   : 1090
No. of slope eval.      :  500
No. of bisections       :   55
Necessary list length   :   17
Run-time (h:m:sec)      : 0:00:00,780
```

```
Global minimizer in     : [      1.08726645E+000,      1.08726647E+000 ]
                          [      9.12733531E-001,      9.12733539E-001 ]
Global minimizer in     : [      9.1273353E-001,       9.1273355E-001 ]
                          [      1.08726645E+000,      1.08726647E+000 ]
Global minimizer in     : [      9.12733536E-001,      9.12733539E-001 ]
                          [      9.12733535E-001,      9.12733541E-001 ]
Global minimizer in     : [      1.08726646E+000,      1.08726647E+000 ]
                          [      1.087266461E+000,     1.087266464E+000 ]
Global minimum value in : [      1.17453292E+000,      1.17453294E+000 ]
```

Example 15. We minimize the function

$$f(x) = \sum_{i=1}^{n-1} |x_i - 1|(1 + |\sin(3\pi x_{i+1})|)$$
$$+ |x_n - 1|(1 + |\sin(2\pi x_n)|) + |\sin(3\pi x_1)| + 1.$$

With Algorithm 4.2 we obtain for $x \in I\!\!R^9$ and $n = 9$

```
Search interval         : [ -10, 10 ]
                          [ -10, 10 ]
                          [ -10, 10 ]
                          [ -10, 10 ]
                          [ -10, 10 ]
                          [ -10, 10 ]
                          [ -10, 10 ]
                          [ -10, 10 ]
                          [ -10, 10 ]
Tolerance (relative)    : 1E-8
No. of function eval.   : 18192
No. of slope eval.      :  8984
No. of bisections       :   609
Necessary list length   :   134
Run-time (h:m:sec)      : 0:00:45,920
Global minimizer in     : [      9.99999997E-001,      1.00000001E+000 ]
                          [      9.99999991E-001,      1.00000001E+000 ]
                          [      9.99999991E-001,      1.00000001E+000 ]
                          [      9.99999991E-001,      1.00000001E+000 ]
                          [      9.99999991E-001,      1.00000001E+000 ]
                          [      9.99999991E-001,      1.00000001E+000 ]
                          [      9.99999991E-001,      1.00000001E+000 ]
                          [      9.99999991E-001,      1.00000001E+000 ]
                          [      9.99999991E-001,      1.00000001E+000 ]
Global minimum value in : [      1.00000000E+000,      1.00000001E+000 ]
```

Example 16. We minimize the function (its plot is turned upside down)

$$f(x) = \frac{|x_1| + |x_2 - 2|}{200} + |\cos(x_1)\cos(\frac{x_2 - 2}{\sqrt{2}})| + 2.$$

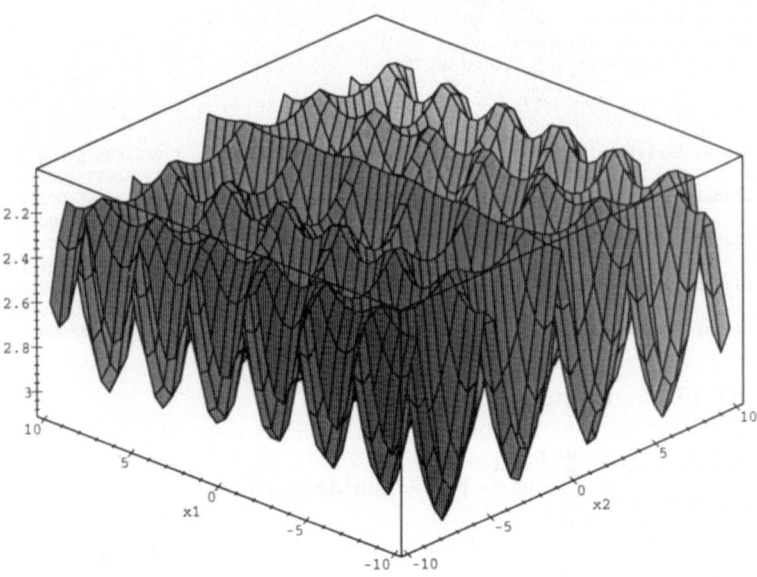

With Algorithm 4.2 we obtain for $x \in I\!R^2$

```
Search interval          : [ -100, 100 ]
                           [ -100, 100 ]
Tolerance (relative)     : 1E-8
No. of function eval.    : 804
No. of slope eval.       : 372
No. of bisections        :  60
Necessary list length    :  26
Run-time (h:m:sec)       : 0:00:00,900
Global minimizer in      : [        1.57079632E+000,        1.57079633E+000 ]
                           [         1.999999E+000,         2.000001E+000 ]
Global minimizer in      : [     -1.5707963269E+000,     -1.5707963267E+000 ]
                           [        1.9999999E+000,        2.0000003E+000 ]
Global minimum value in  : [     2.00785398162E+000,     2.00785398164E+000 ]
```

Example 17. We minimize the function (its plots are turned upside down)

$$f(x) = -\min\{|3\sin(\pi(\frac{x_1 - x_2^2}{4} - 2))|, |\sin(\pi(\frac{x_2}{4} + 0.5))|, |\sin(5\pi(\frac{x_2 - x_1^2}{4} - 1.5))|,$$
$$|\cos(\pi(1 - \frac{x_1}{4}))|, |\cos(\frac{5\pi x_1 x_2}{4})| - |\frac{x_1}{4}| - |\frac{x_2}{4}|\} - 1$$

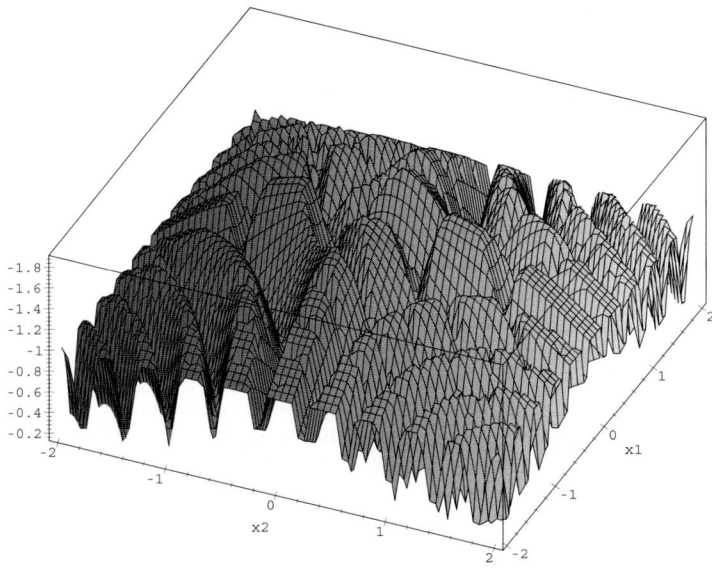

With Algorithm 4.2 we obtain for $x \in I\!\!R^2$

```
Search interval        : [ -10, 10 ]
                         [ -10, 10 ]
Tolerance (relative)   : 1E-8
No. of function eval.  : 1351
No. of slope eval.     :  626
No. of bisections      :   86
Necessary list length  :   26
Run-time (h:m:sec)     : 0:00:01,800
Global minimizer in    : [      -3.8484984E-001,      -3.8484983E-001 ]
                         [       3.043775E-002,       3.043777E-002 ]
Global minimum value in : [    -1.89512027E+000,    -1.89512025E+000 ]
```

References

1. Alefeld, G. and Herzberger, J. (1983): *Introduction to Interval Computations*, Academic Press, New York.
2. Bomze, I. M., Csendes, T., Horst, R., and Pardalos, P. M. (Eds.) (1996): *Developments in Global Optimization*, Kluwer Academic Publishers, Dordrecht.
3. Clarke, F. H. (1983): *Optimization and Nonsmooth Analysis*, Wiley, New York.
4. Csendes, T. (1989): *An Interval Method for Bounding Level Sets of Parameter Estimation Problems*, Computing, **41**, 75–86.
5. Csendes, T. and Pintér, J. (1993): *The Impact of Accelerating Tools on the Interval Subdivision Algorithm for Global Optimization*, European Journal of Operational Research, **65**, 314–320.
6. Csendes, T. and Ratz, D. (1997): *Subdivision Direction Selection in Interval Methods for Global Optimization*, SIAM Journal on Numerical Analysis, **34**, 922–938.
7. Floudas, C. A. and Pardalos, P. M. (Eds.) (1996): *State of the Art in Global Optimization*, Kluwer Academic Publishers, Dordrecht.
8. Grossmann, I. E. (Ed.) (1995): *Global Optimization in Engineering Design*, Kluwer Academic Publishers, Dordrecht.

9. Hammer, R., Hocks, M., Kulisch, U. and Ratz, D. (1993): *Numerical Toolbox for Verified Computing I*, Springer-Verlag, Berlin.

10. Hammer, R., Hocks, M., Kulisch, U. and Ratz, D. (1995): *C++ Toolbox for Verified Computing I*, Springer-Verlag, Berlin.

11. Hansen, E. (1979): *Global Optimization Using Interval Analysis – The One-Dimensional Case*, Journal of Optimization Theory and Applications, **29**, 331–344.

12. Hansen, E. (1980): *Global Optimization Using Interval Analysis – The Multi-Dimensional Case*, Numerische Mathematik, **34**, 247–270.

13. Hansen, E. (1992): *Global Optimization Using Interval Analysis*, Marcel Dekker, New York.

14. Hansen, P., Jaumard, B., and Xiong, J. (1994): *Cord-Slope Form of Taylor's Expansion in Univariate Global Optimization*, Journal of Optimization Theory and Applications, **80**, 441–464.

15. Hofschuster, W. and Krämer, W. (1997): *A Computer Oriented Approach to Get Sharp Reliable Error Bounds*, Reliable Computing, **3**, 239–248.

16. Hofschuster, W. and Krämer, W. (1997): *A Fast Public Domain Interval Library in ANSI C*, To appear in: Proceedings of the 15th IMACS World Congress 1997.

17. Horst, R. and Pardalos, P. M. (Eds.) (1995): *Handbook of Global Optimization*, Kluwer Academic Publishers, Dordrecht.

18. Ichida, K. and Fujii, Y. (1979): *An Interval Arithmetic Method for Global Optimization*, Computing, **23**, 85–97.

19. Jansson, C. (1994): *On Self-Validating Methods for Optimization Problems*, In: Herzberger, J.: *Topics in Validated Computations*, North-Holland, Amsterdam.

20. Kearfott, R. B. (1996): *Rigorous Global Search: Continuous Problems*, Kluwer Academic Publishers, Boston.

21. Klatte, R., Kulisch, U., Neaga, M., Ullrich, Ch. and Ratz, D. (1992): *PASCAL–XSC – Language Description with Examples*, Springer-Verlag, Berlin.

22. Krawczyk, R. and Nickel, K. (1982): *The Centered Form in Interval Arithmetics: Quadratic Convergence and Inclusion Isotonicity*, Computing, **28**, 117–137.

23. Krawczyk, R. and Neumaier, A. (1985): *Interval Slopes for Rational Functions and Associated Centered Forms*, SIAM Journal on Numerical Analysis, **22**, 604–616.

24. Kulisch, U. and Miranker, W. L. (1981): *Computer Arithmetic in Theory and Practice*, Academic Press, New York.

25. Lemarechal, C. and Mifflin, R. (Eds.) (1978): *Nonsmooth Optimization*, Proceedings of a IIASA Workshop, held at Institute for Research in Applied Mathematics and Systems workshop, 1977. IIASA Proceeding Series, Vol. 3, Pergamon Press, Oxford.

26. Levy, A. L., Montalvo, A., Gomez, S., and Calderon, A. (1981): *Topics in Global Optimization*, In: Hennart, J.-P. (Ed.): *Numerical Analysis*. Proceedings of the 3rd IIMAS Institute for Research in Applied Mathematics and Systems workshop, held at Cocoyoc, Mexico, 1981. Lecture notes in mathematics, No. 909, Springer-Verlag, Berlin.

27. Moore, R. E. (1966): *Interval Analysis*, Prentice-Hall, Englewood Cliffs, New Jersey.

28. Murty, K. G. and Kabadi S. N. (1987): *Some NP-Complete Problems in Quadratic and Nonlinear Programming*, Mathematical Programming, **39**, 117–130.

29. Neumaier, A. (1990): *Interval Methods for Systems of Equations*, Cambridge University Press, Cambridge.
30. Rall, L. B. (1981): *Automatic Differentiation, Techniques and Applications*, Lecture Notes in Computer Science, No. 120, Springer-Verlag, Berlin.
31. Ratschek, H. and Rokne, J. (1988): *New Computer Methods for Global Optimization*, Ellis Horwood, Chichester.
32. Ratz, D. (1992): *Automatische Ergebnisverifikation bei globalen Optimierungsproblemen*, Dissertation, Universität Karlsruhe.
33. Ratz, D. (1994): *Box-Splitting Strategies for the Interval Gauss-Seidel Step in a Global Optimization Method*, Computing, **53**, 337–353, Springer-Verlag, Wien.
34. Ratz, D. and Csendes, T. (1995): *On the Selection of Subdivision Directions in Interval Branch-and-Bound Methods for Global Optimization*, Journal of Global Optimization, **7**, 183–207.
35. Ratz, D. (1996): *An Optimized Interval Slope Arithmetic and its Application*, Forschungsschwerpunkt Computerarithmetik, Intervallrechnung und Numerische Algorithmen mit Ergebnisverifikation, Bericht **4/1996**.
36. Ratz, D. (1996): *On Branching Rules in Second-Order Branch-and-Bound Methods for Global Optimization*, In: Alefeld, G., Frommer, A. und Lang, B. (Eds.), *Scientific Computing and Validated Numerics*, 221–227, Akademie-Verlag, Berlin.
37. Ratz, D. (1997): *A Nonsmooth Global Optimization Technique Using Slopes – The One-Dimensional Case*, Accepted for publication in Journal of Global Optimization.
38. Ratz, D. (1998): *Automatic Slope Computation and its Application in Nonsmooth Global Optimization*. Shaker-Verlag, Aachen.
39. Skelboe, S. (1974): *Computation of Rational Interval Functions*, BIT, **4**, 87–95.
40. Törn, A. and Žilinskas, A. (1989): *Global Optimization*, Lecture Notes in Computer Science, No. 350, Springer-Verlag, Berlin.

Index

SpringerMathematics

Götz Alefeld et al. (eds.)

Symbolic Algebraic Methods and Verification Methods

2001. IX, 266 pages. 40 figures.
Softcover DM 129,90, öS 909,–
(recommended retail price)
ISBN 3-211-83593-8

The usual "implementation" of real numbers as floating point numbers on existing computers has the well-known disadvantage that most of the real numbers are not exactly representable in floating point. Also the four basic arithmetic operations can usually not be performed exactly. During the last years research in different areas has been intensified in order to overcome these problems. (LEDA-Library by K. Mehlhorn et al., "Exact arithmetic with real numbers" by A. Edalat et al., Symbolic algebraic methods, verification methods).

The latest development is the combination of symbolic-algebraic methods and verification methods to so-called hybrid methods. – This book contains a collection of worked out talks on these subjects given during a Dagstuhl seminar at the Forschungszentrum für Informatik, Schloß Dagstuhl, Germany, presenting the state of the art.

Please visit our new website: **www.springer.at**

SpringerWienNewYork

A-1201 Wien, Sachsenplatz 4–6, P.O. Box 89, Fax +43.1.330 24 26, e-mail: books@springer.at, Internet: **www.springer.at**
D-69126 Heidelberg, Haberstraße 7, Fax +49.6221.345-229, e-mail: orders@springer.de
USA, Secaucus, NJ 07096-2485, P.O. Box 2485, Fax +1.201.348-4505, e-mail: orders@springer-ny.com
Eastern Book Service, Japan, Tokyo 113, 3–13, Hongo 3-chome, Bunkyo-ku, Fax +81.3.38 18 08 64, e-mail: orders@svt-ebs.co.jp

SpringerMathematics

Dongming Wang

Elimination Methods

2001. XIII. 244 pages. 12 figures.
Softcover DM 108,–, öS 756,–
(recommended retail price)
ISBN 3-211-83241-6
Texts and Monographs in Symbolic Computation

This book provides a systematic and uniform presentation of elimination methods and the underlying theories, along the central line of decomposing arbitrary systems of polynomials into triangular systems of various kinds. Highlighting methods based on triangular sets, the book also covers the theory and techniques of resultants and Gröbner bases.

The methods and their efficiency are illustrated by fully worked out examples and their applications to selected problems such as from polynomial ideal theory, automated theorem proving in geometry and the qualitative study of differential equations. The reader will find the formally described algorithms ready for immediate implementation and applicable to many other problems.

Suitable as a graduate text, this book offers an indispensable reference for everyone interested in mathematical computation, computer algebra (software), and systems of algebraic equations.

 SpringerWienNewYork

A-1201 Wien, Sachsenplatz 4–6, P.O. Box 89, Fax +43.1.330 24 26, e-mail: books@springer.at, Internet: **www.springer.at**
D-69126 Heidelberg, Haberstraße 7, Fax +49.6221.345-229, e-mail: orders@springer.de
USA, Secaucus, NJ 07096-2485, P.O. Box 2485, Fax +1.201.348-4505, e-mail: orders@springer-ny.com
Eastern Book Service, Japan, Tokyo 113, 3–13, Hongo 3-chome, Bunkyo-ku, Fax +81.3.38 18 08 64, e-mail: orders@svt-ebs.co.jp

SpringerMathematics

David Colton et al. (eds.)

Surveys on Solution Methods
for Inverse Problems

2000. V, 275 pages. 41 figures.
Softcover DM 98,–, öS 686,–
(recommended retail price)
ISBN 3-211-83470-2

Inverse problems are concerned with determining causes for observed or desired effects. Problems of this type appear in many application fields both in science and in engineering. The mathematical modelling of inverse problems usually leads to ill-posed problems, i.e., problems where solutions need not exist, need not be unique or may depend discontinuously on the data. For this reason, numerical methods for solving inverse problems are especially difficult, special methods have to be developed which are known under the term "regularization methods".

This volume contains twelve survey papers about solution methods for inverse and ill-posed problems and about their application to specific types of inverse problems, e.g., in scattering theory, in tomography and medical applications, in geophysics and in image processing. The papers have been written by leading experts in the field and provide an up-to-date account of solution methods for inverse problems.

SpringerWienNewYork

A-1201 Wien, Sachsenplatz 4–6, P.O. Box 89, Fax +43.1.330 24 26, e-mail: books@springer.at, Internet: **www.springer.at**
D-69126 Heidelberg, Haberstraße 7, Fax +49.6221.345-229, e-mail: orders@springer.de
USA, Secaucus, NJ 07096-2485, P.O. Box 2485, Fax +1.201.348-4505, e-mail: orders@springer-ny.com
Eastern Book Service, Japan, Tokyo 113, 3–13, Hongo 3-chome, Bunkyo-ku, Fax +81.3.38 18 08 64, e-mail: orders@svt-ebs.co.jp

SpringerComputerScience

Guido Brunnett,
Hanspeter Bieri, Gerald Farin (eds.)

Geometric Modeling

Dagstuhl 1999

2001. Approx. 380 pages.
Softcover DM 220,–, öS 1540,–
Reduced price for subscribers to "Computing":
Softcover DM 198,–, öS 1386,–
(recommended retail price)

From the contents

- Converting Orthogonal Polyhedra from Extreme Vertices Model to B-Rep and to Alternating Sum of Volumes (A. Aguilera, D. Ayala)
- Smooth Shell Construction with Mixed Prism Fat Surfaces (C. L. Bajaj, G. Xu)
- Geometric Modeling of Parallel Curves on Surfaces (G. Brunnett)
- Computing Volume Properties Using Low-discrepancy Sequences (T. J. G. Davies, R. R. Martin, A. Bowyer)
- Bisectors and α-Sectors of Rational Varieties (G. Elber, G. Barquet, M.-S. Kim)
- Piecewise Linear Wavelets Over Type-2 Triangulations (M. S. Floater, E. G. Quak)
- Feature-based Matching of Triangular Meshes (M. Fröhlich, H. Müller, C. Pillokat, F. Weller)
- C^4 Interpolatory Shape-Preserving Polynomial Splines of Variable Degree (N. C. Gabrielides, P. D. Kaklis)
- Blossoming and Divided Difference (R. Goldman)
- Localizing the 4-Split Method for G^1 Free-Form Surface Fitting (S. Hahmann, G.-P. Bonneau, R. Taleb)

 SpringerWienNewYork

A-1201 Wien, Sachsenplatz 4–6, P.O. Box 89, Fax +43.1.330 24 26, e-mail: books@springer.at, Internet: **www.springer.at**
D-69126 Heidelberg, Haberstraße 7, Fax +49.6221.345-229, e-mail: orders@springer.de
USA, Secaucus, NJ 07096-2485, P.O. Box 2485, Fax +1.201.348-4505, e-mail: orders@springer-ny.com
Eastern Book Service, Japan, Tokyo 113, 3–13, Hongo 3-chome, Bunkyo-ku, Fax +81.3.38 18 08 64, e-mail: orders@svt-ebs.co.jp